# Lecture Notes in Engineering

Edited by C. A. Brebbia and S. A. Orszag



58

S. Naomis, P. C. M. Lau



# Computational Tensor Analysis of Shell Structures

Springer-Verlag
Berlin Heidelberg New York London
Paris Tokyo Hong Kong Barcelona

**Authors**
Dr. Steve Naomis

Dr. Paul C. M. Lau
Civil Engineering Department
University of Western Australia
Nedlands, WA 6009
Australia



ISBN 3-540-52852-0 Springer-Verlag Berlin Heidelberg New York
ISBN 0-387-52852-0 Springer-Verlag New York Berlin Heidelberg



Printing: Mercedes-Druck, Berlin
Binding: B. Helm, Berlin
2161/3020-543210 Printed on acid-free paper.

# PREFACE

This book presents a method which is capable of evaluating the deformation characteristics of thin shell structures A free vibration analysis is chosen as a convenient means of studying the displacement behaviour of the shell, enabling it to deform naturally without imposing any particular loading conditions. The strain-displacement equations for thin shells of arbitrary geometry are developed. These relationships are expressed in general curvilinear coordinates and are formulated entirely in the framework of tensor calculus. The resulting theory is not restricted to shell structures characterized by any particular geometric form, loading or boundary conditions.

The complete displacement and strain equations developed by Flugge are approximated by the curvilinear finite difference method and are applied to computing the natural frequencies and mode shapes of general thin shells. This approach enables both the displacement components and geometric properties of the shell to be approximated numerically and accurately.

The selection of an appropriate displacement field to approximate the deformation of the shell within each finite difference mesh is discussed in detail. In addition, comparisons are made between the use of second and third-order finite difference interpolation meshes.

The method is applied to a series of flat plate, cylindrical, spherical and conical shell structures to verify its accuracy. The computed natural frequencies and their associated mode shapes are documented and compared with published results.

The authors wish to acknowledge the supports provided by the Civil Engineering Department, University of Western Australia to the development of the computational tensor analysis of shell structures.

## TABLE OF CONTENTS

# NOTATION

## General

| | |
|---|---|
| [ ] | Matrix notation. |
| { } | Vector notation. |
| $(\ )_{,\alpha}$ | Partial differentiation with respect to the surface coordinates. |
| $(\ )\vert_{\alpha}$ | Covariant differentiation with respect to the coordinates $x^i$. |
| $(\ )\Vert_{\alpha}$ | Covariant differentiation with respect to the surface coordinates $x^{\alpha}$. |
| $(\dot{\ })$ | Time derivative. |
| $[\ ]^t$ | Transpose of a matrix. |
| $[\ ]^{-1}$ | Inverse of a matrix. |
| [K] | Structure's stiffness matrix |
| [M] | Structure's mass matrix. |
| $\mathbb{R}$ | Two dimensional region. |
| $\mathbb{S}$ | Boundary curve. |
| $r_i$ | Integration point for a Gauss-Legendre numerical integration scheme. |
| $w_i$ | Weight factors for a Gauss-Legendre numerical integration scheme. |
| $x^i$ | Coordinates of a general three-dimensional curvilinear coordinate system. |
| $x^{\alpha}$ | Coordinates of a general two-dimensional curvilinear coordinate system on the shell's middle surface. |
| $x^3$ | Coordinate axis normal to $x^{\alpha}$. |
| $\delta_i^j$ | Kronecker delta. |
| $\lambda$ | Eigenvalue. |

| | |
|---|---|
| $\psi$ | General two-dimensional scalar function. |
| $\Psi_i$ | Scalar quantities at nodal points. |

## Material Properties

| | |
|---|---|
| K | Bending stiffness: $Et^3/(12(1-\nu^2)$. |
| C | Damping constant. |
| E | Young's modulus of elasticity. |
| t | Shell's thickness. |
| $\nu$ | Poisson's ratio. |
| $\rho$ | Material density. |

## Geometric Properties

| | |
|---|---|
| $\mathbf{a}_\alpha$, $\mathbf{a}^\alpha$ | Covariant and contravariant base vectors at a point $x^\alpha$ on the surface $z = 0$. |
| $\mathbf{a}_3$, $\mathbf{a}^3$ | Covariant and contravariant unit normal base vectors. |
| $a_{\alpha\beta}$, $a^{\alpha\beta}$ | Covariant and contravariant components of the metric tensor on the surface $z = 0$. |
| $b_{\alpha\beta}$, $b^{\alpha\beta}$, $b^\beta_\alpha$ | Covariant, contravariant and mixed components of the curvature tensor. |
| dA | Area element. |
| ds | Line element. |
| dV | Volume element. |
| $\mathbf{g}_i$, $\mathbf{g}^i$ | Covariant and contravariant base vectors at a point $x^\alpha$ on the surface $z \neq 0$. |

| | |
|---|---|
| $g_{ij}$ $g^{ij}$ | Covariant and contravariant components of the metric tensors at a point $x^{\alpha}$ on the surface $z \neq 0$. |
| g | determinant of the metric tensor component matrix. |
| **i, j, k** | Base vectors in the cartesian coordinate system. |
| [J] | Jacobian matrix. |
| **n, t, s** | Base vectors associated with the boundary coordinate system. |
| **r** | Position vector of a point on the surface $z \neq 0$ before deformation. |
| **s** | Position vector of a point on the surface $z=0$ before deformation. |
| x, y, z | Cartesian coordinates. |
| z | A coordinate along the $x^3$ axis. |
| $\beta_i^j$ | Transformation coefficients. |
| $\in_{\alpha\beta}$, $\in^{\alpha\beta}$ | permutation tensors. |
| $\Gamma_{ij}^{k}$ $\Gamma_{ijk}$ | Christoffel symbols. |
| $\mu_{\alpha}^{\beta}$, $\lambda_{\alpha}^{\beta}$ | Shift tensors. |

**Deformation**

| | |
|---|---|
| **f** | A vector defining the extension and rotation of the middle surface resulting from relaxing the conservation of normals assumption. |
| **u** | Middle surface displacement vector. |
| **v** | General surface displacement vector. |
| w | Displacement component in the direction normal to the shell surface. |
| $\varepsilon_{\alpha\beta}$, $\varepsilon^{\alpha\beta}$, $\varepsilon_{\alpha}^{\beta}$ | Covariant, contravariant and mixed components of the middle surface in-plane strain tensor. |

| | |
|---|---|
| $\eta_{\alpha\beta}$, $\eta^{\alpha\beta}$, $\eta^{\beta}_{\alpha}$ | Covariant, contravariant and mixed components of the general surface in-plane strain tensor. |
| $\gamma_{ij}$ | General three-dimensional strain tensor. |
| $\kappa_{\alpha\beta}$ $\kappa^{\alpha\beta}$ | Covariant and contravariant components of the change in curvature tensor. |

## Stresses

| | |
|---|---|
| $\sigma^{\alpha\beta}$, $\sigma^{\beta}_{\alpha}$ | Contravariant and mixed components of the stress tensor. |
| $M^{\alpha\beta}$ | Contravariant components of the bending stress tensor. |
| $N^{\alpha\beta}$ | Contravariant components of the membrane stress tensor. |
| $\{P\}$ | External force vector. |

# 1. INTRODUCTION

Shell structures are widely used in a variety of engineering applications ranging from domes for major buildings and components of flight structures to liquid storage containers. These structures often have arbitrary shape and support conditions to meet functional and manufacturing requirements.

In modelling a shell numerically, an attempt is made to summarize its three dimensional behaviour in terms of the deformation of its middle surface. In order to achieve this, a number of approximations must be introduced into the formulation. Unlike the theory of plates where the differential equation of motion is universally agreed upon, the introduction of various geometric and displacement approximations give rise to a large number of shell theories.

The works of Flugge [1], Donnell [2], Love [3], Mushtari [4], Naghdi [5], Timoshenko [6], Reissner [7] and Vlasov [8] apply to shells of arbitrary curvature and are based on Love's 'first approximation' [3]. Essentially, differences between theories arise from simplifications involving the thickness to radius of curvature terms.

Comparisons of the various thin shell theories are documented in the works of Leissa [9] and Kraus [10]. A brief discussion on their suitability to modern computerized structural analysis is given by Bushnell [11].

Historically the thin shell equations of motion were first applied to structures characterized by a prescribed analytical geometry and constrained by ideal boundary conditions. This enabled the governing differential equations to be simplified and analytical solutions to be obtained for the dynamic characteristics of the structure. References [12] - [20] are typical of the approach used by early researchers. In addition, many authors (e.g. references [21] - [27]) have combined various numerical techniques with simplified versions of the shells governing strain-displacement equations. The literature in this area is extensive. Leissa [9] in his monograph 'Vibration of Shells' presents a comprehensive survey and discussion of the vibration of cylindrical, spherical and conical based shells. Since its release in 1973 new formulations, particularly in the area of vibrating shell panels, have been developed [24] - [27].

The requirement to analyse structures of arbitrary shape with irregular loading and support conditions limits the scope of the available analytical solutions. One technique which has been used successfully for the analysis of general shell structures since its introduction in the early 1960's is the Finite Element Method.

The literature detailing the application of the method to the dynamic analysis of shell structures is extensive. Only a brief overview is given here and references cited are indicative of the work carried out in the field. A detailed list of references has been compiled by Norrie and deVries [28].

One of the first applications to appear in the literature was the free vibration of shells of revolution [29] - [41]. Theoretically, the geometric nature of the problem allows the complex strain-displacement equations to be simplified. In early work [30], the shell was represented by a stacked assembly of elements which geometrically duplicated the frustum of a cone. Each element was permitted to deform in membrane and bending states and the continuity of displacement and slope was enforced on nodal circles along which neighbouring elements are connected. Continuing refinement in element technology has seen the introduction of general curved elements to improve the geometric representation of the surface[36].

The development of the flat plate bending element enabled the curved shell surface to be replaced by an assembly of flat triangular or rectangular elements [42] - [46]. The elements are formed by the superposition of stretching and bending behaviour which are uncoupled provided deformations are small. Many of the flat plate elements [48] differ from each other by the choice of shape functions and by the connections imposed between adjacent elements.

A number of difficulties and shortcomings of the methodology have been discussed in the literature [49,50]. These include the presence of discontinuity moments at the junction of adjacent elements and the difficulty of treating the junction where all elements are coplanar. As noted by Zienkiewicz [46] and Gallagher [50], the discontinuity moments can in many cases be suppressed by using a large number of elements.

The need to represent the surface geometry accurately has led to the development of a number of curved thin shell elements [51] - [56]. Many elements (e.g. Olsen & Lindberg [52]) are formulated for specific geometric configurations thus eliminating complex nodal parameter transformations.

The use of 'shallow shell theory' has enabled the development of a number of shallow curved elements [53, 54]. In particular, Olsen and Lindberg [54] reported the results of a high order conforming element which utilized first and second order derivatives at each corner of the element. Since the resultant nodal derivative parameters include only those derivatives with respect to the x-y coordinate system, the formulation is restricted to analysing shell structures which have the correct geometric properties and can be orientated in space to account for the implementation's shallow shell assumptions.

Zienkiewicz et al [57] presented an isoparametric shell element based on three dimensional theory. This element, formed by degenerating the three dimensional solid isoparametric element, was used successfully for a variety of relatively thick shell problems. As reported by Gill and Ucmaklioglu [58], the element was found to be over stiff when applied to thin shell situations. A modified version of the element using a reduced integration technique for the determination of the element stiffness matrix was applied to vibration problems by Hofmeister and Evensen [59] and by Bossak and Zienkiewicz [60]. The technique was found to produce acceptable results for the analysis of flat, spherical and cylindrical shells. However, the element performed poorly when applied to the case of a turbine blade. The inaccurate results were attributed to the failure of the reduced integration scheme to capture the rapid local change in curvature along part of the blade.

Generally, the main research objective behind the development of many shell elements has been to provide an element which is versatile, robust and reliable for all possible analysis conditions. In addition, some authors (e.g. Bathe and Dvorkin [61]) have expressed the additional constraint of the element fitting into the existing framework of finite element methodology (i.e. the use of six degrees of freedom per node) .

In the general shell problem, the element can be arbitrarily orientated in space and be geometrically representing a curved surface. In addition, the first derivative of deflection across the boundary of adjacent elements should be continuous. The bending components of the general shell equations require at least first order continuity in the out of plane displacements and, to maintain slope compatibility, higher second order derivatives must be introduced at the nodes. These derivatives must then be transformed to a common coordinate system so that the stiffness and mass matrices of the structure can be assembled. Having transformed the local partial derivatives, it is possible that some of them are zero as a result of the 'conservation of normals' assumption. For example, consider a part of a shell structure lying in the X-Y plane. The nodal parameter $U_{Z,Z}$ is zero and must be accounted for in the assembly of the structure's stiffness and mass matrices. In addition, a compatible element of this type has eighteen degrees of freedom per node which is far in excess of the six favoured by many researchers.

For many elements the conventional practice of ignoring the in-plane rotation when describing the shell kinematics, results in a poorly conditioned problem if the elements are nearly coplanar. A number of numerical techniques have been introduced to overcome this problem [ 45,46,62]. However, in the cases referenced, there are no explanations for the techniques and their performance is both problem and computer dependant.

The Finite Difference Method provides an alternative methodology which avoids the problems associated with computing accurately the geometric properties of the surface and transforming the unknown nodal derivative parameters to a common coordinate system.

In essence, expressions defining the local partial derivatives are written in terms of the adjacent nodal displacements. These are then combined with the pertinent surface geometric properties and substituted into the governing differential equations of motion. The resultant system of equations is then solved to yield the dynamic characteristics of the structure.

Although theoretically viable, the technique described above does have a number of problems associated with its implementation. Firstly, the differential equations describing the dynamic behaviour of general thin shells are extremely complex. They contain fourth order partial derivatives and require the algebraic derivation and expansion of such quantities as the second covariant derivative of the moment tensor in terms of the unknown nodal displacements. Secondly, the resultant stiffness and mass matrices will have a high bandwidth due to the problems fourth order nature and in the general case be non-symmetric. Numerical solution techniques which solve non-symmetric eigenvalue problems and incorporate a pivoting algorithm to overcome possible ill-conditioning associated with the boundary conditions would therefore need to be employed.

The application of the finite difference method to the analysis of plates and shells has been under investigation by a number of researchers [63] - [69]. In general, the work has been restricted to the static analysis of structures having a predefined geometry . This enables the governing differential equations to be simplified.

Morino, Leech and Witmer [63] - [65] have developed a tensor based methodology for the calculation of large elastic-plastic dynamically-induced deformations of general thin shells. Their program, PETROS 2, uses a combination of forward, backward and central finite difference equations which are derived from Taylor-series expansions using constant grid spacings. In addition the program is restricted to shells whose boundary conditions are uniform along each side and lie parallel to the global coordinate axes.

The Finite Difference Energy Method combines some of the advantages of both the finite element and finite difference techniques. In this approach, the surface is discretized by establishing a unique set of parametric curves and assigning node numbers to their intersecting points. In many formulations [70] - [76] these curves need not be restricted to orthogonal cases. Associated with each node is a non-overlapping subdomain which, when combined with adjacent nodes, represents the shell surface numerically. Within a subdomain, the finite difference approximations for the displacements and corresponding derivatives are substituted into the integral form of the differential equations of motion. The total potential energy of the system is computed as the sum of the potential energies summed over the set of nodal subdomains. The resulting system of equations are then solved to yield the shell's dynamic characteristics.

There are a number of advantages in formulating shell problems based on the above approach. Firstly, the integral form of the differential equation of motion contains only first and second order displacement derivatives. Therefore, a second order finite differencing scheme can be used, enabling a reduction in the bandwidths associated with the stiffness and mass matrices of the structure. Secondly, the resultant system of equations describing the dynamic behaviour of the shell will be symmetric and therefore allow the use of efficient eigenvalue/eigenvector solution techniques.

A number of shell analysis programs based on the above method have been developed. The earliest examples appearing within the literature were applied to the analysis of axisymmetric shells and are detailed in references [77] - [80]. In particular, the BOSOR series of programs developed by Bushnell [79,80] were developed with the capacity to perform linear, nonlinear, stability and vibration analyses of complex segmented, ring-stiffened shells of revolution with various wall constructions.

The application of the finite difference energy method to the analysis of shells of arbitrary geometry with the use of irregular grids was first published by Johnson [70]. In his approach, Greens Theorem is used to obtain finite difference expressions for the displacement derivatives which are required in computing the energy function. However, this application was restricted to the static analysis of shells only.

The STAGS computer program developed by Bushnell [79] provides the capability of performing a free vibration analysis on shells of arbitrary geometry. Although the program permits variable mesh spacing, the mesh lines are required to be parallel to the coordinate lines thereby restricting its use to a particular range of problems.

The development of a technique which is capable to express the partial derviatives as finite differences based on a curvilinear coordinate system and to make the finite difference method readily programmable was suggested by Lau in 1977[74].This technique was initially applied to develope the Cell collocation method for solving continuum mechanics problems by Lau and Brebbia [112].In the process of extending the applications to three dimensional potential equations[113], biharmonic equations[114] and plate bending problems[74], the name Curvilinear Finite Differences was used by one of the authors to denote such a finite difference technique.

Recently, Kwok [71,72] combined the Curvilinear Finite Difference (CFD) method introduced by Lau [74] with the general strain-displacement equations of Flugge [1]. The resulting methodology was capable of performing a linear and non-linear static analysis of general thin shells. Its application to non-shallow shells highlighted the importance of determining accurately the geometric properties of a surface and applying the complete strain-displacement equations to model the shell's stress-strain characteristics.However, the application examples covered caese in which the curvilinear grids were nearly regular.

Knowledge of the dynamic characteristics of thin shell structures becomes an important consideration in establishing their performance under various loading conditions. In applying numerical procedures to structures, it has been common practice by many researchers [42,54,56] to present a set of eigenvalues as a means of verifying a techniques accuracy. However, as illustrated in examples by Martins and Owen [81] and Weingarten et al [82] the corresponding eigenvectors may not necessarily be accurately represented.

# 2. GENERAL THEORY

This chapter deals with the mathematical formulation of the theory of general thin shells.

A shell may be defined as a three dimensional body that is bounded by two closely spaced surfaces. The shell theories of Flugge [1], Donnell [2], Love [3], Mushtari [4], Naghdi [5], Timoshenko [6], Reissner [7] and Vlasov [8] attempt to summarize the three dimensional behaviour of the shell in terms of the deformation characteristics of its middle surface. Differences arising from each of the theories are the result of the approximations employed to render the governing three dimensional equations into a two dimensional form. A discussion of the various approximations employed and their effects are detailed in the works of Leissa [9] and Kraus [10].

Recently, a number of authors [83,84] have highlighted the importance of formulating the governing strain-displacement equations within the framework of tensor calculus. Based on such an approach, the governing equations would be valid for any coordinate system within the reference surface of the shell. Furthermore, the use of tensors enables many long and complicated formulae to be written in a simple and compact form which is ideal for numerical computation work.

The following chapter begins by defining the tensor notation and a number of fundamental relationships which are essential in the development of the shell theory presented in the proposed method. In section 2.2, these expressions are applied to the theory of curved surfaces.

Sections 2.3 and 2.4 deal with the strain-displacement equations and the definition of stress within a continuous three dimensional body. For materials which are isotropic in nature, section 2.5 contains the development of the stress-strain relationships in tensor form.

Finally, the strain-displacement equations of general thin shells are derived by considering the deformation of a surface which is located a distance z along a normal coordinate above the middle surface of the shell. This development is presented in section 2.6.2. The resultant expressions are initially three dimensional in nature and by applying the approximations introduced by Love [3] it becomes possible to reduce them to a two dimensional form. Interpretation of these results are presented in section 2.6.5 and the final equations are found to be identical to those derived by Flugge[1].

Throughout the development of the proposed method, all Latin indices (subscripts or superscripts) take the values 1, 2, 3 and Greek indices (subscripts or superscripts) have the range 1, 2. Over a repeated index, (one superscript and one subscript) the usual summation convention is employed. Furthermore, a comma denotes partial derivatives with respect to the coordinates and a single vertical line ( | ) is used to symbolize the covariant derivative.

## 2.1 A SUMMARY OF THE TENSORIAL QUANTITIES REQUIRED IN THE FORMULATION OF A SHELL THEORY

### 2.1.1 Base Vectors

Consider a point P and its associated coordinate system as illustrated in figure 2.1 below.

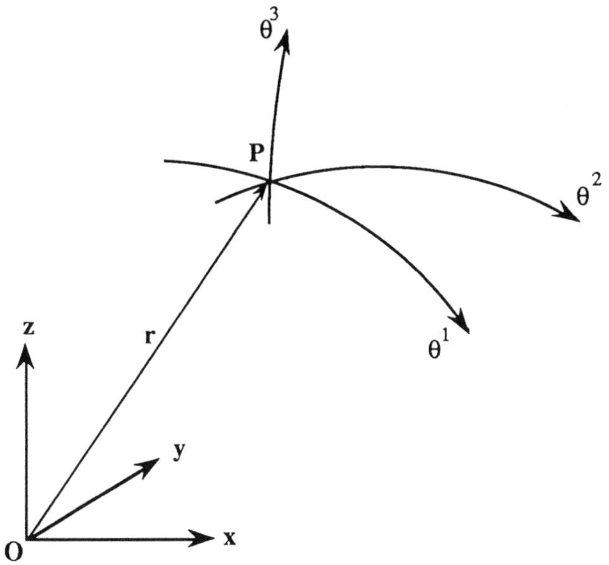


Figure 2.1: The curvilinear coordinate system.

The position vector **r** leading from a fixed point O to the point P can be expressed as a function of the general coordinates $\theta^i$.

$$\mathbf{r} = \mathbf{r}(\theta^1, \theta^2, \theta^3) \tag{2.1.1}$$

The covariant base vectors $\mathbf{g}_i$ characterize the change in the position vector along one of the coordinate curves from the given point P. Mathematically, they can be defined by the relationship:

$$g_i = \frac{\partial r}{\partial \theta^i} \quad (2.1.2)$$

These vectors are directed tangentially along the coordinate curves and may be applied to all vectors associated with the point P.

Equation 2.1.2 assumes the position vector to be differentiable and uniquely defined at each point. This condition is of particular importance in the numerical analysis of shells where the continuum is discretized into a finite number of elements and the displacement behaviour of each element is specified in terms of a local coordinate system.

A set of vectors, referred to as contravariant base vectors, $g^i$, can be defined such that they are normal to all the covariant base vectors.

i.e.
$$g_i \cdot g^j = \delta^j_i \quad (2.1.3)$$

where $\delta^j_i$ is termed the Kronecker delta and has the following values.

$$\delta^j_i = \begin{cases} 0 & (i \neq j) \\ 1 & (i = j) \end{cases} \quad (2.1.4)$$

### 2.1.2 Metric Tensors

The metric tensor can be defined by considering the elementary distance between two adjacent points. Let P be a point whose coordinates are $x^a$ and Q a neighbouring point with coordinates $x^a + dx^a$ (figure 2.2).

If dr defines the infinitesimal vector PQ, then

$$dr = g_i \, dx^i \quad (2.1.5)$$

The magnitude of the vector **dr**, conventionally referred to as the line element ds, can be computed by dot multiplying equation 2.1.5 by itself.

i.e.
$$ds^2 = \mathbf{dr} \cdot \mathbf{dr} = \mathbf{g}_i \, dx^i \cdot \mathbf{g}_j \, dx^j \qquad (2.1.6)$$

Defining
$$g_{ij} = \mathbf{g}_i \cdot \mathbf{g}_j \qquad (2.1.7)$$

then equation 2.1.6 becomes

$$ds^2 = g_{ij} \, dx^i \, dx^j \qquad (2.1.8)$$

The quantities $g_{ij}$ are the components of a symmetric tensor referred to as the metric or fundamental tensor.

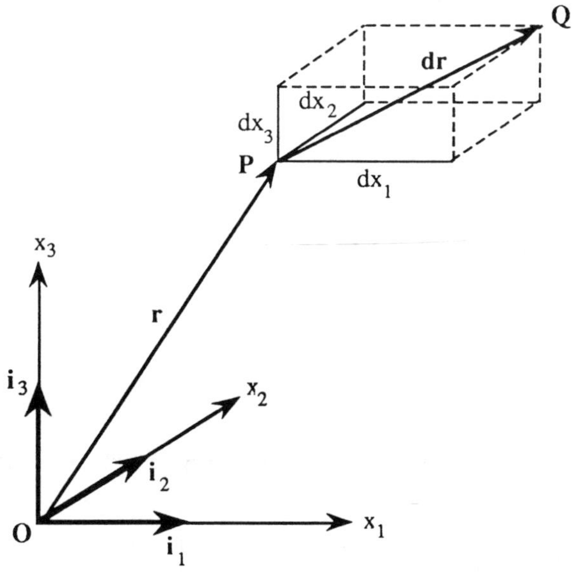


Figure 2.2: Position vector and its differential.

By dot multiplying both sides of equation 2.1.7 with the base vector $\mathbf{g}^j$ the metric tensor can be used to relate the covariant and contravariant base vectors.

$$\mathbf{g}_i = g_{ij}\,\mathbf{g}^j \tag{2.1.9}$$

Similarly, the contravariant components of the metric tensor can be defined by the dot product of two contravariant base vectors and expressions analogous to equations 2.1.7 and 2.1.9 can be constructed:

$$g^{ij} = \mathbf{g}^i \cdot \mathbf{g}^j \tag{2.1.10}$$

$$\mathbf{g}^i = g^{ij}\,\mathbf{g}_j \tag{2.1.11}$$

An important relationship which is often used in the numerical computation of the contravariant components of the metric tensor can be derived by dot multiplying both sides of equation 2.1.11 by $\mathbf{g}_k$:

$$\mathbf{g}^i \cdot \mathbf{g}_k = g^{ij}\,\mathbf{g}_j \cdot \mathbf{g}_k \tag{2.1.12}$$

Substituting equations 2.1.3 and 2.1.7 into the above relationship yields

$$g^{ik}\,g_{kj} = \delta^i_j \tag{2.1.13}$$

which can be written in matrix form to represent a system of nine equations:

$$\begin{bmatrix} g_{11} & g_{12} & g_{13} \\ g_{21} & g_{22} & g_{23} \\ g_{31} & g_{32} & g_{33} \end{bmatrix} \begin{bmatrix} g^{11} & g^{12} & g^{13} \\ g^{21} & g^{22} & g^{23} \\ g^{31} & g^{32} & g^{33} \end{bmatrix} = \begin{bmatrix} 1 & 0 & 0 \\ 0 & 1 & 0 \\ 0 & 0 & 1 \end{bmatrix} \tag{2.1.14}$$

Having computed the covariant components of the metric tensor and assembled them in the form required by equation 2.1.14 it becomes possible to compute the contravariant components by simply inverting the covariant component matrix. Provided two independent base vectors are selected an inverse will always exists.

### 2.1.3 Coordinate Transformations

The ability to establish a set of transformation equations between two coordinate systems provides a valuable tool in the numerical analysis of shell structures. Consider a set of two different curvilinear coordinate systems $\theta^i$, $\theta^{i'}$ and assume a relationship between them exists in the form

$$\theta^i = \theta^{i'}(\theta^{1'}, \theta^{2'}, \theta^{3'}) \tag{2.1.15}$$

If the above transformation is reversible and possesses as many derivatives as required, the inverse can be written in the form

$$\theta^{i'} = \theta^i(\theta^1, \theta^2, \theta^3). \tag{2.1.16}$$

From equations 2.1.15 and 2.1.16, the transformation derivatives $d\theta^i$ and $d\theta^{i'}$ are

$$d\theta^i = \frac{\partial\theta^i}{\partial\theta^{j'}} d\theta^{j'} \tag{2.1.17}$$

$$d\theta^{i'} = \frac{\partial\theta^{i'}}{\partial\theta^j} d\theta^j \tag{2.1.18}$$

which can be combined to form the relationship

$$\delta^i_j = \frac{\partial\theta^i}{\partial\theta^{k'}} \frac{\partial\theta^{k'}}{\partial\theta^j} \tag{2.1.19}$$

In the proposed method, the transformation coefficients introduced in equations 2.1.17 and 2.1.18 are denoted by the symbol β. Hence,

$$\beta^{i}_{j'} = \frac{\partial \theta^{i}}{\partial \theta^{j'}} \tag{2.1.20}$$

$$\beta^{j'}_{i} = \frac{\partial \theta^{j'}}{\partial \theta^{i}} \tag{2.1.21}$$

Equation 2.1.19 can therefore be written in the form:

$$\beta^{k'}_{i}\,\beta^{j}_{k'} = \delta^{j}_{i} \tag{2.1.22}$$

Consider an arbitrary vector **v** at a point P and define $v^{i} = (v^{1}, v^{2}, v^{3})$ and $v^{i'} = (v^{1'}, v^{2'}, v^{3'})$ as the contravariant components of **v** with coordinates $\theta^{i}$ and $\theta^{i'}$ respectively. Using the definitions introduced by equations 2.1.20 and 2.1.21, the following transformations can be written:

$$v^{i} = \beta^{i}_{j'}\, v^{j'} \tag{2.1.23}$$

$$v^{j'} = \beta^{j'}_{i}\, v^{i} \tag{2.1.24}$$

Similarly, the relationships between the covariant components are:

$$v_{i} = \beta^{j'}_{i}\, v_{j'} \tag{2.1.25}$$

$$v_{j'} = \beta^{i}_{j'}\, v_{i} \tag{2.1.26}$$

Now, the displacement vector v, can be defined with respect to either of the coordinate systems. i.e.,

$$\mathbf{v} = v^{i}\,\mathbf{g}_{i} = v^{i'}\,\mathbf{g}_{i'} \qquad (2.1.27)$$

Dot multiplying both sides of equation 2.1.27 by $\mathbf{g}^{k}$,

$$v^{k} = v^{i'}\,\mathbf{g}_{i'} \cdot \mathbf{g}^{k} \qquad (2.1.28)$$

Comparing equations 2.1.23 and 2.1.28 enables the transformation tensor $\beta^{i}_{j'}$ to be defined alternatively as:

$$\beta^{i}_{j'} = \mathbf{g}_{j'} \cdot \mathbf{g}^{i} \qquad (2.1.29)$$

By applying the chain rule for partial differentiation, the concepts discussed above can be extended to the transformation of partial derivatives. Therefore, if the transformation $\beta^{j'}_{i}$ can be defined explicitly, equation 2.1.25 can be differentiated to form the following transformations.

$$v_{i',j} = \beta^{m'}_{i\ ,j}\, v_{m'} + \beta^{m'}_{i}\, v_{m',j} \qquad (2.1.30)$$

$$v_{i',jk} = \beta^{m'}_{i\ ,jk}\, v_{m'} + \beta^{m'}_{i\ ,j}\, v_{m',k} + \beta^{m'}_{i\ ,k}\, v_{m',j} + \beta^{m'}_{i}\, v_{m',jk} \qquad (2.1.31)$$

### 2.1.4 Christoffel Symbols

When a base vector is differentiated with respect to the coordinates $\theta^i$, the resulting expressions can be written as components of either a contravariant or covariant base vector. i.e.,

$$\mathbf{g}_{i,j} = \Gamma^k_{ij}\, \mathbf{g}_k = \Gamma_{ijk}\, \mathbf{g}^k \qquad (2.1.32)$$

By dot multiplying equation 2.1.32 with $\mathbf{g}^l$ or $\mathbf{g}_l$, the symbols $\Gamma^k_{ij}$ and $\Gamma_{ijk}$ can be isolated.

$$\Gamma_{ijk} = \mathbf{g}_{i,j} \cdot \mathbf{g}_k \qquad (2.1.33)$$

$$\Gamma^k_{ij} = \mathbf{g}_{i,j} \cdot \mathbf{g}^k \qquad (2.1.34)$$

Equations 2.1.33 and 2.1.34 define the Christoffel symbols of the first and second kind respectively.

The third index of the Christoffel symbol can be raised and lowered by using the metric tensor:

$$\Gamma_{ijk} = \Gamma^l_{ij}\, g_{lk} \qquad (2.1.35)$$

$$\Gamma^k_{ij} = \Gamma_{ijl}\, g^{lk} \qquad (2.1.36)$$

In addition, differentiating equation 2.1.1, yields

$$\mathbf{r}_{,ij} = \mathbf{g}_{i,j} = \mathbf{r}_{,ji} = \mathbf{g}_{j,i} \qquad (2.1.37)$$

which, combined with equation 2.1.32 shows that the Christoffel symbols are symmetric with respect to the first two subscripts.

$$\Gamma_{ijk} = \Gamma_{jik} \tag{2.1.38}$$

$$\Gamma^{k}_{ij} = \Gamma^{k}_{ji} \tag{2.1.39}$$

In determining the tensorial quantities associated with a coordinate system it is usually possible to obtain explicit expressions for the covariant components of the metric tensor. These expressions can be differentiated and combined with equation 2.1.32 to yield an alternative form for the definition of the Christoffel symbol:

$$\Gamma_{ijk} = \frac{1}{2}(g_{jk},_{i} + g_{ki},_{j} - g_{ij},_{k}) \tag{2.1.40}$$

### 2.1.5 Covariant Derivatives

Let **v** be an arbitrary vector defined in terms of its contravariant components:

$$\mathbf{v} = v^{i}\mathbf{g}_{i} \tag{2.1.41}$$

Differentiating equation 2.1.41 yields

$$\mathbf{v},_{j} = v^{i},_{j}\mathbf{g}_{i} + v^{i}\mathbf{g}_{i},_{j}$$

which can be combined with equation 2.1.32 to form

$$\mathbf{v}_{,j} = (v^i_{,j} + v^k\Gamma^i_{kj})\,\mathbf{g}_i \tag{2.1.42}$$

Defining
$$v^i|_j = (v^i_{,j} + v^k\Gamma^i_{kj}) \tag{2.1.43}$$

enables equation 2.1.42 to be simplified to

$$\mathbf{v}_{,j} = v^i|_j\,\mathbf{g}_i \tag{2.1.44}$$

The quantity $v^i|_j$ is called the covariant derivative of the vector $v^i$ and is a second order tensor. By adopting a similar approach, the covariant derivative of the vector $v_i$ can be expressed in the form

$$v_i|_j = (v_{i,j} - v_k\Gamma^k_{ij}) \tag{2.1.45}$$

Since $v^i|_j$ and $v_i|_j$ are the components of a tensor, the index i can be raised and lowered in the conventional manner. i.e.,

$$v_i|_j = v^k|_j\,g_{ki} \tag{2.1.46}$$

$$v^i|_j = v_k|_j\,g^{ki} \tag{2.1.47}$$

The above concepts can be extended to obtain expressions for the covariant derivatives of higher order tensors. Of particular interest are the equations describing the covariant derivatives of second order tensors. These are used in the formulation of a number of geometric properties related to a surface's curvature.

Let $\phi$ be a scalar expressed as the product of a second order tensor $A_{ij}$ with two arbitrary vectors $u^i$ and $v^j$:

$$\phi = A_{ij}\, u^i\, v^j \qquad (2.1.48)$$

Differentiating $\phi$ with respect to $x^k$ yields

$$\phi,_k = A_{ij},_k\, u^i\, v^j + A_{ij}\, u^i,_k\, v^j + A_{ij}\, u^i\, v^j,_k \qquad (2.1.49)$$

Similarly the covariant derivative of equation 2.1.48 may be written in the form

$$\phi|_k = A_{ij}|_k\, u^i\, v^j + A_{ij}\, u^i|_k\, v^j + A_{ij}\, u^i\, v^j|_k \qquad (2.1.50)$$

Using equation 2.1.43,

$$\phi,_k = A_{ij},_k\, u^i\, v^j + A_{ij}\, u^i|_k\, v^j + A_{ij}\, u^i\, v^j|_k - A_{ij}\, u^l\, \Gamma^i_{kl} v^j - A_{ij}\, u^i\, v^l\, \Gamma^j_{kl}$$

(2.1.51)

and since $\phi$ is a scalar , $\phi,_k = \phi|_k$ , enabling equations 2.1.50 and 2.1.51 to be used to define the covariant derivative of $A_{ij}$ as

$$A_{ij}|_k = A_{ij},_k - A_{lj}\, \Gamma^l_{ik} - A_{il}\, \Gamma^l_{kj} \qquad (2.1.52a)$$

In a similar way, relationships can be derived for the contravariant and mixed components of the tensor A:

$$A^i_{\ j}|_k = A^i_{\ j},_k + A^l_{\ j}\, \Gamma^i_{kl} - A^i_{\ l}\, \Gamma^l_{jk} \qquad (2.1.52b)$$

$$A_i^{\ j}|_k = A_i^{\ j},_k - A_l^{\ j}\, \Gamma^l_{ik} + A_i^{\ l}\, \Gamma^j_{kl} \qquad (2.1.52c)$$

$$A^{ij}|_k = A^{ij},_k + A^{lj}\, \Gamma^i_{kl} + A^{il}\, \Gamma^j_{kl} \qquad (2.1.52d)$$

## 2.2 SURFACE GEOMETRY

### 2.2.1 Curvilinear Coordinates On a Surface

A surface can be defined as the locus of a point whose coordinates are functions of two independent parameters. Its equation can be expressed in the form:

$$x^i = x^i(x^\alpha, x^\beta) \tag{2.2.1}$$

The variables $x^\alpha$ constitute a system of curvilinear coordinates while the curves on a surface along which one parameter remains constant are called parametric curves.

In developing a shell theory, an attempt is made to describe the overall behaviour of the shell in terms of the deformation of its middle surface. Consequently a three dimensional coordinate system consisting of two curvilinear coordinates $x^\alpha$ on the middle surface and a third coordinate $x^3$ which is normal to $x^\alpha$ is introduced. It also becomes advantageous to adopt separate notations for tensorial quantities related to points lying on the middle surface and those which are within its immediate vicinity. The notations are summarised in Table 2.1 and figure 2.3 .

### 2.2.2 Geometry of a Curved Surface

In previous sections, quantities often used in the study of elasticity problems have been described in tensor notation. By adopting the monoclinic reference frame introduced above, these relationships can be simplified and written in a form which relates the tensorial quantities of points adjacent to the surface to the corresponding quantities associated with points on the surface.

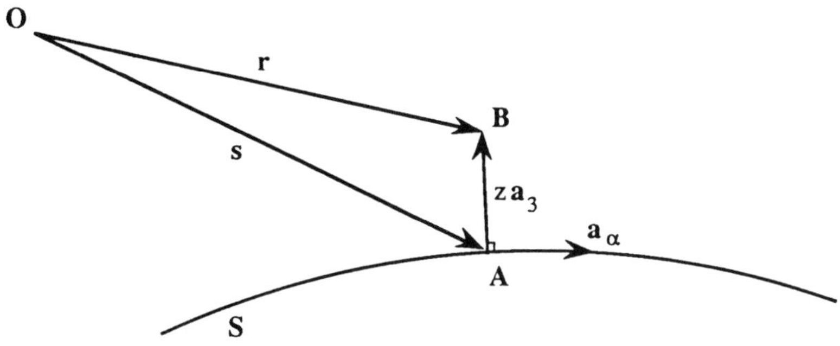


Figure 2.3: Position vector of a surface.

| | Middle Surface | General Surface |
|---|---|---|
| *Geometry* | | |
| position vector | $\mathbf{s}$ | $\mathbf{r}$ |
| line element | $d\mathbf{s}$ | $d\mathbf{r}$ |
| base vectors | $\mathbf{a}_\alpha, \mathbf{a}^\alpha, \mathbf{a}_3 = \mathbf{a}^3$ | $\mathbf{g}_\alpha, \mathbf{g}^\alpha, \mathbf{g}_3 = \mathbf{g}^3$ |
| metric tensors | $a_{\alpha\beta}, a^{\alpha\beta}$ | $g_{\alpha\beta}, g^{\alpha\beta}$ |
| Christoffel symbols | $\Gamma^\gamma_{\beta\alpha}$ | $\bar{\Gamma}^\gamma_{\beta\alpha}$ |
| permutation tensor | $\in_{\alpha\beta}, \in^{\alpha\beta}$ | $\bar{\in}_{\alpha\beta}, \bar{\in}^{\alpha\beta}$ |
| | | |
| *Deformation* | | |
| displacement | $\mathbf{u} = u_\alpha \mathbf{a}^\alpha + u_3 \mathbf{a}^3$ | $\mathbf{v} = v_\alpha \mathbf{g}^\alpha + v_3 \mathbf{g}^3$ |
| strain tensor | $\varepsilon_{ij}, \varepsilon_{\alpha\beta}$ | $\eta_{ij}, \eta_{\alpha\beta}$ |

Table 2.1: Middle and general surface notation.

From figure 2.3, the position vector **r** of a point B can be expressed in the form

$$\mathbf{r} = \mathbf{s} + z\mathbf{a}_3 \tag{2.2.2}$$

where **s** is a function of $x^\alpha$ and $\mathbf{a}_3$ is a unit vector normal to the surface S.

Using equations 2.1.2 and 2.2.2 the in-surface base vectors may be defined as

$$\mathbf{a}_\alpha = \mathbf{s},_\alpha \tag{2.2.3}$$

$$\mathbf{g}_\alpha = \mathbf{r},_\alpha = \mathbf{a}_\alpha + z\mathbf{a}_{3,\alpha} \tag{2.2.4}$$

Since $\mathbf{g}_3$ is assumed to be normal to the vectors $\mathbf{g}_\alpha$ and $\mathbf{g}_\beta$ it follows that

$$\mathbf{g}_3 = \frac{\mathbf{g}_\alpha \times \mathbf{g}_\beta}{\| \mathbf{g}_\alpha \times \mathbf{g}_\beta \|} \tag{2.2.5}$$

and

$$\mathbf{g}_3 \cdot \mathbf{g}_3 = 1 \tag{2.2.6a}$$

$$\mathbf{g}_\alpha \cdot \mathbf{g}_3 = g_{\alpha 3} = 0 \tag{2.2.6b}$$

$$\mathbf{g}_{3,\alpha} \cdot \mathbf{g}_3 = 0 \tag{2.2.6c}$$

$$\mathbf{g}_{3,3} \cdot \mathbf{g}_3 = 0 \tag{2.2.6d}$$

As a result, the metric tensor components expressed in matrix form are:

$$g_{ij} = \begin{bmatrix} g_{11} & g_{12} & 0 \\ g_{21} & g_{22} & 0 \\ 0 & 0 & 1 \end{bmatrix} . \tag{2.2.7}$$

The Christoffel symbol $\Gamma_{ijk}$ is defined previously by equation 2.1.33. It follows from equations 2.2.6 c & d that

$$\bar{\Gamma}_{3\alpha 3} = \bar{\Gamma}_{\alpha 33} = \bar{\Gamma}_{33\alpha} = \bar{\Gamma}_{333} = 0 \tag{2.2.8}$$

In addition, by differentiating equation 2.2.6b

$$\mathbf{g}_{\alpha,\beta} \cdot \mathbf{g}_3 + \mathbf{g}_{3,\beta} \cdot \mathbf{g}_\alpha = 0 \tag{2.2.9}$$

which after combining with 2.1.33 yields

$$\bar{\Gamma}_{\alpha\beta 3} = -\bar{\Gamma}_{3\beta\alpha} = -\bar{\Gamma}_{\beta 3\alpha} = \bar{\Gamma}_{\beta\alpha 3} = -\bar{\Gamma}_{3\alpha\beta} = -\bar{\Gamma}_{\alpha 3\beta} \tag{2.2.10}$$

In addition, equation 2.1.36 may be combined with equations 2.2.7 and 2.2.8 to simplify a number of the Christoffel symbols of the second kind:

$$\bar{\Gamma}^3_{3\alpha} = \bar{\Gamma}^\alpha_{33} = \bar{\Gamma}^3_{33} = 0 \tag{2.2.11}$$

These results can be substituted into equations 2.1.43 and 2.1.45 to arrive at the following relationships for the covariant derivatives of a vector component $v_\alpha$:

$$v_\alpha|_\beta = v_{\alpha,\beta} - v_\gamma \bar{\Gamma}^\gamma_{\alpha\beta} - v_3 \bar{\Gamma}^3_{\alpha\beta} \tag{2.2.12a}$$

$$v_\alpha|_3 = v_{\alpha,3} - v_\gamma \bar{\Gamma}^\gamma_{\alpha 3} \tag{2.2.12b}$$

$$v_3|_\alpha = v_{3,\alpha} - v_\gamma \bar{\Gamma}^\gamma_{3\alpha} \tag{2.2.12c}$$

$$v^\alpha|_\beta = v^\alpha{}_{,\beta} + v^\gamma \bar{\Gamma}^\alpha_{\beta\gamma} + v^3 \bar{\Gamma}^\alpha_{\beta 3} \tag{2.2.13a}$$

$$v^{\alpha}|_3 = v^{\alpha}{}_{,3} + v^{\gamma}\overline{\Gamma}^{\alpha}_{\gamma 3} \tag{2.2.13b}$$

$$v^{3}|_{\alpha} = v^{3}{}_{,\alpha} + v^{\gamma}\overline{\Gamma}^{3}_{\alpha\gamma} \tag{2.2.13c}$$

In section 2.1.4, the Christoffel symbol is introduced to represent the components of the derivative of a base vector. For the middle surface z = 0, the third component of equation 2.1.32 may be expanded and written in the form

$$\mathbf{a}_{3,\alpha} = \Gamma_{3\alpha\beta}\,\mathbf{a}^{\beta} + \Gamma_{3\alpha 3}\,\mathbf{a}^{3}$$

Using equations 2.2.8 and 2.2.10, the above expression can be simplified to

$$\mathbf{a}_{3,\alpha} = -\Gamma_{\alpha\beta 3}\,\mathbf{a}^{\beta} \tag{2.2.14}$$

The term $\Gamma_{\alpha\beta 3}$ is a measure of the rate of change of the base vector $\mathbf{a}_3$ along a parametric curve. $\Gamma_{\alpha\beta 3}$ is a plane symmetric tensor and is expressed by the symbol $b_{\alpha\beta}$ which is referred to as the curvature tensor. Thus equation 2.2.14 can be rewritten as

$$b_{\alpha\beta} = -\,\mathbf{a}_{3,\alpha} \cdot \mathbf{a}_{\beta} \tag{2.2.15}$$

Alternatively, if equation 2.2.9 is restricted to the surface z = 0, the curvature tensor may be defined by the expression

$$\mathbf{a}_{\alpha,\beta} \cdot \mathbf{a}_3 = -\,\mathbf{a}_{3,\beta} \cdot \mathbf{a}_{\alpha} = b_{\alpha\beta} \tag{2.2.16}$$

The above formula is simple and particularly useful in the numerical computation of the curvature tensor where it is often possible to express the base vectors $\mathbf{a}_{\alpha}$ as functions of the curvilinear coordinates $x^{\alpha}$.

The mixed and contravariant components of the curvature tensor can be derived using the metric tensor:

$$b^{\beta}_{\alpha} = b_{\alpha\gamma}\, a^{\gamma\beta} \tag{2.2.17a}$$

$$b^{\alpha\beta} = b^{\alpha}_{\gamma}\, a^{\gamma\beta} \tag{2.2.17b}$$

Consider a vector $\mathbf{v}$ which is defined in the form

$$\mathbf{v} = v_{\alpha}\, \mathbf{a}^{\alpha} + v_3\, \mathbf{a}^3 \tag{2.2.18}$$

Taking the derivative of $\mathbf{v}$ with respect to $x^{\beta}$ and making use of equations 2.1.44 and 2.2.12

$$\begin{aligned} \mathbf{v}_{,\beta} &= v_{\alpha}|_{\beta}\, \mathbf{a}^{\alpha} + v_3|_{\beta}\, \mathbf{a}^3 \\ &= ( v_{\alpha,\beta} - v_{\gamma}\Gamma^{\gamma}_{\alpha\beta} - v_3\, b_{\alpha\beta} )\, \mathbf{a}^{\alpha} + ( v_{3,\beta} + v_{\gamma} b^{\gamma}_{\beta} )\, \mathbf{a}^3 \end{aligned} \tag{2.2.19}$$

Introducing the two dimensional counterpart to the covariant derivative defined by equation 2.1.45

$$v_{\alpha}\|_{\beta} = v_{\alpha,\beta} - v_{\gamma}\Gamma^{\gamma}_{\alpha\beta} \tag{2.2.20}$$

This enables equation 2.2.19 to be simplified yielding

$$\mathbf{v}_{,\beta} = ( v_{\alpha}\|_{\beta} - v_3\, b_{\alpha\beta} )\, \mathbf{a}^{\alpha} + ( v_{3,\beta} + v_{\gamma} b^{\gamma}_{\beta} )\, \mathbf{a}^3 \tag{2.2.21}$$

The covariant derivative of $v_{\alpha}$ can therefore be written in the form

$$v_{\alpha}|_{\beta} = v_{\alpha}\|_{\beta} - v_3\, b_{\alpha\beta} \tag{2.2.22}$$

Similarly, if the vector **v** is defined in terms of its contravariant components the following relationships can be derived:

$$v^{\alpha}\|_{\beta} = v^{\alpha}{}_{,\beta} + v^{\gamma}\Gamma^{\alpha}_{\beta\gamma} \qquad (2.2.23)$$

$$v^{\alpha}|_{\beta} = v^{\alpha}\|_{\beta} - v^{3}b^{\alpha}_{\beta} \qquad (2.2.24)$$

$$\mathbf{v}_{,\beta} = (v^{\alpha}\|_{\beta} - v^{3}b^{\alpha}_{\beta})\,\mathbf{a}_{\alpha} + (v^{3}{}_{,\beta} + v^{\gamma}b_{\gamma\beta})\,\mathbf{a}_{3} \qquad (2.2.25)$$

Generally, **v** is also defined for points adjacent to the surface. Hence, the partial derivative of **v** with respect to $x^3$ exists and can be expressed as a function of the contravariant or covariant base vectors as detailed below:

$$\mathbf{v}_{,3} = v_{\alpha}|_{3}\,\mathbf{a}^{\alpha} + v_{3}|_{3}\,\mathbf{a}^{3} = (v_{\alpha,3} + v_{\gamma}b^{\gamma}_{\alpha})\,\mathbf{a}^{\alpha} + v_{3,3}\,\mathbf{a}^{3} \qquad (2.2.26)$$

$$\mathbf{v}_{,3} = v^{\alpha}|_{3}\,\mathbf{a}_{\alpha} + v^{3}|_{3}\,\mathbf{a}_{3} = (v^{\alpha}{}_{,3} - v^{\gamma}b^{\alpha}_{\gamma})\,\mathbf{a}_{\alpha} + v^{3}{}_{,3}\,\mathbf{a}_{3} \qquad (2.2.27)$$

The idea of a two dimensional counterpart for the covariant derivative of surface vectors can be extended to surface tensors of any order. Of particular interest is its application to plane tensors such as the curvature tensor. Since $b_{3\alpha} = b_{\alpha 3} = b_{33} = 0$, equation 2.1.52a can be expanded to yield

$$b_{\alpha\beta}|_{\gamma} = b_{\alpha\beta}\|_{\gamma} = b_{\alpha\beta,\gamma} - b_{\delta\beta}\Gamma^{\delta}_{\alpha\gamma} - b_{\alpha\delta}\Gamma^{\delta}_{\beta\gamma} \qquad (2.2.28)$$

An important group of relationships required in developing a general shell theory in tensor form are the Gauss - Codazzi equations of differential geometry. Briefly, these equations allow the terms $\beta$ and $\gamma$ within equation 2.2.28 to be interchanged and can be derived by comparing the components of the partial derivatives $\mathbf{a}_{\alpha',\beta\gamma}$ and $\mathbf{a}_{\alpha',\gamma\beta}$. Restating equation 2.1.32 in planar form:

$$\mathbf{a}_{\alpha',\beta} = \Gamma^{\delta}_{\alpha\beta}\,\mathbf{a}_{\delta} + b_{\alpha\beta}\,\mathbf{a}_{3} \qquad (2.2.29)$$

Differentiating 2.2.29 with respect to $x^{\gamma}$ and collecting terms,

$$\mathbf{a}_{\alpha',\beta\gamma} = (\Gamma^{\delta}_{\alpha\beta',\gamma} + \Gamma^{\xi}_{\alpha\beta}\,\Gamma^{\delta}_{\xi\gamma} - b_{\alpha\beta}\,b^{\delta}_{\gamma})\,\mathbf{a}_{\delta} + (\Gamma^{\xi}_{\alpha\beta}\,b_{\xi\gamma} + b_{\alpha\beta',\gamma})\,\mathbf{a}_{3}$$

(2.2.30)

Now $\mathbf{a}_{\alpha',\gamma\beta}$ can be obtained by interchanging the indices $\beta$ and $\gamma$. Since $\mathbf{a}_{\alpha',\beta\gamma}$ and $\mathbf{a}_{\alpha',\gamma\beta}$ are equivalent, the $\mathbf{a}_{3}$ components may be compared to yield the relationship

$$b_{\alpha\beta',\gamma} - \Gamma^{\xi}_{\alpha\gamma}\,b_{\xi\beta} = b_{\alpha\gamma',\beta} - \Gamma^{\xi}_{\alpha\beta}\,b_{\xi\gamma} \qquad (2.2.31)$$

Substituting 2.2.31 into equation 2.2.28 enables the standard form of the Gauss - Codazzi equations to be written. i.e.,

$$b_{\alpha\beta}\|_{\gamma} = b_{\alpha\gamma}\|_{\beta} \qquad (2.2.32a)$$

Adopting a similar line of reasoning it may also be concluded that

$$b^{\alpha}_{\beta}\|_{\gamma} = b^{\alpha}_{\gamma}\|_{\beta} \qquad (2.2.32b)$$

## 2.3 THE STRAIN TENSOR

Consider two adjacent points A and B which are within a continuous three dimensional body illustrated by figure 2.4. While the body is in an undeformed state, a curvilinear coordinate system $x^i$ is established and the points A and B are assumed to denote the end points of an element vector **ds**. Using equations 2.1.5 and 2.1.6, the vector **ds** may be written in the form

$$\mathbf{ds} = \mathbf{g}_i \, dx^i$$

and the square of the line element, ds, computed from the relationship

$$ds^2 = \mathbf{ds} \cdot \mathbf{ds} = g_{ij} \, dx^i \, dx^j \qquad (2.3.1)$$

During deformation, point A undergoes the displacement **u** and moves to $\tilde{A}$, while point B experiences a slightly different displacement **u** + **du** when moving to $\tilde{B}$. If the coordinate system is allowed to undergo the same deformation as the body, the deformed vector $\mathbf{d\tilde{s}}$ can be written as

$$\mathbf{d\tilde{s}} = \tilde{\mathbf{g}}_i \, dx^i$$

Similarly, the square of the line element $\tilde{s}$ is

$$d\tilde{s}^2 = \tilde{g}_{ij} \, dx^i \, dx^j \qquad (2.3.2)$$

The degree of deformation is characterized by the difference in the squares of the line elements ds and $d\tilde{s}$. i.e.,

$$d\tilde{s}^2 - ds^2 = ( \tilde{g}_{ij} - g_{ij} ) \, dx^i \, dx^j \qquad (2.3.3)$$

Defining $$\gamma_{ij} = \frac{1}{2}( \tilde{g}_{ij} - g_{ij} ) \qquad (2.3.4)$$

equation 2.3.3 becomes

$$d\tilde{s}^2 - ds^2 = 2\gamma_{ij}\, dx^i\, dx^j \qquad (2.3.5)$$

The quantities $\gamma_{ij}$ are the symmetrical components of a covariant tensor which is called the strain tensor.

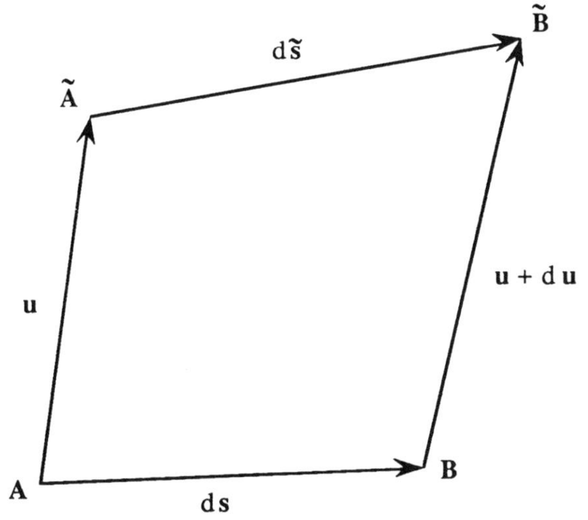


Figure 2.4: Displacement and strain.

It is possible to express the strain components in terms of the displacement vector **u** by examining the deformation of two adjacent points within a body. From figure 2.4, the following vector equation can be constructed:

$$d\tilde{\mathbf{s}} + \mathbf{u} = d\mathbf{s} + \mathbf{u} + d\mathbf{u} \tag{2.3.6}$$

Using equation 2.1.44, the elemental change in displacement, **du**, can be written as

$$d\mathbf{u} = u_k|_j \, \mathbf{g}^k \, dx^j \tag{2.3.7}$$

Substituting 2.3.7 into 2.3.6 yields

$$d\tilde{\mathbf{s}} = d\mathbf{s} + u_k|_j \, \mathbf{g}^k \, dx^j \tag{2.3.8}$$

which allows the deformed line element to be expressed as

$$d\tilde{\mathbf{s}} \cdot d\tilde{\mathbf{s}} = ( \mathbf{g}_i \, dx^i + u_k|_i \, \mathbf{g}^k \, dx^i ) \cdot ( \mathbf{g}_j \, dx^j + u_l|_j \, \mathbf{g}^l \, dx^j )$$

$$= ( g_{ij} + u_i|_j + u_j|_i + u^k|_i \, u_k|_j ) \, dx^i \, dx^j \tag{2.3.9}$$

Through equations 2.3.1, 2.3.3 and 2.3.6, the above expression simplifies to

$$2\gamma_{ij} \, dx^i \, dx^j = ( u_i|_j + u_j|_i + u^k|_i \, u_k|_j ) \, dx^i \, dx^j \tag{2.3.10}$$

The factors $dx^i$ and $dx^j$ cannot simply be cancelled on both sides since these expressions represent the sums of products containing the factors $\gamma_{11}$, $\gamma_{12}$, ... etc. However, equation 2.3.10 holds for the components of any line element vector **ds**. By selecting a line element for which only $dx^1 \neq 0$, it can be shown that

$$\gamma_{ij} = \tfrac{1}{2}( u_i|_j + u_j|_i + u^k|_i \, u_k|_j ) \tag{2.3.11}$$

This process may then be repeated for i, j = 2, 3. For mixed values of the indices, a line element in the form

$$ds = g_1\, dx^1 + g_2\, dx^2$$

is selected and a similar process is adopted to show equation 2.3.11 is valid for all i,j.

The strain - displacement equation defined above is nonlinear as a result of the quadratic term $u^k|_i\, u_k|_j$. It is based solely on the assumption that the body undergoing deformation is continuous. For small displacements, the strains associated with the quadratic terms can be assumed negligible in comparison to the linear ones thereby reducing equation 2.3.11 to

$$\gamma_{ij} = \frac{1}{2}( u_i|_j + u_j|_i ) \qquad (2.3.12)$$

## 2.4 THE STRESS TENSOR

If a three dimensional body is in equilibrium under a system of forces, then the stresses within the body can be studied by considering an elemental area dA of arbitrary size and orientation. At any point within the body, a reference frame and its associated base vectors $g_i$ may be established. Together with the three curves AB, AC and CB illustrated in figure 2.5a, the base vectors may be used to form the edges of a tetrahedron.

The area of the triangle ABC can be computed via the vector cross product:

$$dA = \frac{1}{2} dr \times ds \qquad (2.4.1)$$

It follows from figure 2.5a that

$$dr = db - da \quad \text{and} \quad ds = dc - da.$$

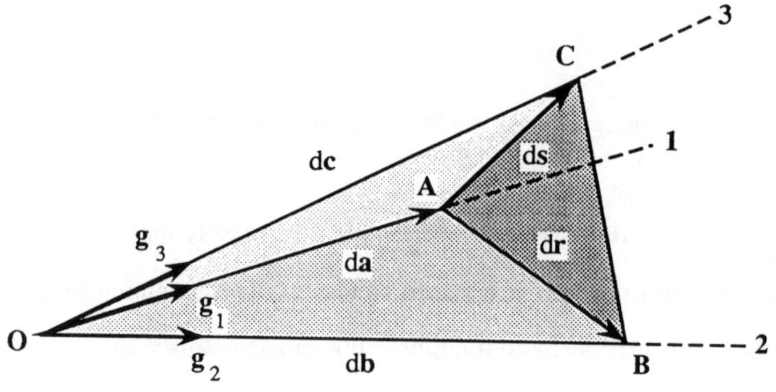


(a)

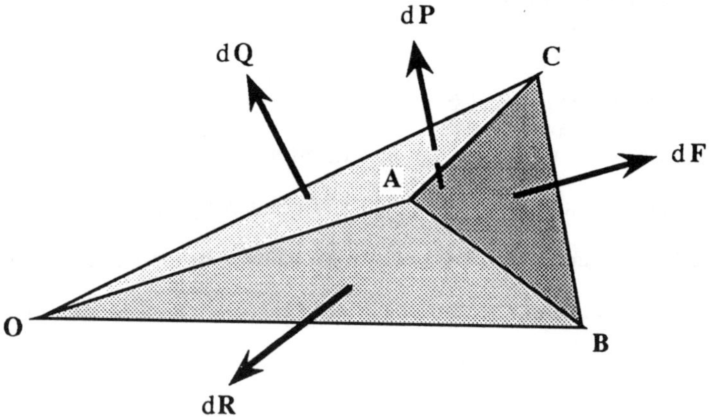


(b)

Figure 2.5: Definition of stress.

Therefore equation 2.4.1 may be expanded to yield

$$dA = \frac{1}{2}(\mathbf{db} - \mathbf{da}) \times (\mathbf{dc} - \mathbf{da})$$

$$= \frac{1}{2}(\mathbf{db} \times \mathbf{dc} + \mathbf{dc} \times \mathbf{da} + \mathbf{da} \times \mathbf{db}) \qquad (2.4.2)$$

The vectors **da**, **db** and **dc** have only one non zero component and can be written in the form

$$\mathbf{da} = da^1\,\mathbf{g}_1\,, \qquad \mathbf{db} = db^2\,\mathbf{g}_2\,, \qquad \mathbf{dc} = dc^3\,\mathbf{g}_3 \qquad (2.4.3)$$

Expanding 2.4.2,

$$dA = \frac{1}{2}(db^2\,dc^3\,\epsilon_{231}\,\mathbf{g}^1 + dc^3\,da^1\,\epsilon_{312}\,\mathbf{g}^2 + da^1\,db^2\,\epsilon_{123}\,\mathbf{g}^3)$$

$$= dA_i\,\mathbf{g}^i \qquad (2.4.4)$$

with $dA_1 = \frac{1}{2}db^2\,dc^3\,\epsilon_{231}$, $dA_2 = \frac{1}{2}dc^3\,da^1\,\epsilon_{312}$, $dA_3 = \frac{1}{2}da^1\,db^2\,\epsilon_{123}$.

In equation 2.4.4, the quantities $\epsilon_{ijk}$ are referred to as the permutation symbols and are defined by the following set of rules:

$\epsilon_{ijk} = \sqrt{g}$ if i, j, k is a cyclic sequence.

$\epsilon_{ijk} = -\sqrt{g}$ if i, j, k is an anticyclic sequence.

$\epsilon_{ijk} = 0$ if i, j, k is an acyclic sequence. (2.4.5)

The term g is determined by arranging the nine components of the metric tensor in a square matrix and then calculating its determinant.

Using a similar procedure, it is also possible to compute the areas associated with the triangles OAB, OAC and OCB:

$$dA_{OAB} = \tfrac{1}{2}(\mathbf{db} \times \mathbf{da}) = \tfrac{1}{2}\, db^2\, da^1\, \epsilon_{213}\, \mathbf{g}^3 = -dA_3\, \mathbf{g}^3 \qquad (2.4.6a)$$

$$dA_{OAC} = \tfrac{1}{2}(\mathbf{da} \times \mathbf{dc}) = \tfrac{1}{2}\, da^1\, dc^3\, \epsilon_{132}\, \mathbf{g}^2 = -dA_2\, \mathbf{g}^2 \qquad (2.4.6b)$$

$$dA_{OCB} = \tfrac{1}{2}(\mathbf{dc} \times \mathbf{db}) = \tfrac{1}{2}\, dc^3\, db^2\, \epsilon_{321}\, \mathbf{g}^1 = -dA_1\, \mathbf{g}^1 \qquad (2.4.6c)$$

Comparing equations 2.4.4 and 2.4.6 it is evident that the normal area associated with the sides of the tetrahedron joined at the point O are the covariant components of **dA**. Having derived these components, the forces **dP**, **dQ** and **dR** may be defined in the form

$$\mathbf{dP} = -\sigma^{1j}\, dA_1\, \mathbf{g}_j \qquad (2.4.7a)$$

$$\mathbf{dQ} = -\sigma^{2j}\, dA_2\, \mathbf{g}_j \qquad (2.4.7b)$$

$$\mathbf{dR} = -\sigma^{3j}\, dA_3\, \mathbf{g}_j \qquad (2.4.7c)$$

The quantities $\sigma^{ij}$ are the contravariant components of a symmetric second order tensor called the stress tensor. It is important to note that the quantities $\sigma^{1j}\, dA_1$, $\sigma^{2j}\, dA_2$ and $\sigma^{3j}\, dA_3$ are not exactly forces since $\mathbf{g}_i$ is generally not a dimensionless unit vector.

From figure 2.5, the equilibrium equation

$$\mathbf{dF} = -\mathbf{dP} - \mathbf{dQ} - \mathbf{dR} \qquad (2.4.8)$$

can be written and combined with equations 2.4.7a - c to yield

$$\mathbf{dF} = (\sigma^{1j}\, dA_1 + \sigma^{2j}\, dA_2 + \sigma^{3j}\, dA_3)\, \mathbf{g}_j = \sigma^{ij}\, dA_i\, \mathbf{g}_j \qquad (2.4.9)$$

This relationship allows the force components $dF^j$ to be expressed in the form

$$dF^j = \sigma^{ij} \, dA_i \tag{2.4.10}$$

## 2.5 THE CONSTITUTIVE EQUATIONS

For an elastic material whose stress-strain behaviour is essentially linear, Hooke's law may be written in the form

$$\sigma^{ij} = E^{ijkl} \varepsilon_{kl} \tag{2.5.1}$$

where $E^{ijkl}$ is a fourth order tensor referred to as the elastic modulus. This system of equations is applicable to a variety of materials possessing either orthotropic, isotropic or anisotropic properties.

In the proposed method, it is assumed that the material has isotropic properties. As a result, many of the components of the elastic modulus are identical and equation 2.5.1 may be written in a more convenient form. To achieve this, the stress-strain properties of an isotropic material expressed with reference to a Cartesian coordinate system are examined. In matrix form,

$$\begin{bmatrix} \sigma^{11} \\ \sigma^{22} \\ \sigma^{33} \\ \sigma^{12} \\ \sigma^{23} \\ \sigma^{31} \end{bmatrix} = \frac{E}{(1+\nu)\,(1-2\nu)} \begin{bmatrix} (1-\nu) & \nu & \nu & 0 & 0 & 0 \\ \nu & (1-\nu) & \nu & 0 & 0 & 0 \\ \nu & \nu & (1-\nu) & 0 & 0 & 0 \\ 0 & 0 & 0 & (1-2\nu) & 0 & 0 \\ 0 & 0 & 0 & 0 & (1-2\nu) & 0 \\ 0 & 0 & 0 & 0 & 0 & (1-2\nu) \end{bmatrix} \begin{bmatrix} \varepsilon_{11} \\ \varepsilon_{22} \\ \varepsilon_{33} \\ \varepsilon_{12} \\ \varepsilon_{23} \\ \varepsilon_{31} \end{bmatrix} \tag{2.5.2}$$

where E and ν refer to Young's modulus and Poisson's ratio respectively.

By using the symmetric properties associated with the stress and strain tensors it becomes possible to summarize the 81 components of the tensor $E^{ijkl}$ in the following form:

$$E^{ijkl} = \frac{E\nu}{(1+\nu)(1-2\nu)}\delta^{ij}\delta^{kl} + \frac{E}{2(1+\nu)}(\delta^{ik}\delta^{jl} + \delta^{il}\delta^{jk}) \qquad (2.5.3)$$

Although this equation is valid only within a Cartesian coordinate system, it can be generalized by replacing the Kronecker deltas with the corresponding components of the metric tensor:

$$E^{ijkl} = \frac{E}{2(1-\nu)}\left(\frac{2\nu}{(1-2\nu)}g^{ij}g^{kl} + g^{ik}g^{jl} + g^{il}g^{jk}\right) \qquad (2.5.4)$$

Equation 2.5.4 can now be combined with 2.5.1 to yield the constitutive equations for a linear elastic material.

$$\sigma^{ij} = \frac{E}{2(1-\nu)}\left(\frac{2\nu}{(1-2\nu)}g^{ij}g^{lm} + g^{il}g^{jm} + g^{im}g^{jl}\right)\varepsilon_{lm} \qquad (2.5.5)$$

The inverse relationship may be computed by firstly expressing equation 2.5.5 in terms of the mixed components $\varepsilon^i_j$ and $\sigma^i_j$. i.e.,

$$\sigma^i_j = \frac{E}{(1+\nu)}\left(\varepsilon^i_j + \frac{\nu}{(1-2\nu)}\varepsilon^m_m\delta^i_j\right) \qquad (2.5.6)$$

Secondly, an expression for $\varepsilon^m_m$ is obtained by letting i = j. Equation 2.5.6 then becomes

$$\sigma^i_i = \frac{E}{(1+\nu)}\left(\varepsilon^i_i + \frac{3\nu}{(1-2\nu)}\varepsilon^m_m\right) = \frac{E}{(1-2\nu)}\varepsilon^m_m$$

which can be substituted into 2.5.6 to yield

$$\varepsilon^i_j = \frac{1}{E}[\,(1+\nu)\,\sigma^i_j \;-\; \nu\,\sigma^m_m\,\delta^i_j\,] \qquad (2.5.7)$$

The application of equations 2.5.5 and 2.5.7 to plane stress systems is of particular importance to the development of a shell theory. When a body is in a state of plane stress, the stresses $\sigma^\alpha_3$ , $\sigma^3_\alpha$ and $\sigma^3_3$ are assumed to be zero. Equation 2.5.7 can therefore be written as

$$E\,\varepsilon^\alpha_\beta = (1+\nu)\,\sigma^\alpha_\beta \;-\; \nu\,\sigma^\xi_\xi\,\delta^\alpha_\beta$$

$$\varepsilon^\alpha_3 = \varepsilon^3_\alpha = 0$$

$$E\varepsilon^3_3 = -\nu\,\sigma^\xi_\xi \qquad (2.5.8)$$

Similarly, the procedure used to invert equation 2.5.5 can be applied to the above expression. The following relationship defining the stress-strain properties of an isotropic material may be developed.

$$\sigma^\alpha_\beta = \frac{E}{(1+\nu)}[\,\varepsilon^\alpha_\beta + \frac{\nu}{(1-\nu)}\varepsilon^\xi_\xi\,\delta^\alpha_\beta\,] \qquad (2.5.9)$$

## 2.6 THE THEORY OF SHELLS

### 2.6.1 Shell Geometry

A shell structure may be defined as a body consisting of an inner surface, called the middle surface, and two outer surfaces which are parallel to each other. The normal distances from the middle to the outer surfaces are assumed to be the same although they need not be constant.

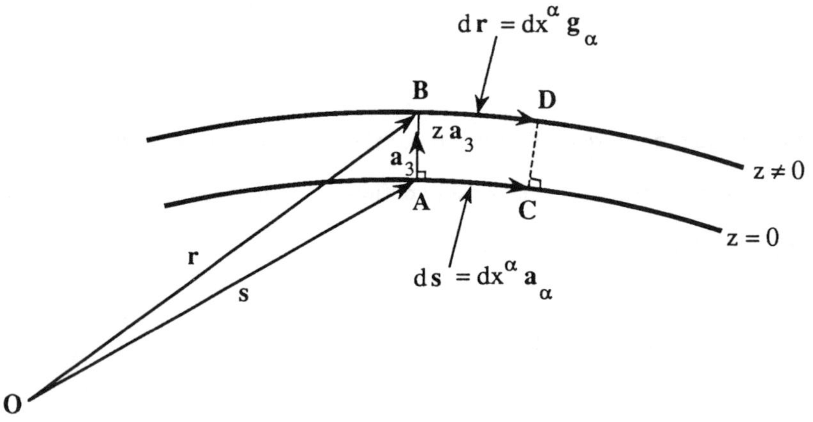


Figure 2.6 Position vector of a point on a shell.

Consider a shell of uniform thickness t as illustrated in figure 2.6 above. Let **s** and **r** denote the position vectors of two points A and B which are located at the same coordinates $x^\alpha$ but are separated by a distance z normal to the middle surface.

From figure 2.6, the vector equation

$$\mathbf{r} = \mathbf{s} + z\,\mathbf{a}_3 \tag{2.6.1}$$

may be written and differentiated to form the expression

$$\mathbf{r}_{,\alpha} = \mathbf{s}_{,\alpha} + z\,\mathbf{a}_{3,\alpha} \tag{2.6.2}$$

By using the definition of a base vector introduced in section 2.2, equation 2.6.2 can be expressed as

$$\mathbf{g}_\alpha = \mathbf{a}_\alpha - z\,b^\beta_\alpha\,\mathbf{a}_\beta = (\,\delta^\beta_\alpha - z\,b^\beta_\alpha\,)\,\mathbf{a}_\beta \tag{2.6.3}$$

The coefficient ( $\delta^\beta_\alpha - z\,b^\beta_\alpha$ ) relates the covariant base vectors associated with the general surface $x^3 = z$ to the middle surface base vectors $\mathbf{a}_\alpha$. It is called the shift tensor and is denoted by the symbol $\mu^\beta_\alpha$. Hence,

$$\mu^\beta_\alpha = (\,\delta^\beta_\alpha - z\,b^\beta_\alpha\,) \tag{2.6.4}$$

and equation 2.6.3 becomes

$$\mathbf{g}_\alpha = \mu^\beta_\alpha\,\mathbf{a}_\beta \tag{2.6.5}$$

Similarly, by introducing a tensor $\lambda^\beta_\alpha$, which relates the contravariant base vectors:

$$\mathbf{g}^\beta = \lambda^\beta_\alpha\,\mathbf{a}^\alpha \tag{2.6.6}$$

Applying equation 2.6.5 and 2.6.6 to 2.1.3 yields

$$\mathbf{g}^\alpha \cdot \mathbf{g}_\beta = \lambda^\alpha_\gamma\,\mathbf{a}^\gamma \cdot \mu^\delta_\beta\,\mathbf{a}_\delta = \lambda^\alpha_\gamma\,\mu^\gamma_\beta$$

from which it can be concluded that

$$\delta^{\alpha}_{\beta} = \lambda^{\alpha}_{\gamma}\mu^{\gamma}_{\beta} \tag{2.6.7}$$

The approach adopted in section 2.1.2 for the computation of the contravariant components of the metric tensor can also be applied to the above expression. Equation 2.6.7 represents a system of four equations and if expanded, may be written in the form

$$\begin{bmatrix} \lambda^{\alpha}_{\alpha} & \lambda^{\alpha}_{\beta} \\ \lambda^{\beta}_{\alpha} & \lambda^{\beta}_{\beta} \end{bmatrix} \begin{bmatrix} \mu^{\alpha}_{\alpha} & \mu^{\alpha}_{\beta} \\ \mu^{\beta}_{\alpha} & \mu^{\beta}_{\beta} \end{bmatrix} = \begin{bmatrix} 1 & 0 \\ 0 & 1 \end{bmatrix} \tag{2.6.8}$$

Having computed the tensor components $\mu^{\alpha}_{\beta}$ and assembled them in the form required by 2.6.8, $\lambda^{\alpha}_{\beta}$ can be computed by simply inverting the matrix containing the shift tensor components $\mu^{\alpha}_{\beta}$.

Similarly, the metric tensors $g_{\alpha\beta}$ and $g^{\alpha\beta}$ at the point B (see figure 2.6) may be expressed in terms of the corresponding quantities $a_{\alpha\beta}$ and $a^{\alpha\beta}$ at the point A. Starting with the definition of the metric tensor given by equation 2.1.7 and combining it with 2.6.5 and 2.6.6:

$$g_{\alpha\beta} = \mathbf{g}_{\alpha}\cdot\mathbf{g}_{\beta} = \mu^{\gamma}_{\alpha}\mathbf{a}_{\gamma}\cdot\mu^{\delta}_{\beta}\mathbf{a}_{\delta} = \mu^{\gamma}_{\alpha}\mu^{\delta}_{\beta}a_{\gamma\delta} \tag{2.6.9a}$$

$$g^{\alpha\beta} = \mathbf{g}^{\alpha}\cdot\mathbf{g}^{\beta} = \lambda^{\alpha}_{\gamma}\mathbf{a}^{\gamma}\cdot\lambda^{\beta}_{\delta}\mathbf{a}^{\delta} = \lambda^{\alpha}_{\gamma}\lambda^{\beta}_{\delta}a^{\gamma\delta} \tag{2.6.9b}$$

Alternatively, if the product of two base vectors from opposite systems, $\mathbf{g}_{\alpha}$ and $\mathbf{a}^{\beta}$, are considered then the shift tensors may be computed from the relationships

$$\mathbf{g}_{\alpha}\cdot\mathbf{a}^{\beta} = \mu^{\gamma}_{\alpha}\mathbf{a}_{\gamma}\cdot\mathbf{a}^{\beta} = \mu^{\beta}_{\alpha} \tag{2.6.10a}$$

$$\mathbf{g}^{\alpha} \cdot \mathbf{a}_{\beta} = \lambda^{\alpha}_{\gamma}\, \mathbf{a}^{\gamma} \cdot \mathbf{a}_{\beta} = \lambda^{\alpha}_{\beta} \qquad (2.6.10b)$$

From the definition of the Christoffel symbol given in section 2.1.4

$$\bar{\Gamma}^{\gamma}_{\alpha\beta} = \mathbf{g}_{\alpha,\beta} \cdot \mathbf{g}^{\gamma} = (\mu^{\delta}_{\alpha}\, \mathbf{a}_{\delta})_{,\beta} \cdot \lambda^{\gamma}_{\xi}\, \mathbf{a}^{\xi} = (\mu^{\xi}_{\alpha,\beta} + \mu^{\delta}_{\alpha}\, \Gamma^{\xi}_{\delta\beta})\, \lambda^{\gamma}_{\xi} \qquad (2.6.11)$$

However, the planar derivative of the shift tensor is

$$\mu^{\xi}_{\alpha}\|_{\beta} = \mu^{\xi}_{\alpha,\beta} + \mu^{\delta}_{\alpha}\, \Gamma^{\xi}_{\delta\beta} - \mu^{\xi}_{\delta}\, \Gamma^{\delta}_{\alpha\beta} \qquad (2.6.12)$$

which enables equation 2.6.11 to be simplified yielding

$$\bar{\Gamma}^{\gamma}_{\alpha\beta} = \Gamma^{\gamma}_{\alpha\beta} + \lambda^{\gamma}_{\xi}\, \mu^{\xi}_{\alpha}\|_{\beta} \qquad (2.6.13)$$

Similarly, by applying equations 2.1.33 and 2.1.34

$$\bar{\Gamma}^{3}_{\alpha\beta} = \mathbf{g}_{\alpha,\beta} \cdot \mathbf{g}^{3} = (\mu^{\delta}_{\alpha}\, \mathbf{a}_{\delta})_{,\beta} \cdot \mathbf{a}^{3} = \mu^{\delta}_{\alpha}\, b_{\delta\beta} \qquad (2.6.14)$$

$$\bar{\Gamma}^{\alpha}_{3\beta} = \mathbf{g}_{3,\beta} \cdot \mathbf{g}^{\alpha} = -b_{\beta\gamma}\, \mathbf{a}^{\gamma} \cdot \lambda^{\alpha}_{\delta}\, \mathbf{a}^{\delta} = -\lambda^{\alpha}_{\delta}\, b^{\delta}_{\beta} \qquad (2.6.15)$$

### 2.6.2 Deformation Characteristics

Figure 2.7 illustrates two points A and B lying within a shell as it undergoes deformation. In the undeformed state, A lies on the shell's middle surface and B lies a distance z from the point A along the normal vector $\mathbf{a}_3$. After the shell undergoes deformation, point A moves to $\hat{A}$ and point B to $\hat{B}$ . The displacement vector connecting the points B and $\hat{B}$ can be written as

$$\mathbf{v} = \mathbf{u} - z\,\mathbf{a}_3 + \bar{\mathbf{a}} \tag{2.6.16}$$

The vector $\bar{\mathbf{a}}$ is expressed in the following form

$$\bar{\mathbf{a}} = z\,\hat{\mathbf{a}}_3 + z\,\mathbf{f} \tag{2.6.17}$$

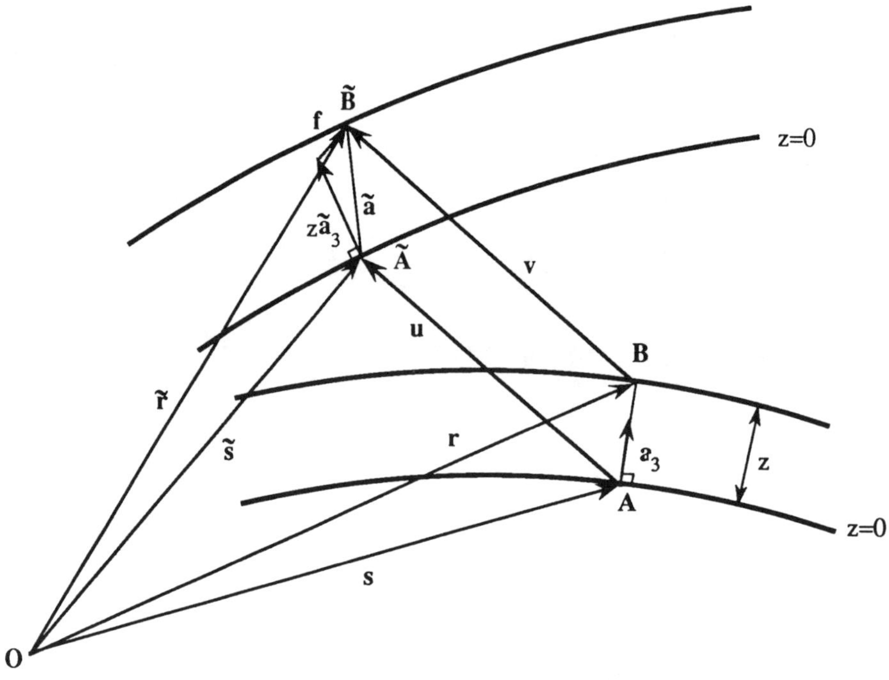


Figure 2.7: Section through a shell before and after deformation.

The components of the vector **f** describe a combination of rotational and stretching modes which allow the general surface to deform freely. Substituting equation 2.6.17 into 2.6.16 yields

$$\mathbf{v} = \mathbf{u} + z(\hat{\mathbf{a}}_3 - \mathbf{a}_3) + z\mathbf{f} \tag{2.6.18}$$

Referring to figure 2.7, the vector equation

$$\hat{\mathbf{s}} = \mathbf{s} + \mathbf{u} \tag{2.6.19}$$

may be written and differentiated with respect to $x^3$ to yield

$$\hat{\mathbf{s}}_{,3} = \mathbf{s}_{,3} + \mathbf{u}_{,3} \tag{2.6.20}$$

From equation 2.1.2,

$$\mathbf{s}_{,3} = \mathbf{a}_3 \quad \text{and} \quad \hat{\mathbf{s}}_{,3} = \hat{\mathbf{a}}_3 \tag{2.6.21}$$

enabling equation 2.6.20 to be written in the form

$$\hat{\mathbf{a}}_3 - \mathbf{a}_3 = \mathbf{u}_{,3} \tag{2.6.22}$$

For any point on the middle surface, the displacement vector **u** can be specified as a function of the coordinates $x^\alpha$. An expression defining $\mathbf{u}_{,3}$ can therefore be formed using equation 2.2.26:

$$\mathbf{u}_{,3} = (u_{\alpha'3} + u_\gamma b^\gamma_\alpha)\,\mathbf{a}^\alpha + u_{3'3}\,\mathbf{a}^3 \tag{2.6.23}$$

Substituting the above relationship into equations 2.6.22 and 2.6.18 yields

$$v = [u_\alpha + z(u_{\alpha,3} + u_\gamma b^\gamma_\alpha + f_\alpha)]\mathbf{a}^\alpha + [w + z(w_{,3} + f_3)]\mathbf{a}^3 \tag{2.6.24}$$

where $w = u_3$.

Alternatively, the general surface displacement vector $\mathbf{v}$ can be expanded into its normal and inplane components:

$$\mathbf{v} = v_\alpha \mathbf{g}^\alpha + v_3 \mathbf{g}^3 \tag{2.6.25}$$

The displacement components $v_\alpha$ can be expressed in terms of the middle surface displacement components $u_\alpha$ and $f_\alpha$ by dot multiplying both sides of equation 2.6.24 with the base vector $\mathbf{g}_\alpha$:

$$v_\alpha = [u_\delta + z(u_{\delta,3} + u_\gamma b^\gamma_\delta + f_\delta)]\mathbf{a}^\delta \cdot \mathbf{g}_\alpha$$

This expression can be further simplified through the use of equation 2.6.10a:

$$v_\alpha = [u_\delta + z(u_{\delta,3} + u_\gamma b^\gamma_\delta + f_\delta)]\mu^\delta_\alpha \tag{2.6.26}$$

Similarly, the displacement components $v_3$ are determined by dot multiplying both sides of equation 2.6.24 with the base vector $\mathbf{g}^3$:

$$v_3 = w + z(w_{,3} + f_3) \tag{2.6.27}$$

### 2.6.3 The Change in Curvature Tensor

The change of curvature in a shell's middle surface may be defined as the difference between the curvature before and after deformation. This change is denoted by the tensor $\kappa_\alpha^\gamma$ and can be expressed in the form

$$\kappa_\alpha^\gamma = \hat{b}_\alpha^\gamma - b_\alpha^\gamma \tag{2.6.28}$$

where $b_\alpha^\gamma$ and $\hat{b}_\alpha^\gamma$ are the curvature tensors before and after deformation respectively. The covariant components are obtained by applying the metric tensor. i.e.,

$$\kappa_{\alpha\beta} = \hat{b}_\alpha^\gamma a_{\gamma\beta} - b_{\alpha\beta} \tag{2.6.29}$$

From equation 2.3.4 the middle surface strain tensor $\varepsilon_{\alpha\beta}$ can be written in the form

$$\varepsilon_{\alpha\beta} = \frac{1}{2}[\hat{a}_{\alpha\beta} - a_{\alpha\beta}] \tag{2.6.30}$$

Re-arranging the above expression to isolate $a_{\alpha\beta}$ and then substituting into 2.6.29 yields:

$$\kappa_{\alpha\beta} = \hat{b}_{\alpha\beta} - b_{\alpha\beta} - 2\hat{b}_\alpha^\gamma \varepsilon_{\gamma\beta} \tag{2.6.31}$$

The metric tensor for the deformed middle surface can be described by the equation

$$\hat{a}^{\delta\gamma} = a^{\delta\gamma} - 2\varepsilon^{\delta\gamma}$$

which can be combined with the relationship

$$\hat{b}^{\gamma}_{\alpha} = \hat{b}_{\delta\alpha}\, \hat{a}^{\delta\gamma} \tag{2.6.32}$$

to yield

$$\hat{b}^{\gamma}_{\alpha} = \hat{b}_{\delta\alpha}\, [a^{\delta\gamma} - 2\,\varepsilon^{\delta\gamma}] \tag{2.6.33}$$

Multiplying both sides of the above expression by $2\,\varepsilon_{\gamma\beta}$ and neglecting quadratic terms in strain, the equation

$$2\,\hat{b}^{\gamma}_{\alpha}\,\varepsilon_{\gamma\beta} = 2\,\hat{b}_{\delta\alpha}\, a^{\delta\gamma}\, \varepsilon_{\gamma\beta}$$

is obtained. The change in curvature tensor can now be expressed in the form

$$\kappa_{\alpha\beta} = \hat{b}_{\alpha\beta} - b_{\alpha\beta} - 2\,\hat{b}_{\delta\alpha}\, a^{\delta\gamma}\, \varepsilon_{\gamma\beta} \tag{2.6.34}$$

Defining the curvature tensor $\hat{b}_{\alpha\beta}$ through the use of equation 2.2.15:

$$\hat{b}_{\alpha\beta} = - \hat{\mathbf{a}}_{3,\alpha} \cdot \hat{\mathbf{a}}_{\beta} \tag{2.6.35}$$

Differentiating equation 2.6.22 with respect to $x^{\alpha}$ yields

$$\hat{\mathbf{a}}_{3,\alpha} = \mathbf{a}_{3,\alpha} + \mathbf{u}_{,3\alpha} \tag{2.6.36}$$

which can be combined with equation 2.6.23 to form

$$\hat{\mathbf{a}}_{3,\alpha} = \mathbf{a}_{3,\alpha} + (u_{\delta,3} + u_{\gamma}\, b^{\gamma}_{\delta})_{,\alpha}\, \mathbf{a}^{\delta} + (u_{\delta,3} + u_{\gamma}\, b^{\gamma}_{\delta})\, \mathbf{a}^{\delta}_{,\alpha}$$
$$+ w_{,3\alpha}\, \mathbf{a}^{3} + w_{,3}\, \mathbf{a}^{3}_{,\alpha} \tag{2.6.37}$$

In addition, equation 2.6.19 may be differentiated with respect to $x^\beta$ to yield

$$\hat{\mathbf{a}}_\beta = \mathbf{a}_\beta + \mathbf{u}_{,\beta} \qquad (2.6.38)$$

The displacement derivative $\mathbf{u}_{,\beta}$ may also be defined using equation 2.2.21:

$$\mathbf{u}_{,\beta} = ( u_\delta \|_\beta - w\, b_{\delta\beta} )\, \mathbf{a}^\delta + ( w_{,\beta} + u_\gamma b^\gamma_\beta )\, \mathbf{a}^3 \qquad (2.6.39)$$

Equations 2.6.37 - 2.6.39 can now be substituted into 2.6.35. Keeping only the terms linear in displacement results in the expression

$$\begin{aligned} -\hat{b}_{\alpha\beta} = {} & [\, \mathbf{a}_{3,\alpha} + (u_{\delta,3} + u_\gamma b^\gamma_\delta)_{,\alpha}\, \mathbf{a}^\delta + ( u_{\delta,3} + u_\gamma b^\gamma_\delta )\, \mathbf{a}^\delta{}_{,\alpha} + w_{,3\alpha}\, \mathbf{a}^3 + \\ & w_{,3}\, \mathbf{a}^3{}_{,\alpha} \,] \cdot [\, \mathbf{a}_\beta + ( u_\rho \|_\beta - w\, b_{\rho\beta} )\, \mathbf{a}^\rho + ( w_{,\beta} + u_\xi b^\xi_\beta )\, \mathbf{a}^3 \,] \end{aligned}$$

$$\begin{aligned} = {} & -b_{\alpha\beta} + (u_{\beta,3} + u_\gamma b^\gamma_\beta)_{,\alpha} + ( u_{\delta,3} + u_\gamma b^\gamma_\delta )\, \mathbf{a}^\delta{}_{,\alpha} \cdot \mathbf{a}_\beta - w_{,3}\, b_{\alpha\beta} \\ & - ( u_\rho \|_\beta - w\, b_{\rho\beta} )\, b^\rho_\alpha \end{aligned}$$

$$= -b_{\alpha\beta} - ( u_\rho \|_\beta - w\, b_{\rho\beta} )\, b^\rho_\alpha + (u_{\beta,3} + u_\gamma b^\gamma_\beta) \|_\alpha - w_{,3}\, b_{\alpha\beta} \qquad (2.6.40)$$

The above equation is combined with 2.6.34 to arrive at an expression which describes the change of curvature of the middle surface in terms of the displacements $u_\alpha$ and $w$. Once again, only linear terms are retained within the formulation:

$$\kappa_{\alpha\beta} = ( u_\rho \|_\beta - w\, b_{\rho\beta} )\, b^\rho_\alpha - (u_{\beta,3} + u_\gamma b^\gamma_\beta) \|_\alpha + w_{,3}\, b_{\alpha\beta} - 2\, b^\gamma_\alpha\, \varepsilon_{\gamma\beta} \qquad (2.6.41)$$

### 2.6.4 The Strain-Displacement Equations

In section 2.3 the deformation characteristic of a continuous three dimensional body were examined. For small displacements the general strain tensor $\eta_{ij}$ can be expressed in the form

$$\eta_{ij} = \frac{1}{2}( v_i|_j + v_j|_i ) \tag{2.6.42}$$

The aim in this section is to express the general strain components in terms of the middle surface displacement quantities $u_\alpha$, $u_3$, $f_\alpha$ and $f_3$. An expression defining the inplane strain $\eta_{\alpha\beta}$ may be obtained by letting the indices $i = \alpha$ and $j = \beta$. i.e.,

$$\eta_{\alpha\beta} = \frac{1}{2}( v_\alpha|_\beta + v_\beta|_\alpha ) \tag{2.6.43}$$

The covariant derivative of the displacement components $v_\alpha$ can be derived by combining equations 2.1.45, 2.6.13 and 2.6.14:

$$v_\alpha|_\beta = v_{\alpha,\beta} - v_\rho \Gamma^\rho_{\alpha\beta} - v_3 \Gamma^3_{\alpha\beta}$$

$$= v_{\alpha,\beta} - v_\rho \Gamma^\rho_{\alpha\beta} - v_\rho \lambda^\rho_\xi \mu^\xi_\alpha\|_\beta - v_3 \mu^\delta_\alpha b_{\delta\beta}$$

$$= v_\alpha\|_\beta - v_\rho \lambda^\rho_\xi \mu^\xi_\alpha\|_\beta - v_3 \mu^\delta_\alpha b_{\delta\beta} \tag{2.6.44}$$

For clarity, each term in equation 2.6.44 will be considered individually. The quantity $v_\alpha\|_\beta$ is defined by differentiating the inplane displacement vector $v_\alpha$:

$$v_\alpha\|_\beta = [ ( u_\delta + z ( u_{\delta,3} + u_\gamma b^\gamma_\delta + f_\delta ) ) \mu^\delta_\alpha ] \|_\beta$$

$$= [ u_\delta + z ( u_{\delta,3} + u_\gamma b^\gamma_\delta + f_\delta ) ] \mu^\delta_\alpha\|_\beta +$$

$$[\, u_\delta||_\beta + z\,(u_{\delta,3}||_\beta + u_\gamma||_\beta\, b^\gamma_\delta + u_\gamma\, b^\gamma_\delta||_\beta + f_\delta||_\beta)\,]\,\mu^\delta_\alpha$$

$$= u_\delta\, \mu^\delta_\alpha||_\beta + u_\delta||_\beta\, \mu^\delta_\alpha + z\,[\,(u_{\delta,3} + u_\gamma\, b^\gamma_\delta + f_\delta)\, \mu^\delta_\alpha||_\beta + (u_{\delta,3}||_\beta + u_\gamma||_\beta\, b^\gamma_\delta + u_\gamma\, b^\gamma_\delta||_\beta + f_\delta||_\beta)\, \mu^\delta_\alpha] \quad (2.6.45)$$

Equation 2.6.7 is combined with 2.6.26 when expanding the second term:

$$v_\rho\, \lambda^\rho_\xi\, \mu^\xi_\alpha||_\beta = [\, u_\delta + z\,(u_{\delta,3} + u_\gamma\, b^\gamma_\delta + f_\delta)\,]\, \mu^\delta_\rho\, \lambda^\rho_\xi\, \mu^\xi_\alpha||_\beta$$

$$= [\, u_\delta + z\,(u_{\delta,3} + u_\gamma\, b^\gamma_\delta + f_\delta)\,]\, \mu^\delta_\alpha||_\beta \quad (2.6.46)$$

and equation 2.6.27 is utilized to form

$$v_3\, \mu^\delta_\alpha\, b_{\delta\beta} = [\, w + z\,(w_{,3} + f_3)\,]\, \mu^\delta_\alpha\, b_{\delta\beta} \quad (2.6.47)$$

The three equations, 2.6.45, 2.6.46 and 2.6.47 are substituted into 2.6.44 to form

$$v_\alpha|_\beta = u_\delta||_\beta\, \mu^\delta_\alpha - w\, \mu^\delta_\alpha\, b_{\delta\beta} + z\,[\, u_{\delta,3}||_\beta + u_\gamma||_\beta\, b^\gamma_\delta + u_\gamma\, b^\gamma_\delta||_\beta + f_\delta||_\beta\,]\, \mu^\delta_\alpha - z\,[w_{,3} + f_3]\, \mu^\delta_\alpha\, b_{\delta\beta} \quad (2.6.48)$$

From 2.6.48, $v_\beta|_\alpha$ is obtained by interchanging the indices $\alpha$ and $\beta$. An expression for the general strain tensor $\eta_{\alpha\beta}$ can now be formulated by substituting equation 2.6.48 into 2.6.43:

$$\begin{aligned} 2\eta_{\alpha\beta} = {} & u_\delta||_\beta\, \mu^\delta_\alpha - w\, \mu^\delta_\alpha\, b_{\delta\beta} + z\,[\, u_{\delta,3}||_\beta + u_\gamma||_\beta\, b^\gamma_\delta + u_\gamma\, b^\gamma_\delta||_\beta + f_\delta||_\beta\,]\, \mu^\delta_\alpha \\ & + u_\delta||_\alpha\, \mu^\delta_\beta - w\, \mu^\delta_\beta\, b_{\delta\alpha} + z\,[\, u_{\delta,3}||_\alpha + u_\gamma||_\alpha\, b^\gamma_\delta + u_\gamma\, b^\gamma_\delta||_\alpha + f_\delta||_\alpha\,]\, \mu^\delta_\beta \\ & - z\,[w_{,3} + f_3]\, \mu^\delta_\alpha\, b_{\delta\beta} - z\,[w_{,3} + f_3]\, \mu^\delta_\beta\, b_{\delta\alpha} \end{aligned} \quad (2.6.49)$$

The relationship describing the strain of the middle surface, denoted by the symbol $\varepsilon_{\alpha\beta}$, is obtained by letting $z = 0$:

$$\varepsilon_{\alpha\beta} = \tfrac{1}{2}[\, u_\alpha\|_\beta + u_\beta\|_\alpha - 2\, w\, b_{\alpha\beta}\,] \tag{2.6.50}$$

This equation can be substituted into 2.6.41 to yield the following expression for the change in curvature of the middle surface:

$$\kappa_{\alpha\beta} = -u_\gamma\|_\alpha\, b^\gamma_\beta - u_\beta\|_\gamma\, b^\gamma_\alpha - u_{\beta,3}\|_\alpha - u_\gamma\, b^\gamma_\beta\|_\alpha + w_{,3}\, b_{\alpha\beta} + w\, b_{\gamma\beta}\, b^\gamma_\alpha \tag{2.6.51}$$

Equations 2.6.50 and 2.6.51 are relationships which can be used to simplify the general strain-displacement equations (2.6.49). This process begins by considering the difference $2(\eta_{\alpha\beta} - \varepsilon_{\gamma\delta}\,\mu^\gamma_\alpha\,\mu^\delta_\beta)$. i.e.,

$$2(\eta_{\alpha\beta} - \varepsilon_{\gamma\delta}\,\mu^\gamma_\alpha\,\mu^\delta_\beta) = \underbrace{u_\delta\|_\beta\,\mu^\delta_\alpha + u_\delta\|_\alpha\,\mu^\delta_\beta - u_\gamma\|_\delta\,\mu^\delta_\beta\,\mu^\gamma_\alpha - u_\delta\|_\gamma\,\mu^\gamma_\alpha\,\mu^\delta_\beta}_{[1]} +$$

$$\underbrace{w\,(2\, b_{\gamma\delta}\,\mu^\gamma_\alpha\,\mu^\delta_\beta - \mu^\delta_\alpha\, b_{\delta\beta} - \mu^\delta_\beta\, b_{\delta\alpha})}_{[2]} +$$

$$\underbrace{\begin{aligned} z\,(\, & u_{\delta,3}\|_\beta\,\mu^\delta_\alpha + u_{\delta,3}\|_\alpha\,\mu^\delta_\beta + u_\gamma\|_\beta\, b^\gamma_\delta\,\mu^\delta_\alpha + u_\gamma\|_\alpha\, b^\gamma_\delta\,\mu^\delta_\beta \\ & + u_\gamma\, b^\gamma_\delta\|_\beta\,\mu^\delta_\alpha + u_\gamma\, b^\gamma_\delta\|_\alpha\,\mu^\delta_\beta + f_\delta\|_\beta\,\mu^\delta_\alpha + f_\delta\|_\alpha\,\mu^\delta_\beta\,)\end{aligned}}_{[3]} -$$

$$\underbrace{z\,(\, w_{,3} + f_3\,)\,(\,\mu^\delta_\alpha\, b_{\delta\beta} + \mu^\delta_\beta\, b_{\delta\alpha}\,)}_{[4]} \tag{2.6.52}$$

Next, using the definition for $\mu^\beta_\alpha$ given by equation 2.6.4, each of the four designated groups is considered individually. Starting with group 1:

$$[1] = u_\delta\|_\beta (\delta^\delta_\alpha - z b^\delta_\alpha) + u_\delta\|_\alpha (\delta^\delta_\beta - z b^\delta_\beta) - u_\gamma\|_\delta (\delta^\delta_\beta \delta^\gamma_\alpha - z b^\gamma_\alpha \delta^\delta_\beta - z b^\delta_\beta \delta^\gamma_\alpha + z^2 b^\gamma_\alpha b^\delta_\beta) - u_\delta\|_\gamma (\delta^\delta_\beta \delta^\gamma_\alpha - z b^\gamma_\alpha \delta^\delta_\beta - z b^\delta_\beta \delta^\gamma_\alpha + z^2 b^\gamma_\alpha b^\delta_\beta)$$

$$= z [ u_\gamma\|_\delta b^\delta_\beta (\delta^\gamma_\alpha - z b^\gamma_\alpha) + u_\delta\|_\gamma b^\gamma_\alpha (\delta^\delta_\beta - z b^\delta_\beta) ]$$

$$= z\, u_\delta\|_\gamma [ b^\gamma_\beta \mu^\delta_\alpha + b^\gamma_\alpha \mu^\delta_\beta ]$$

$$= z\, u_\delta\|_\gamma b^\gamma_\nu [ \delta^\nu_\beta \mu^\delta_\alpha + \delta^\nu_\alpha \mu^\delta_\beta ]$$

Similarly, group 2 is expanded term by term:

$$[2] = w\, b_{\gamma\delta} ( 2 \mu^\gamma_\alpha \mu^\delta_\beta - \mu^\delta_\alpha \delta^\gamma_\beta - \mu^\delta_\beta \delta^\gamma_\alpha )$$

$$= w\, b_{\gamma\delta} ( 2 \delta^\gamma_\alpha \delta^\delta_\beta - 2 z \delta^\gamma_\alpha b^\delta_\beta - 2 z b^\gamma_\alpha \delta^\delta_\beta + 2 z^2 b^\gamma_\alpha b^\delta_\beta - \delta^\gamma_\alpha \delta^\delta_\beta + z b^\delta_\alpha \delta^\gamma_\beta - \delta^\gamma_\alpha \delta^\delta_\beta + z b^\delta_\beta \delta^\gamma_\alpha )$$

$$= - z w [ b_{\gamma\delta} b^\delta_\beta (\delta^\gamma_\alpha - z b^\gamma_\alpha) + b_{\gamma\delta} b^\gamma_\alpha (\delta^\delta_\beta - z b^\delta_\beta) ]$$

$$= - z w\, b_{\gamma\delta} [ b^\delta_\beta \mu^\gamma_\alpha + b^\delta_\alpha \mu^\gamma_\beta ]$$

$$= - z w\, b_{\gamma\delta} b^\delta_\nu [ \delta^\nu_\beta \mu^\gamma_\alpha + \delta^\nu_\alpha \mu^\gamma_\beta ]$$

Examining the third group,

$$[3] = z [ u_{\delta,3}\|_\nu (\delta^\nu_\beta \mu^\delta_\alpha + \delta^\nu_\alpha \mu^\delta_\beta) + u_\gamma\|_\nu b^\gamma_\delta (\delta^\nu_\beta \mu^\delta_\alpha + \delta^\nu_\alpha \mu^\delta_\beta) + u_\gamma b^\gamma_\delta\|_\nu (\delta^\nu_\beta \mu^\delta_\alpha + \delta^\nu_\alpha \mu^\delta_\beta) + f_\delta\|_\nu (\delta^\nu_\beta \mu^\delta_\alpha + \delta^\nu_\alpha \mu^\delta_\beta) ]$$

$$= z\,(u_{\delta,3}\|_\nu + u_\gamma\|_\nu\, b^\gamma_\delta + u_\gamma\, b^\gamma_\delta\|_\nu + f_\delta\|_\nu)\,[\,\delta^\nu_\beta\,\mu^\delta_\alpha + \delta^\nu_\alpha\,\mu^\delta_\beta\,]$$

and finally, the fourth group simplifies to

$$[4] \quad = -z\,(w_{,3} + f_3)\,b_{\delta\nu}\,[\,\mu^\delta_\alpha\,\delta^\nu_\beta + \mu^\delta_\beta\,\delta^\nu_\alpha\,]$$

Substituting groups [1] - [4] into equation 2.6.52 yields

$$\eta_{\alpha\beta} = \varepsilon_{\delta\nu}\,\mu^\delta_\alpha\,\mu^\nu_\beta + \tfrac{1}{2}z\,(u_{\delta,3}\|_\nu + u_\gamma\|_\nu\, b^\gamma_\delta + u_\gamma\, b^\gamma_\delta\|_\nu + u_\delta\|_\gamma\, b^\gamma_\nu + f_\delta\|_\nu - w\, b_{\gamma\delta}\, b^\delta_\nu - (w_{,3} + f_3)\, b_{\delta\nu})\,[\,\mu^\delta_\alpha\,\delta^\nu_\beta + \mu^\delta_\beta\,\delta^\nu_\alpha\,]$$

which combined with equation 2.6.51 becomes

$$\eta_{\alpha\beta} = \varepsilon_{\delta\nu}\,\mu^\delta_\alpha\,\mu^\nu_\beta - \tfrac{1}{2}z\,(\kappa_{\delta\nu} + f_3\, b_{\delta\nu} - f_\delta\|_\nu)\,[\,\mu^\delta_\alpha\,\delta^\nu_\beta + \mu^\delta_\beta\,\delta^\nu_\alpha\,] \qquad (2.6.53)$$

By introducing the notation

$$\phi_{\delta\nu} = f_3\, b_{\delta\nu} - f_\delta\|_\nu \qquad (2.6.54)$$

equation 2.6.53 becomes

$$\eta_{\alpha\beta} = \varepsilon_{\delta\nu}\,\mu^\delta_\alpha\,\mu^\nu_\beta - \tfrac{1}{2}z\,(\kappa_{\delta\nu} + \phi_{\delta\nu})\,[\,\mu^\delta_\alpha\,\delta^\nu_\beta + \mu^\delta_\beta\,\delta^\nu_\alpha\,] \qquad (2.6.55)$$

For the strain components $\eta_{\alpha 3}$ the indices i and j in equation 2.6.42 are set to $\alpha$ and 3 respectively:

$$\eta_{\alpha 3} = \tfrac{1}{2}(v_\alpha|_3 + v_3|_\alpha) \qquad (2.6.56)$$

Expressions for the covariant derivatives of the displacement vector **v** can be obtained from equation 2.1.45. These are simplified using equations 2.2.11 and 2.6.15 enabling the following relationship to be written:

$$\eta_{\alpha 3} = \frac{1}{2}( v_{\alpha ' 3} + v_{3 ' \alpha} + 2\, v_{\gamma} \lambda^{\gamma}_{\delta} b^{\delta}_{\alpha} ) \tag{2.6.57}$$

Expressing $\eta_{\alpha 3}$ in terms of the middle surface quantities **u** and **f** is achieved by expanding each term individually. Differentiating 2.6.26 with respect to $x^3$:

$$\begin{aligned} v_{\alpha ' 3} = {} & [\, u_{\delta} + z ( u_{\delta ' 3} + u_{\gamma} b^{\gamma}_{\delta} + f_{\delta} ) ]_{, 3}\, \mu^{\delta}_{\alpha} + \\ & [\, u_{\delta} + z ( u_{\delta ' 3} + u_{\gamma} b^{\gamma}_{\delta} + f_{\delta} ) ]\, \mu^{\delta}_{\alpha ' 3} \end{aligned}$$

The thickness of a shell is assumed to be small when compared with the other dimensions of the structure, in particular, with its radius of curvature. Consequently, variations in the curvature tensor along the $x^3$ axis are assumed to be negligible. Hence, the above expression can be simplified to

$$v_{\alpha ' 3} = [\, u_{\delta ' 3} + z ( u_{\delta ' 33} + u_{\gamma ' 3} b^{\gamma}_{\delta} + f_{\delta ' 3} ) ]\, \mu^{\delta}_{\alpha} \tag{2.6.58}$$

Similarly, equation 2.6.27 is differentiated with respect to $x^{\alpha}$ to form the expression

$$v_{3 ' \alpha} = w_{, \alpha} + z ( w_{, 3\alpha} + f_{3 ' \alpha} ) \tag{2.6.59}$$

which can be combined with equations 2.6.26, 2.6.57 and 2.6.58 to yield

$$\begin{aligned} 2\eta_{\alpha 3} = {} & u_{\alpha ' 3} + 2\, u_{\delta} b^{\delta}_{\alpha} + w_{, \alpha} + z [\, 2\, u_{\gamma} b^{\gamma}_{\delta} b^{\delta}_{\alpha} + u_{\delta ' 3} b^{\delta}_{\nu} ( \delta^{\nu}_{\alpha} + \mu^{\nu}_{\alpha} ) + \\ & u_{\delta ' 33} \mu^{\delta}_{\alpha} + w_{, 3\alpha} + 2\, f_{\delta} b^{\delta}_{\alpha} + f_{\delta ' 3} \mu^{\delta}_{\alpha} + f_{3 ' \alpha} ] \end{aligned} \tag{2.6.60}$$

The middle surface strain tensor $\varepsilon_{\alpha 3}$ can be easily obtained by letting z = 0. i.e.,

$$\varepsilon_{\alpha 3} = \frac{1}{2}[\, u_{\alpha ' 3} + 2\, u_{\delta}\, b^{\delta}_{\alpha} + w_{,\alpha} \,] \tag{2.6.61}$$

Using this expression, the strain tensor $\eta_{\alpha 3}$ can be written in the form:

$$\eta_{\alpha 3} = \varepsilon_{\alpha 3} + \frac{1}{2} z\, [\, 2\, u_{\gamma}\, b^{\gamma}_{\delta}\, b^{\delta}_{\alpha} + u_{\delta ' 3}\, b^{\delta}_{\nu} (\, \delta^{\nu}_{\alpha} + \mu^{\nu}_{\alpha}\,) + u_{\delta ' 33}\, \mu^{\delta}_{\alpha} + w_{,3\alpha} + 2\, f_{\delta}\, b^{\delta}_{\alpha} + f_{\delta ' 3}\, \mu^{\delta}_{\alpha} + f_{3 ' \alpha} \,] \tag{2.6.62}$$

Finally, the strain components $\eta_{33}$ are formulated by letting i = j = 3 in equation 2.6.42. This leads to the relationship

$$\eta_{33} = \varepsilon_{33} + z\, (\, w_{,33} + f_{3 ' 3}) \tag{2.6.63}$$

where $$\varepsilon_{33} = w_{,3} \tag{2.6.64}$$

## 2.6.5 Interpretation and Discussion of the Strain-Displacement Equations

Equations 2.6.55, 2.6.62 and 2.6.63 describe the strain-displacement behaviour of a shell structure. They are essentially three dimensional in nature and allow the general surface to deform freely. From a computational point of view, the middle surface quantities **u** and **f** require the evaluation of first and second order derivatives with respect to the normal coordinate $x^3$.

Any numerical implementation capable of analysing shells of arbitrary geometry such as the finite element or finite difference methods would therefore need to introduce the derivatives as nodal unknowns. In order to maintain consistency in the assembly of the structure stiffness matrix each unknown derivative must be transformed to a common coordinate system. However, the orientation of the shell's middle surface is arbitrary. Hence, eighteen derivative components ( nine for each $u_{i,j}$ and $f_{i,j}$ ) would be required at each node. Assuming the transformation equations and their inverse exist (see equations 2.1.30-2.1.31 and Appendix A), then the number of nodal degrees of freedom would be twenty four. This severely restricts the application of the theory to the dynamic analysis of shell structures where fine meshes are normally required to produce mode shapes accurately.

In contrast, many shell theories [1] - [8] are two dimensional in nature and are based on a series of assumptions which allow the middle surface displacements to be approximated by functions of only two independent variables.

In his classical linear theory of thin shells, Love [3] made the following four assumptions:

(1) The thickness of the shell is small compared with the smallest radius of curvature of its reference surface.

(2) Strains and displacements are small enough to allow second-order terms in the strain-displacement equations to be neglected in comparison with the first-order terms.

(3) The transverse normal stress is small compared with the other normal stress components and may be neglected.

(4) Normals to the undeformed middle surface remain straight and normal to the deformed middle surface and do not change in length.

The first two assumptions have already been incorporated in the derivation of equations 2.6.55, 2.6.62 and 2.6.63. The third, restricts the application of shell theory to situations where the work done by the normal stresses is negligible when compared to the total change in strain energy during deformation. The fourth, known as Kirchhoff's hypothesis, allows the transverse shearing strains to be shearing strains to be neglected and permits the reduction of a three dimensional problem to a two dimensional bending and stretching problem.

However, if Hooke's law is considered, a number of inconsistencies exist in the four assumptions outlined above. For example, if a material is isotropic in nature and the length of the normals remain unchanged during deformation then a substantial normal stress will arise from the Poisson effect. As discussed by a number of authors (e.g. Leissa [9] and Bushnell [11]) this normal stress would be comparable to the inplane stresses. Similarly, Kirchhoff's hypothesis implies zero transverse shear stress. However, the shell element cannot in general remain in equilibrium without the transverse shear forces acting along its edges.

Bushnell [11] points out that these inconsistencies do not diminish the value of shell theory as an engineering tool. In his discussion, he introduces a corollary rule which states the average normal stresses and transverse shear stress resultants must be calculated from considerations of equilibrium rather than directly from the kinematics embodied in Kirchhoff's hypothesis.

The development of the shell theory derived above is further refined by applying the assumption of conservation of normals. This amounts to assuming that the strain components $\eta_{\alpha 3}$ and the middle surface quantities $f^{\alpha}$ and $f^{3}$ are zero. When these conditions are applied to equation 2.6.61 the following relationship may be obtained:

$$u_{\alpha',3} = -\left( w_{,\alpha} + 2\, u_{\delta}\, b^{\delta}_{\alpha} \right) \qquad (2.6.65)$$

Furthermore if the length of the normals does not change during deformation one may assume $\eta_{33}$ is zero. From equation 2.6.64 it can be concluded that

$$w_{,3} = 0 \qquad (2.6.66)$$

Combining equations 2.6.65 with 2.6.26 yields the following expression for the covariant components of the displacement $v_\alpha$:

$$v_\alpha = [\, u_\delta - z\,( w_{,\delta} + u_\gamma b^\gamma_\delta )\,]\, \mu^\delta_\alpha \qquad (2.6.67)$$

Substituting the relationships 2.6.65 and 2.6.66 into equation 2.6.51, the change of curvature tensor $\kappa_{\alpha\beta}$ becomes

$$\kappa_{\alpha\beta} = u_\gamma b^\gamma_\beta\|_\alpha + u_\gamma\|_\alpha b^\gamma_\beta - u_\beta\|_\gamma b^\gamma_\alpha + w\|_{\alpha\beta} + w\, b^\gamma_\alpha b_{\gamma\beta} \qquad (2.6.68)$$

Similarly, the general strain tensor $\eta_{\alpha\beta}$ defined by equation 2.6.55 reduces to

$$\eta_{\alpha\beta} = \varepsilon_{\delta\nu}\, \mu^\delta_\alpha \mu^\nu_\beta + \tfrac{1}{2} z\, \kappa_{\delta\nu}\, [\, \mu^\delta_\alpha \delta^\nu_\beta + \mu^\delta_\beta \mu^\nu_\alpha \,] \qquad (2.6.69)$$

This expression shows that the strain $\eta_{\alpha\beta}$ depends only on the deformation of the middle surface. The equations are therefore two dimensional and identical in form to the ones developed by Flugge [1].

# 3. NUMERICAL FUNDAMENTALS

In the proposed method, the curvilinear finite difference (CFD) method [74] is combined with the strain-displacement equations developed by Flugge [1] and the Principle of Virtual Displacements to formulate a technique capable of computing the dynamic characteristics of thin shells. This chapter presents and discusses the numerical techniques utilized in the formulation.

Section 3.1 introduces the CFD method with particular reference to the two dimensional second-order strain-displacement equations described in the previous chapter. The method is presented in its general form to enable both the displacement and geometric characteristics to be approximated by either a second or higher order interpolation function.

In section 3.2, expressions defining the geometric properties of a shell's middle surface in terms of the surface local partial derivatives are developed. These quantities are then combined with the strain-displacement and constitutive relationships in section 3.5 to form numerical approximations for a shell's stress-strain characteristics.

In sections 3.3 and 3.4, the Principle of Virtual Displacements is applied to formulate the stiffness and mass characteristics of the shell structure. The selection of an appropriate displacement field to approximate the deformation of the shell within each mesh is therefore a critical step and is discussed in detail within section 3.4.

Prior to solving the governing system of equations, displacement and force boundary conditions are applied. The details of a procedure capable of enforcing the constraints associated with simply supported, fixed and free boundary conditions are presented in section 3.6.

Section 3.7 contains a brief summary of the 'subspace iteration' technique which was utilized in the proposed method to solve the system of equations describing the free vibration characteristics of shell structures.

Essentially, the computational procedure may be summarized in the following manner:

a) Discretize the middle surface of the shell into computational grids using an assembly of curvilinear meshes.

b) To each node, assign the values of its x, y and z coordinates and establish three global translational components of displacement. The coordinates of the nodes defining the mesh are used to approximate the surface of the shell and provide a basis for computing the transformation coefficients, surface metric tensors, curvature tensors, Christoffel symbols and their pertinent derivatives. These quantities are required in the application of Flugge's strain-displacement equations.

c) Transform the nodal displacement components and their associated derivatives to the local coordinate system defined by the curvilinear mesh.

d) Evaluate the local stiffness and mass matrices and assemble them into the structure's stiffness and mass matrices respectively.

e) Solve the resulting system of equations to obtain the natural frequencies and corresponding mode shapes of the structure.

## 3.1 THE CURVILINEAR FINITE DIFFERENCE METHOD [74]

Consider the two dimensional region $\mathbb{R}$ illustrated in figure 3.1 which is bounded by the curve $\mathbb{S}$. Suppose within this region a function $\psi(\alpha,\beta)$ is required to satisfy the following second order partial differential equation

$$A\psi + B^i\psi,_i + C^{ij}\psi,_{ij} = F \tag{3.1.1a}$$

subject to the boundary conditions

$$D^i\psi,_i + E\psi = G \tag{3.1.1b}$$

Essentially, these equations are the generalized form of the shell strain-displacement relationships derived in the previous chapter.

The region $\mathbb{R}$ is discretized to form a computational mesh by defining a local curvilinear coordinate system and computational grid at each node. For the system of equations described above, the local function $\psi(\alpha,\beta)$ is required to be approximated by at least a complete 2nd-order polynomial to ensure continuity of the function up to and including its 2nd-order partial derivatives. A curvilinear finite difference mesh can therefore be constructed from nine nodes bounded by the coordinates $\alpha = \pm 1$ and $\beta = \pm 1$ as illustrated in figure 3.2a. Alternatively, a higher order approximation may be formed by using a sixteen node mesh similar to the one shown in figure 3.2b.

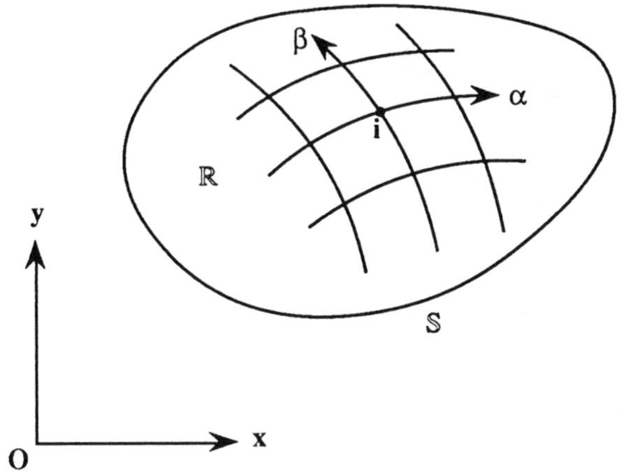


Figure 3.1: Curvilinear mesh discretization.

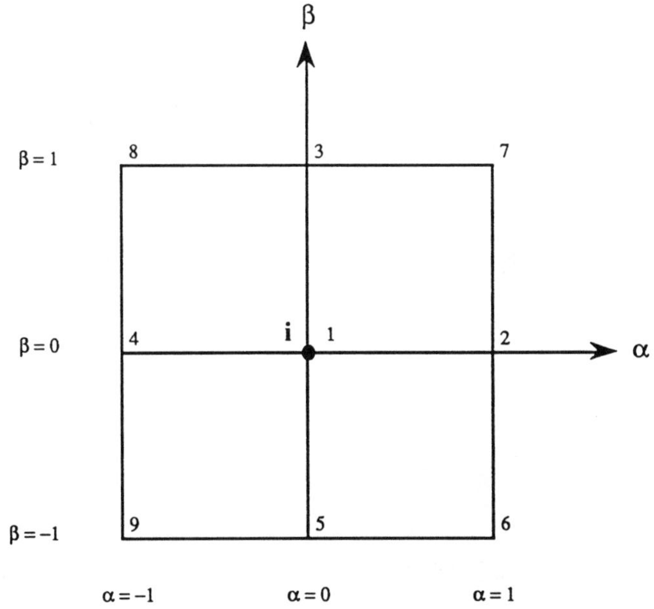


Figure 3.2a: Local 9-node curvilinear grid.

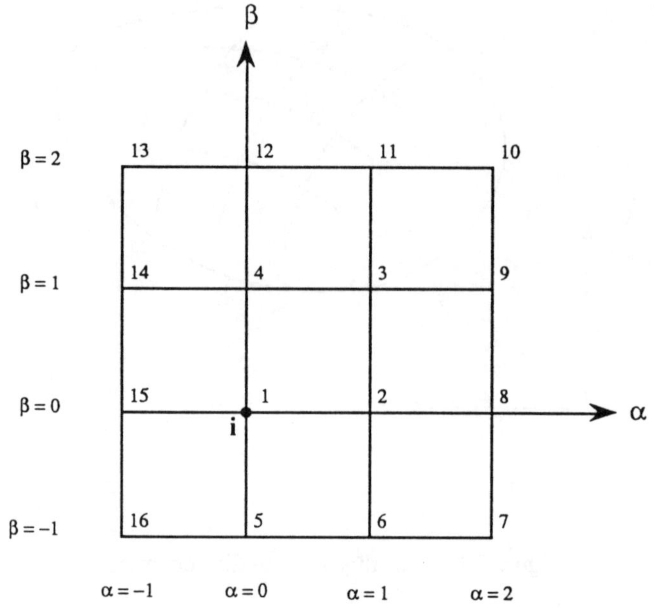


Figure 3.2b: Local 16-node curvilinear grid.

At any point within the mesh, the values of the function $\psi$ may be approximated from the expression

$$\psi = a_1 + a_2\alpha + a_3\beta + a_4\alpha^2 + a_5\beta^2 + a_6\alpha\beta + a_7\alpha^2\beta + a_8\alpha\beta^2 + a_9\alpha^2\beta^2 + \ldots \tag{3.1.2}$$

where $a_1, a_2, a_3, \ldots, a_n$ are constant coefficients. Equation 3.1.2 is more conveniently expressed in the form

$$\psi = [\,\phi\,]^t \{\Psi_p\} \tag{3.1.3}$$

where $[\phi]$ is an $n{\times}1$ matrix of interpolation functions. For a nine node mesh,

$$[\phi] = \begin{bmatrix} 1 - \alpha^2 - \beta^2 + \alpha^2\beta^2 \\ \frac{1}{2}(\alpha + \alpha^2 - \alpha\beta^2 - \alpha^2\beta^2) \\ \frac{1}{2}(\beta + \beta^2 - \alpha^2\beta - \alpha^2\beta^2) \\ \frac{1}{2}(-\alpha + \alpha^2 + \alpha\beta^2 - \alpha^2\beta^2) \\ \frac{1}{2}(-\beta + \beta^2 + \alpha^2\beta - \alpha^2\beta^2) \\ \frac{1}{4}(-\alpha\beta - \alpha^2\beta + \alpha\beta^2 + \alpha^2\beta^2) \\ \frac{1}{4}(\alpha\beta + \alpha^2\beta + \alpha\beta^2 + \alpha^2\beta^2) \\ \frac{1}{4}(-\alpha\beta + \alpha^2\beta - \alpha\beta^2 + \alpha^2\beta^2) \\ \frac{1}{4}(\alpha\beta - \alpha^2\beta - \alpha\beta^2 + \alpha^2\beta^2) \end{bmatrix} \tag{3.1.4}$$

The vector $\{\Psi_p\}$ contains the nodal values of the unknowns in the order $\Psi_1$, $\Psi_2$, ... ,$\Psi_n$. i.e.,

$$\{\Psi_p\}^t = \{\Psi_1 \ \Psi_2 \Psi_3 \ ...... \ \Psi_n\} \tag{3.1.5}$$

Equation 3.1.3 can be differentiated with respect to the coordinates $\alpha$ and $\beta$ to form the finite difference approximations for the partial derivatives in the local coordinate system. For a computational grid based on nine nodes, these are summarized in table 3.1.

Using equation 3.1.3, the geometric coordinates at each of the nodes can be assembled to form a relationship between the local and global coordinates:

$$x = [\phi]^t\{X_p\} \tag{3.1.6a}$$

$$y = [\phi]^t\{Y_p\} \tag{3.1.6b}$$

$$z = [\phi]^t\{Z_p\} \tag{3.1.6c}$$

| $\phi_i$ | $\phi,_{\alpha}$ | $\phi,_{\beta}$ | $\phi,_{\alpha\alpha}$ | $\phi,_{\alpha\beta}$ | $\phi,_{\beta\beta}$ |
|---|---|---|---|---|---|
| $\phi_1$ | $(-2\alpha + 2\alpha\beta^2)$ | $(-2\beta + 2\alpha^2\beta)$ | $-2 + 2\beta^2$ | $4\alpha\beta$ | $-2 + 2\alpha^2$ |
| $\phi_2$ | $(1 + 2\alpha - \beta^2 - 2\alpha\beta^2)/2$ | $(-2\beta\alpha - 2\beta\alpha^2)/2$ | $(2 - 2\beta^2)/2$ | $(-2\beta - 4\alpha\beta)/2$ | $(-2\alpha - 2\alpha^2)/2$ |
| $\phi_3$ | $(-2\alpha\beta - 2\alpha\beta^2)/2$ | $(1 + 2\beta - \alpha^2 - 2\alpha^2\beta)/2$ | $(-2\beta - 2\beta^2)/2$ | $(-2\alpha - 4\alpha\beta)/2$ | $(2 - 2\alpha^2)/2$ |
| $\phi_4$ | $(-1 + 2\alpha + \beta^2 - 2\alpha\beta^2)/2$ | $(2\beta\alpha - 2\beta\alpha^2)/2$ | $(2 - 2\beta^2)/2$ | $(2\beta - 4\alpha\beta)/2$ | $(2\alpha - 2\alpha^2)/2$ |
| $\phi_5$ | $(2\alpha\beta - 2\alpha\beta^2)/2$ | $(-1 + 2\beta + \alpha^2 - 2\alpha^2\beta)/2$ | $(2\beta - 2\beta^2)/2$ | $(2\alpha - 4\alpha\beta)/2$ | $(2 - 2\alpha^2)/2$ |
| $\phi_6$ | $(-\beta - 2\alpha\beta + \beta^2 + 2\alpha\beta^2)/4$ | $(-\alpha - \alpha^2 + 2\beta\alpha + 2\alpha^2\beta)/4$ | $(-2\beta + 2\beta^2)/4$ | $(-1 - 2\alpha + 2\beta + 4\alpha\beta)/4$ | $(2\alpha + 2\alpha^2)/4$ |
| $\phi_7$ | $(\beta + 2\alpha\beta + \beta^2 + 2\alpha\beta^2)/4$ | $(\alpha + \alpha^2 + 2\beta\alpha + 2\alpha^2\beta)/4$ | $(2\beta + 2\beta^2)/4$ | $(1 + 2\alpha + 2\beta + 4\alpha\beta)/4$ | $(2\alpha + 2\alpha^2)/4$ |
| $\phi_8$ | $(-\beta + 2\alpha\beta - \beta^2 + 2\alpha\beta^2)/4$ | $(-\alpha + \alpha^2 - 2\beta\alpha + 2\alpha^2\beta)/4$ | $(2\beta + 2\beta^2)/4$ | $(-1 + 2\alpha - 2\beta + 4\alpha\beta)/4$ | $(-2\alpha + 2\alpha^2)/4$ |
| $\phi_9$ | $(\beta - 2\alpha\beta - \beta^2 + 2\alpha\beta^2)/4$ | $(\alpha - \alpha^2 - 2\beta\alpha + 2\alpha^2\beta)/4$ | $(-2\beta + 2\beta^2)/4$ | $(1 - 2\alpha - 2\beta + 4\alpha\beta)/4$ | $(-2\alpha + 2\alpha^2)/4$ |

| $\phi_i$ | $\phi,_{\alpha\alpha\alpha}$ | $\phi,_{\alpha\beta\alpha}$ | $\phi,_{\beta\beta\alpha}$ | $\phi,_{\beta\beta\beta}$ |
|---|---|---|---|---|
| $\phi_1$ | 0 | $4\beta$ | $4\alpha$ | 0 |
| $\phi_2$ | 0 | $-2\beta$ | $(-1 - 2\alpha)$ | 0 |
| $\phi_3$ | 0 | $(-1 - 2\beta)$ | $-2\alpha$ | 0 |
| $\phi_4$ | 0 | $-2\beta$ | $(1 - 2\alpha)$ | 0 |
| $\phi_5$ | 0 | $(1 - 2\beta)$ | $-2\alpha$ | 0 |
| $\phi_6$ | 0 | $(-1 + 2\beta)/2$ | $(1 + 2\alpha)/2$ | 0 |
| $\phi_7$ | 0 | $(1 + 2\beta)/2$ | $(1 + 2\alpha)/2$ | 0 |
| $\phi_8$ | 0 | $(1 + 2\beta)/2$ | $(-1 + 2\alpha)/2$ | 0 |
| $\phi_9$ | 0 | $(-1 + 2\beta)/2$ | $(-1 + 2\alpha)/2$ | 0 |

Table 3.1: Finite difference approximations for the local partial derivatives based on a 9 node mesh.

The vectors $\{X_p\}$, $\{Y_p\}$ and $\{Z_p\}$ contain the x, y and z coordinates of the nine nodes in the form:

$$\{X_p\}^t = \{ x_1 \ x_2 \ x_3 \ \ldots\ldots\ldots\ldots \ x_n \} \tag{3.1.7a}$$

$$\{Y_p\}^t = \{ y_1 \ y_2 \ y_3 \ \ldots\ldots\ldots\ldots \ y_n \} \tag{3.1.7b}$$

$$\{Z_p\}^t = \{ z_1 \ z_2 \ z_3 \ \ldots\ldots\ldots\ldots \ z_n \} \tag{3.1.7c}$$

In order to solve the system of equations defined by equations 3.1.1a and 3.1.1b, expressions for the partial derivatives of the function $\psi$ with respect to the global x-y coordinate system are required. The CFD method uses the chain rule of partial differentiation to define the relationship between the local and global partial derivatives. If $\psi(\alpha,\beta)$ is differentiated with respect to the x-y coordinate system, then

$$\psi,_i \quad = \beta_i^\gamma \psi,_\gamma \tag{3.1.8}$$

where the indices i and $\gamma$ refer to the global x-y and local curvilinear $\alpha$-$\beta$ coordinates respectively and the coefficient $\beta_i^\gamma$ is defined by the relationship

$$\beta_i^\gamma \quad = \frac{\partial x^\gamma}{\partial x^i} \tag{3.1.9}$$

Similarly the second order partial derivatives $\psi,_{ij}$ can be obtained by operating on equation 3.1.8 as follows:

Let $\qquad \varphi = \psi,_i$

then $\qquad \varphi,_j \quad = \psi,_{ij}$

which can be combined equation 3.1.8 to yield,

$$\psi,_{ij} \;=\; \beta_j^\gamma \varphi,_\gamma \;=\; \beta_j^\gamma (\beta_i^\delta \psi,_\delta ),_\gamma = \beta_j^\gamma \beta_i^\delta,_\gamma \psi,_\delta \;+\; \beta_j^\gamma \beta_i^\delta \psi,_{\delta\gamma} \tag{3.1.10}$$

The transformation coefficients $\beta_j^\gamma$ are computed from the inverse of the Jacobian matrix [J].

Defining

$$[J] = \begin{bmatrix} \frac{\partial x}{\partial \alpha} & \frac{\partial y}{\partial \alpha} \\ \frac{\partial x}{\partial \beta} & \frac{\partial y}{\partial \beta} \end{bmatrix} = \begin{bmatrix} [\phi,_{\alpha}]^t\{X_p\} & [\phi,_{\alpha}]^t\{Y_p\} \\ [\phi,_{\beta}]^t\{X_p\} & [\phi,_{\beta}]^t\{Y_p\} \end{bmatrix} \qquad (3.1.11)$$

then the inverse of the Jacobian becomes

$$[J]^{-1} = \begin{bmatrix} \frac{\partial \alpha}{\partial x} & \frac{\partial \beta}{\partial x} \\ \frac{\partial \alpha}{\partial y} & \frac{\partial \beta}{\partial y} \end{bmatrix} = \begin{bmatrix} \beta_x^{\alpha} & \beta_x^{\beta} \\ \beta_y^{\alpha} & \beta_y^{\beta} \end{bmatrix} \qquad (3.1.12)$$

This matrix is second order, thereby allowing the algebraic expressions for the transformation coefficients to be written in the form

$$\begin{bmatrix} \beta_x^{\alpha} & \beta_x^{\beta} \\ \beta_y^{\alpha} & \beta_y^{\beta} \end{bmatrix} = \frac{1}{C} \begin{bmatrix} y,_{\beta} & -y,_{\alpha} \\ -x,_{\beta} & x,_{\alpha} \end{bmatrix} \qquad (3.1.13)$$

where $C = x,_{\alpha}\, y,_{\beta} - x,_{\beta}\, y,_{\alpha}$

It is now possible to obtain directly from equation 3.1.13 expressions for the coefficients $\beta^{\delta}_{i},_{\gamma}$ :

$$\beta^{\alpha}_{x},_{\alpha} = (y,_{\alpha\beta} C - y,_{\beta} C,_{\alpha}) / C^2$$
$$\beta^{\alpha}_{y},_{\alpha} = (-x,_{\alpha\beta} C + x,_{\beta} C,_{\alpha}) / C^2$$
$$\beta^{\alpha}_{x},_{\beta} = (y,_{\beta\beta} C - y,_{\beta} C,_{\beta}) / C^2$$
$$\beta^{\alpha}_{y},_{\beta} = (-x,_{\beta\beta} C + x,_{\beta} C,_{\beta}) / C^2$$
$$\beta^{\beta}_{x},_{\alpha} = (-y,_{\alpha\alpha} C + y,_{\alpha} C,_{\alpha}) / C^2$$

$$\beta^{\beta}_{y'\alpha} = ( x,_{\alpha\alpha} C - x,_{\alpha} C,_{\alpha} ) / C^2$$

$$\beta^{\beta}_{x'\beta} = ( -y,_{\beta\alpha} C + y,_{\alpha} C,_{\beta} ) / C^2$$

$$\beta^{\beta}_{y'\beta} = ( x,_{\beta\alpha} C - x,_{\alpha} C,_{\beta} ) / C^2 \qquad ...(3.1.14)$$

Equations 3.1.3 and 3.1.10 form the basis of the CFD method which allows the partial derivatives associated with a local region to be transformed to a common global coordinate system. In chapter 2, the general strain-displacement equations for a shell structure were derived in tensor form. These equations are independent of any coordinate system. The ability to relate local displacements and their associated derivatives to a common coordinate system is therefore essential in the numerical analysis of shell structures.

## 3.2 THE NUMERICAL COMPUTATION OF THE SURFACE GEOMETRIC QUANTITIES

In the previous chapter, a surface was defined as the locus of a point whose coordinates were functions of two independent parameters. Its equation was expressed in the form

$$x^i = x^i (\alpha,\beta) \qquad (3.2.1)$$

where the variables $x^\alpha$ and $x^\beta$ established a curvilinear coordinate system. Furthermore, a three dimensional coordinate system consisting of two curvilinear coordinates $x^\alpha$ and $x^\beta$ lying on the surface and a third $x^3$ normal to $x^\alpha$ and $x^\beta$ was introduced.

An important stage in the analysis of shell structures is the computation of a shell's geometric properties. Although many algebraic formulations appear in the literature for surfaces of relatively simple form ( e.g. chapter 13, Flugge [1] ) a general methodology capable of computing quantities such as the metric and curvature tensors for any arbitrary geometry is required.

The curvilinear finite difference method provides a systematic approach for the numerical approximation of the partial derivatives $x,_{\alpha}$ , $y,_{\alpha}$ , $z,_{\alpha}$ , $x,_{\beta}$ , ... etc. This is achieved by firstly discretizing the shell surface to form a computational mesh. At each node, a local curvilinear coordinate system and a computational grid is established. The geometric coordinates of these nodes are assembled to form a relationship between the local and global coordinate system as defined by equation 3.1.6. These expressions are differentiated with respect to the coordinates $x^{\alpha}$ and $x^{\beta}$ to form the finite difference approximations for the partial derivatives.

In this section, expressions defining the geometric quantities (e.g. base vectors, metric tensors and curvature tensors ) in terms of the surface local partial derivatives will be derived.

### 3.2.1 Base Vectors and Metric Tensors

If a surface is defined with respect to a common Cartesian coordinate system, then the position vector **s** of an arbitrary point on the surface can be expressed in the form

$$\mathbf{s} = x\mathbf{i} + y\mathbf{j} + z\mathbf{k} \qquad (3.2.2)$$

The vector components of the covariant base vectors (equation 2.2.3) are obtained by differentiating **s** with respect to the $\alpha$-$\beta$ coordinate system. i.e.,

$$\begin{bmatrix} \mathbf{a}_\alpha \\ \mathbf{a}_\beta \end{bmatrix} = \begin{bmatrix} s,_\alpha \\ s,_\beta \end{bmatrix} = \begin{bmatrix} x,_\alpha & y,_\alpha & z,_\alpha \\ x,_\beta & y,_\beta & z,_\beta \end{bmatrix} \begin{bmatrix} \mathbf{i} \\ \mathbf{j} \\ \mathbf{k} \end{bmatrix} \qquad (3.2.3)$$

To compute the components of the base vector $\mathbf{a}^3$, equation 2.2.5 is written in terms of the surface vectors $\mathbf{a}_\alpha$ and $\mathbf{a}_\beta$:

$$\mathbf{a}_3 = \frac{\mathbf{a}_\alpha \times \mathbf{a}_\beta}{\|\mathbf{a}_\alpha \times \mathbf{a}_\beta\|} \qquad (3.2.4)$$

The vector cross product $\mathbf{a}_\alpha \times \mathbf{a}_\beta$ is expanded by assembling the base vectors $\mathbf{a}_\alpha$ and $\mathbf{a}_\beta$ in the form of a 3x3 determinant:

$$\mathbf{a}_\alpha \times \mathbf{a}_\beta = \begin{vmatrix} \mathbf{i} & \mathbf{j} & \mathbf{k} \\ x,_\alpha & y,_\alpha & z,_\alpha \\ x,_\beta & y,_\beta & z,_\beta \end{vmatrix}$$

$$= [\,(y,_\alpha z,_\beta - z,_\alpha y,_\beta) \;\; (z,_\alpha x,_\beta - x,_\alpha z,_\beta) \;\; (x,_\alpha y,_\beta - y,_\alpha x,_\beta)\,] \begin{bmatrix} \mathbf{i} \\ \mathbf{j} \\ \mathbf{k} \end{bmatrix} \qquad (3.2.5)$$

The direction of the resultant vector is determined by applying the familiar "right-hand" rule.

Defining the quantities

$$\begin{aligned} d_1 &= (y,_\alpha z,_\beta - z,_\alpha y,_\beta) \\ d_2 &= (z,_\alpha x,_\beta - x,_\alpha z,_\beta) \\ d_3 &= (x,_\alpha y,_\beta - y,_\alpha x,_\beta) \end{aligned} \qquad (3.2.6)$$

enables the vector norm $\|\mathbf{a}_\alpha \times \mathbf{a}_\beta\|$ to be expressed as

$$H = \|\mathbf{a}_\alpha \times \mathbf{a}_\beta\| = \sqrt{d_1^2 + d_2^2 + d_3^2} \qquad (3.2.7)$$

Substituting equations 3.2.5, 3.2.6 and 3.2.7 into 3.2.4, allows the components of the base vectors $\mathbf{a}_\alpha$, $\mathbf{a}_\beta$ and $\mathbf{a}_3$ to be summarized in the form:

$$\begin{bmatrix} \mathbf{a}_\alpha \\ \mathbf{a}_\beta \\ \mathbf{a}_3 \end{bmatrix} = \begin{bmatrix} x,_\alpha & y,_\alpha & z,_\alpha \\ x,_\beta & y,_\beta & z,_\beta \\ d_1/H & d_2/H & d_3/H \end{bmatrix} \begin{bmatrix} \mathbf{i} \\ \mathbf{j} \\ \mathbf{k} \end{bmatrix} \qquad (3.2.8)$$

The covariant components of the metric tensor $a_{\alpha\beta}$ are obtained by dot multiplying the base vectors defined in equation 3.2.3. i.e.,

$$a_{\alpha\alpha} = x,_\alpha \, x,_\alpha + y,_\alpha \, y,_\alpha + z,_\alpha \, z,_\alpha \qquad (3.2.9a)$$

$$a_{\alpha\beta} = a_{\beta\alpha} = x,_\alpha \, x,_\beta + y,_\alpha \, y,_\beta + z,_\alpha \, z,_\beta \qquad (3.2.9b)$$

$$a_{\beta\beta} = x,_\beta \, x,_\beta + y,_\beta \, y,_\beta + z,_\beta \, z,_\beta \qquad (3.2.9c)$$

In section 2.1 the following relationship between the covariant and contravariant metric tensors was established.

$$a^{\beta\gamma} a_{\gamma\alpha} = \delta^\beta_\alpha \qquad (3.2.10)$$

This equation is expanded and assembled in matrix form to represent a system of four equations:

$$\begin{bmatrix} a_{\alpha\alpha} & a_{\alpha\beta} \\ a_{\beta\alpha} & a_{\beta\beta} \end{bmatrix} \begin{bmatrix} a^{\alpha\alpha} & a^{\alpha\beta} \\ a^{\beta\alpha} & a^{\beta\beta} \end{bmatrix} = \begin{bmatrix} 1 & 0 \\ 0 & 1 \end{bmatrix} \qquad (3.2.11)$$

Having computed the covariant components of the metric tensor, the contravariant components are determined by simply inverting the matrix containing the covariant components:

$$\begin{bmatrix} a^{\alpha\alpha} & a^{\alpha\beta} \\ a^{\beta\alpha} & a^{\beta\beta} \end{bmatrix} = \frac{1}{(a_{\alpha\alpha}\, a_{\beta\beta} - a_{\alpha\beta}\, a_{\beta\alpha})} \begin{bmatrix} a_{\beta\beta} & -a_{\alpha\beta} \\ -a_{\beta\alpha} & a_{\alpha\alpha} \end{bmatrix} \qquad (3.2.12)$$

Equations 3.2.8 and 3.2.12 enable the contravariant base vectors $\mathbf{a}^{\alpha}$ and $\mathbf{a}^{\beta}$ to be evaluated. It follows from equation 2.1.11 that

$$\mathbf{a}^{\alpha} = a^{\alpha\gamma}\, \mathbf{a}_{\gamma} \qquad (3.2.13)$$

which can be expanded in matrix form to yield

$$\begin{bmatrix} \mathbf{a}^{\alpha} \\ \mathbf{a}^{\beta} \\ \mathbf{a}^{3} \end{bmatrix} = \begin{bmatrix} a^{\alpha\alpha} & a^{\alpha\beta} & 0 \\ a^{\beta\alpha} & a^{\beta\beta} & 0 \\ 0 & 0 & 1 \end{bmatrix} \begin{bmatrix} \mathbf{a}_{\alpha} \\ \mathbf{a}_{\beta} \\ \mathbf{a}_{3} \end{bmatrix} \qquad (3.2.14)$$

### 3.2.2 The Christoffel Symbol

In section 2.1.4, the Christoffel symbol was introduced to represent the components of the derivative of a base vector. The inplane components of the Christoffel symbol of the first kind were defined by the relationship

$$\Gamma_{\alpha\beta\gamma} = \mathbf{a}_{\alpha,\beta} \cdot \mathbf{a}_{\gamma} \qquad (3.2.15)$$

The first term, $\mathbf{a}_{\alpha,\beta}$, is expressed in terms of the unit vectors **i**, **j** and **k** by differentiating equation 3.2.3:

$$\mathbf{a}_{\alpha,\beta} = [\, x_{,\alpha\beta} \;\; y_{,\alpha\beta} \;\; z_{,\alpha\beta} \,] \begin{bmatrix} \mathbf{i} \\ \mathbf{j} \\ \mathbf{k} \end{bmatrix} \qquad (3.2.16)$$

This relationship is combined with equation 3.2.3 to form

$$\Gamma_{\alpha\beta\gamma} = [\, x,_{\alpha\beta} \;\; y,_{\alpha\beta} \;\; z,_{\alpha\beta} \,] \cdot [\, x,_{\gamma} \;\; y,_{\gamma} \;\; z,_{\gamma} \,]$$
$$= x,_{\alpha\beta}\, x,_{\gamma} + y,_{\alpha\beta}\, y,_{\gamma} + z,_{\alpha\beta}\, z,_{\gamma} \qquad (3.2.17)$$

The inplane components of the Christoffel symbols defined above are combined with the contravariant components of the metric tensor to raise the third index. This procedure can be summarized in matrix form and allows the Christoffel symbols of the second kind to be computed:

$$\begin{bmatrix} \Gamma^{\alpha}_{\alpha\alpha} & \Gamma^{\beta}_{\alpha\alpha} \\ \Gamma^{\alpha}_{\alpha\beta} & \Gamma^{\beta}_{\alpha\beta} \\ \Gamma^{\alpha}_{\beta\alpha} & \Gamma^{\beta}_{\beta\alpha} \\ \Gamma^{\alpha}_{\beta\beta} & \Gamma^{\beta}_{\beta\beta} \end{bmatrix} = \begin{bmatrix} \Gamma_{\alpha\alpha\alpha} & \Gamma_{\alpha\alpha\beta} \\ \Gamma_{\alpha\beta\alpha} & \Gamma_{\alpha\beta\beta} \\ \Gamma_{\beta\alpha\alpha} & \Gamma_{\beta\alpha\beta} \\ \Gamma_{\beta\beta\alpha} & \Gamma_{\beta\beta\beta} \end{bmatrix} \begin{bmatrix} a^{\alpha\alpha} & a^{\alpha\beta} \\ a^{\beta\alpha} & a^{\beta\beta} \end{bmatrix} \qquad (3.2.18)$$

### 3.2.3 The Curvature Tensor

The curvature tensor $b_{\alpha\beta}$ may be defined by the scalar product

$$b_{\alpha\beta} = \mathbf{a}_{\alpha'\beta} \cdot \mathbf{a}_3 \qquad (3.2.19)$$

Combining equations 3.2.8, 3.2.16 and 3.2.19 the covariant components are determined from the relationship:

$$b_{\alpha\beta} = [\, x,_{\alpha\beta} \;\; y,_{\alpha\beta} \;\; z,_{\alpha\beta} \,] \cdot [\, d_1/H \;\; d_2/H \;\; d_3/H \,]$$
$$= (\, d_1\, x,_{\alpha\beta} + d_2\, y,_{\alpha\beta} + d_3\, z,_{\alpha\beta} \,)/H \qquad (3.2.20)$$

Adopting the procedure developed for raising the third index of the Christoffel symbol, the mixed and contravariant components of the curvature tensor can be computed from the following set of equations.

$$\begin{bmatrix} b^{\alpha}_{\alpha} & b^{\beta}_{\alpha} \\ b^{\alpha}_{\beta} & b^{\beta}_{\beta} \end{bmatrix} = \begin{bmatrix} b_{\alpha\alpha} & b_{\alpha\beta} \\ b_{\beta\alpha} & b_{\beta\beta} \end{bmatrix} \begin{bmatrix} a^{\alpha\alpha} & a^{\alpha\beta} \\ a^{\beta\alpha} & a^{\beta\beta} \end{bmatrix} \qquad (3.2.21)$$

$$\begin{bmatrix} b^{\alpha\alpha} & b^{\alpha\beta} \\ b^{\beta\alpha} & b^{\beta\beta} \end{bmatrix} = \begin{bmatrix} b^{\alpha}_{\alpha} & b^{\alpha}_{\beta} \\ b^{\beta}_{\alpha} & b^{\beta}_{\beta} \end{bmatrix} \begin{bmatrix} a^{\alpha\alpha} & a^{\alpha\beta} \\ a^{\beta\alpha} & a^{\beta\beta} \end{bmatrix} \qquad (3.2.22)$$

### 3.2.4 Covariant Derivative of the Curvature Tensor

In section 2.2.2 the definition for the two dimensional counterpart to the covariant derivative of a tensor was introduced. Subsequently it was applied to plane tensors and the following relationship for the covariant derivative of the curvature tensor was developed:

$$b_{\alpha\beta}\|_{\gamma} = b_{\alpha\beta,\gamma} - b_{\delta\beta}\,\Gamma^{\delta}_{\alpha\gamma} - b_{\alpha\delta}\,\Gamma^{\delta}_{\beta\gamma} \qquad (3.2.23)$$

The terms $b_{\delta\beta}\,\Gamma^{\delta}_{\alpha\gamma}$ and $b_{\alpha\delta}\,\Gamma^{\delta}_{\beta\gamma}$ can be evaluated from the formulae derived in sections 3.2.2 and 3.2.3. Applying the chain rule for partial differentiation, equation 3.2.20 may be differentiated with respect to $x^{\gamma}$ to yield

$$\begin{aligned} b_{\alpha\beta,\gamma} = \; & (\, d_{1,\gamma}\, x_{,\alpha\beta} + d_{2,\gamma}\, y_{,\alpha\beta} + d_{3,\gamma}\, z_{,\alpha\beta}\,) / H \\ & + (\, d_1\, x_{,\alpha\beta\gamma} + d_2\, y_{,\alpha\beta\gamma} + d_3\, z_{,\alpha\beta\gamma}\,) / H \\ & - (\, d_1\, x_{,\alpha\beta} + d_2\, y_{,\alpha\beta} + d_3\, z_{,\alpha\beta}\,)\, H_{,\gamma} / H^2 \end{aligned} \qquad (3.2.24)$$

The quantities $d_{1,\gamma}$, $d_{2,\gamma}$ and $d_{3,\gamma}$ are obtained by differentiating equations 3.2.6:

$$
\begin{aligned}
d_{1,\gamma} &= y_{,\alpha\gamma}\, z_{,\beta} + y_{,\alpha}\, z_{,\beta\gamma} - ( z_{,\alpha\gamma}\, y_{,\beta} + z_{,\alpha}\, y_{,\beta\gamma} ) \\
d_{2,\gamma} &= z_{,\alpha\gamma}\, x_{,\beta} + z_{,\alpha}\, x_{,\beta\gamma} - ( x_{,\alpha\gamma}\, z_{,\beta} + x_{,\alpha}\, z_{,\beta\gamma} ) \\
d_{3,\gamma} &= x_{,\alpha\gamma}\, y_{,\beta} + x_{,\alpha}\, y_{,\beta\gamma} - ( y_{,\alpha\gamma}\, x_{,\beta} + y_{,\alpha}\, x_{,\beta\gamma} )
\end{aligned}
\qquad (3.2.25)
$$

Similarly, equation 3.2.7 can be differentiated to form

$$
H_{,\gamma} = ( d_1\, d_{1,\gamma} + d_2\, d_{2,\gamma} + d_3\, d_{3,\gamma} ) / H \qquad (3.2.26)
$$

Having computed all the components of equation 3.2.23 the following matrix is assembled for the computation of the covariant derivatives of the curvature tensor.

$$
\begin{bmatrix} b_{\alpha\alpha}\|_{\alpha} & b_{\alpha\alpha}\|_{\beta} \\ b_{\alpha\beta}\|_{\alpha} & b_{\alpha\beta}\|_{\beta} \\ b_{\beta\alpha}\|_{\alpha} & b_{\beta\alpha}\|_{\beta} \\ b_{\beta\beta}\|_{\alpha} & b_{\beta\beta}\|_{\beta} \end{bmatrix}
=
\begin{bmatrix} b_{\alpha\alpha,\alpha} & b_{\alpha\alpha,\beta} \\ b_{\alpha\beta,\alpha} & b_{\alpha\beta,\beta} \\ b_{\beta\alpha,\alpha} & b_{\beta\alpha,\beta} \\ b_{\beta\beta,\alpha} & b_{\beta\beta,\beta} \end{bmatrix}
-
\begin{bmatrix} 2b_{\alpha\alpha} & 0 & 2b_{\beta\alpha} & 0 \\ b_{\alpha\beta} & b_{\alpha\alpha} & b_{\beta\beta} & b_{\alpha\beta} \\ b_{\beta\alpha} & b_{\alpha\alpha} & b_{\beta\beta} & b_{\beta\alpha} \\ 0 & 2b_{\alpha\beta} & 0 & 2b_{\beta\beta} \end{bmatrix}
\begin{bmatrix} \Gamma^{\alpha}_{\alpha\alpha} & \Gamma^{\alpha}_{\alpha\beta} \\ \Gamma^{\alpha}_{\beta\alpha} & \Gamma^{\alpha}_{\beta\beta} \\ \Gamma^{\beta}_{\alpha\alpha} & \Gamma^{\beta}_{\alpha\beta} \\ \Gamma^{\beta}_{\beta\alpha} & \Gamma^{\beta}_{\beta\beta} \end{bmatrix}
\qquad (3.2.27)
$$

The contravariant components of the metric tensor are now used to evaluate the mixed and contravariant components of the covariant derivative of the curvature tensor. i.e.,

$$
b^{\beta}_{\alpha}\|_{\gamma} = b_{\alpha\delta}\|_{\gamma}\, a^{\delta\beta} \qquad (3.2.28)
$$

$$
b^{\alpha\beta}\|_{\gamma} = b^{\alpha}_{\delta}\|_{\gamma}\, a^{\delta\beta} \qquad (3.2.29)
$$

## 3.3 THE PRINCIPLE OF VIRTUAL DISPLACEMENTS

Consider a body which is initially in equilibrium. If the body is assumed to undergo a virtual displacement $\delta \mathbf{u}$ from the equilibrium position, then the Principle of Virtual Displacements states that 'the first-order work $\delta W_e$ done by the external forces is equal to the first-order work $\delta W_d$ done by the internal forces (stresses) during the virtual displacement':

$$\delta W_e = \delta W_d \qquad (3.3.1)$$

Essentially, this statement is an alternative form of the equilibrium equations. It is valid for nonlinear geometries and is independent of material behaviour.

For a three dimensional body in motion, equation 3.3.1 may be expressed in the form

$$\int_V \delta\{\eta\}^t \{\sigma\}\, dV + \int_V C\, \delta\{V\}^t \{\dot{V}\}\, dV + \int_V \rho\, \delta\{V\}^t \{\ddot{V}\}\, dV = \int_S \delta\{V\}^t \{P\}\, dS \qquad (3.3.2)$$

where $\rho$ is simply the mass per unit volume, C is a constant referring to the damping properties of the system and the vectors $\{\sigma\}$ and $\{\eta\}$ reflect the stress and strain characteristics of the body respectively. Since the displacement vector $\{V\}$ and the applied force vector $\{P\}$ are functions of time, the above expression becomes a statement of equilibrium for any specific point in time.

If the body is permitted to vibrate freely, i.e. when $\{P\} = 0$, equation 3.3.2 reduces to

$$\int_V \delta\{\eta\}^t \{\sigma\}\, dV + \int_V C\, \delta\{V\}^t \{\dot{V}\}\, dV + \int_V \rho\, \delta\{V\}^t \{\ddot{V}\}\, dV = 0$$

(3.3.3)

In the proposed method, the strain-displacement and constitutive equations developed in chapter 2 are assumed to represent sufficiently the stress-strain characteristics required by equation 3.3.2. In addition, the technique is restricted to a class of problems for which damping is assumed negligible. Therefore, the equation describing the Principle of Virtual Displacements for a shell undergoing free vibration becomes

$$\int_V \delta\{\eta_{\alpha\beta}\}^t \{\sigma^{\alpha\beta}\}\, dV + \int_V \rho\, \delta\{V^i\}^t \{\ddot{V}_i\}\, dV = 0 \qquad (3.3.4)$$

where $\{\eta_{\alpha\beta}\}^t = \{\eta_{\alpha\alpha} \;\; \eta_{\alpha\beta} \;\; \eta_{\beta\alpha} \;\; \eta_{\beta\beta}\}$

$\{\sigma^{\alpha\beta}\}^t = \{\sigma^{\alpha\alpha} \;\; \sigma^{\alpha\beta} \;\; \sigma^{\beta\alpha} \;\; \sigma^{\beta\beta}\}$

## 3.4 DISCRETIZATION AND DISPLACEMENT FIELDS

In addition to providing a general methodology for the calculation of a shell's geometric properties, the CFD method can also be used to approximate the displacements and their derivatives within the region bounded by the nine node mesh.

At each node, the three components of the displacement vector **u** are assumed. Using equation 3.1.3, each component, $u_i$ , within a mesh can be approximated by the relationship

$$u_i = [\phi]^t \{U_i\} \qquad (3.4.1)$$

where $\{U_i\}$ contains the nodal values of the unknown middle surface displacements.

Since the displacements are associated with one mesh, the Principle of Virtual Displacements for the whole structure can be written as

$$\sum_{i=1}^{n_m} \left[ \int_V \delta\{\eta_{\alpha\beta}\}^t \{\sigma^{\alpha\beta}\}\, dV + \int_V \rho\, \delta\{V^i\}^t \{\ddot{V}_i\}\, dV \right]_i = 0 \quad (3.4.2)$$

From equation 2.6.67 the general displacement vector $\{V^i\}$ can be expressed in the form

$$\{V^i\} = [D^{cn}]\{U_n\} \quad (3.4.3)$$

where $\{U_n\}$ contains the unknown displacement components associated with the curvilinear finite difference mesh and the matrix $[D^{cn}]$ contains a series of expressions related to the shell geometry. Similarly, it will be shown in section 3.5 that the covariant displacement vector $\{V_i\}$, the strain component vector $\{\eta_{\alpha\beta}\}$ and the stress component vector $\{\sigma^{\alpha\beta}\}$ can be represented by the matrix equations

$$\{V_i\} = [D_{cv}]\{U_n\} \quad (3.4.4)$$

$$\{\eta_{\alpha\beta}\} = [H_{\alpha\beta}]\{U_n\} \quad (3.4.5)$$

$$\{\sigma^{\alpha\beta}\} = [\Sigma^{\alpha\beta}]\{U_n\} \quad (3.4.6)$$

Since the matrices $[H_{\alpha\beta}]$ and $[D^{cn}]$ are independent of displacements

$$\delta\{V^\alpha\} = [D^{cn}]\,\delta\{U_n\} \quad (3.4.7)$$

and $$\delta\{\eta_{\alpha\beta}\} = [H_{\alpha\beta}]\,\delta\{U_n\} \tag{3.4.8}$$

Considering the mesh associated with the i[th] node, the above set of equations can be substituted into 3.4.2 to yield:

$$\delta\{U_n\}_i^t\left[\int_V [H_{\alpha\beta}]^t\,[\Sigma^{\alpha\beta}]\,dV\,\{U_n\}_i + \int_V \rho\,[D^{cn}]^t\,[D_{cv}]\,dV\,\{\ddot{U}_n\}_i\right] = 0 \tag{3.4.9}$$

Defining

$$[K]_i = \int_V [H_{\alpha\beta}]^t\,[\Sigma^{\alpha\beta}]\,dV \tag{3.4.10}$$

and $$[M]_i = \int_V \rho\,[D^{cn}]^t\,[D_{cv}]\,dV \tag{3.4.11}$$

then, for an arbitrary virtual displacement, equation 3.4.2 may be written as

$$\sum_{i=1}^{n_m}\left[[M]_i\,\{\ddot{U}_n\}_i + [K]_i\,\{U_n\}_i\right] = \{0\} \tag{3.4.12}$$

The matrices $[K]_i$ and $[M]_i$ are referred to as the stiffness and mass matrices respectively.

One of the most critical steps in applying the curvilinear finite difference method to the analysis of shell structures is the selection of the displacement field. Kwok [71, 72] utilized the invariant properties associated with tensor components to introduce the displacements $u_\alpha$, $u_\beta$, and w as the nodal unknowns in equation 3.4.1. Being invariant, the components are independent of the reference system from which they are referred to and it was concluded by Kwok that any further transformation to a common coordinate system would not be necessary. This proposal was applied successfully to the static analysis of hyperbolic, spherical, cylindrical and conoidal shells.

However, in adopting this approach, Kwok has introduced a number of important theoretical and computational assumptions which limit the techniques generality. By defining the displacement vector **u** in terms of its covariant components, it is assumed that the corresponding base vectors $\mathbf{g}^\alpha$ are continuous and single valued along the coordinate curves. This condition is satisfied when the coordinate curves are exactly defined by a piecewise second order function. Figure 3.3 illustrates some of the admissible mesh discretizations.

In contrast, a distorted mesh such as the one illustrated in figure 3.4 will contain discontinuities along the assumed coordinate curves. The analysis of simple shell and plate structures based on similar mesh discretizations has yielded inaccurate results and will be discussed in further detail in chapter 4.

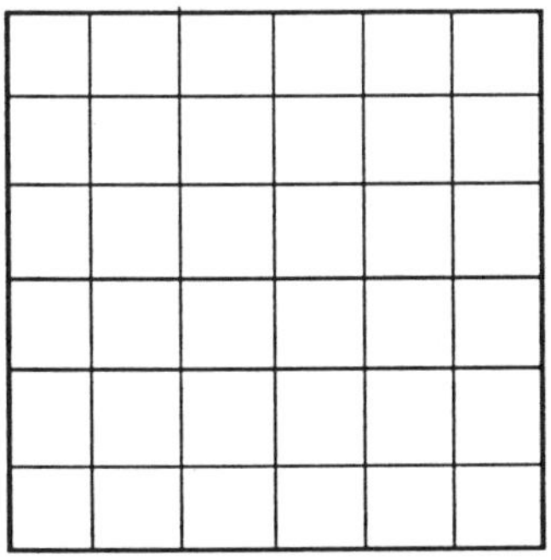

(a) Square Mesh

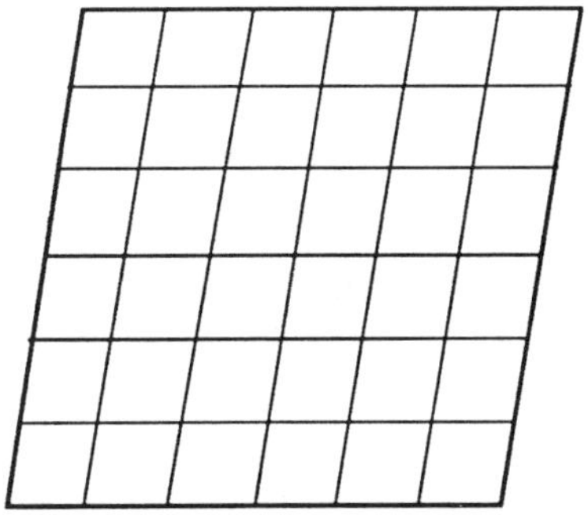

(b) Oblique Mesh

Figures 3.3 (a)-(b): Admissible mesh discretizations.

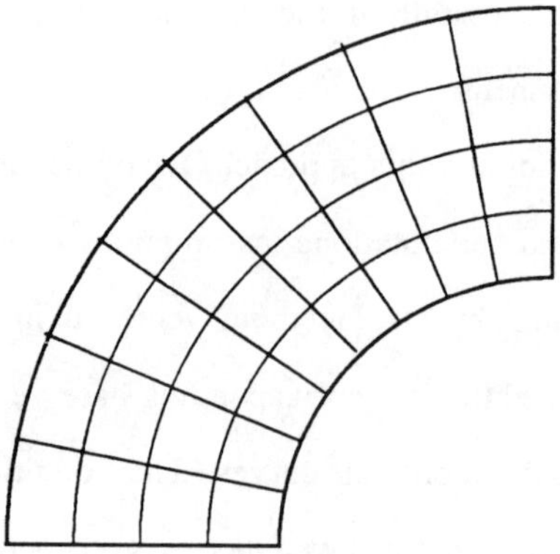

(c) Circular Mesh

Figure 3.3(c): Admissible mesh discretizations.

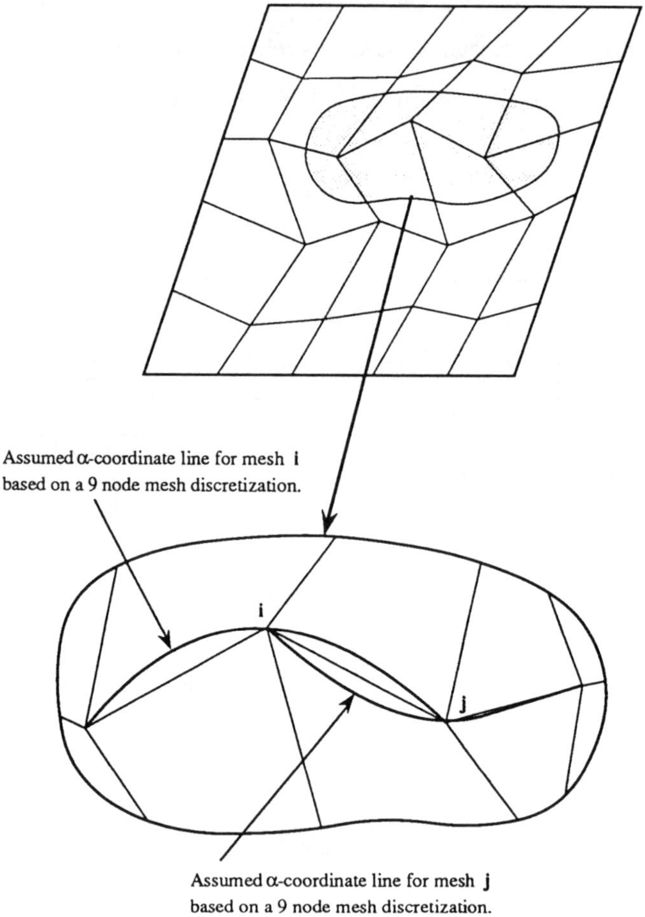


Figure 3.4: A distorted mesh discretization.

To overcome this problem, the displacement field is selected with respect to a common global coordinate system. For an internal mesh, the displacement components $u_x$, $u_y$, and $u_z$ are chosen as nodal unknowns and the vector $\{U_n\}$ is assumed to take the form

$$\{U_n\}^t = \{\ {}^1u_x\ {}^1u_y\ {}^1u_z\ {}^2u_x\ {}^2u_y\ {}^2u_z\ \ldots\ {}^nu_x\ {}^nu_y\ {}^nu_z\ \}^t \qquad (3.4.13)$$

where the quantity ${}^1u_x$ refers to the $u_x$ displacement component associated with the primary local node (see figure 3.5 ).

The covariant components of the local displacements $u_\alpha$, $u_\beta$ and w can be computed by expanding the transformation equations defined in section 2.1.3. i.e.,

$$u_\alpha = \beta^i_\alpha u_i \qquad (3.4.14)$$

$$w = \beta^i_3 u_i \qquad (3.4.15)$$

Similarly, the displacement derivatives required by the general strain-displacement equations are given by the following set of relationships:

$$u_{\alpha,\beta} = \beta^i_{\alpha,\beta} u_i + \beta^i_\alpha u_{i,\beta} \qquad (3.4.16)$$

$$w_{,\alpha} = \beta^i_{3,\alpha} u_i + \beta^i_3 u_{i,\alpha} \qquad (3.4.17)$$

$$u_{\alpha,\beta\gamma} = \beta^i_{\alpha,\beta\gamma} u_i + \beta^i_{\alpha,\beta} u_{i,\gamma} + \beta^i_{\alpha,\gamma} u_{i,\beta} + \beta^i_\alpha u_{i,\beta\gamma} \qquad (3.4.18)$$

$$w_{,\alpha\beta} = \beta^i_{3,\alpha\beta} u_i + \beta^i_{3,\alpha} u_{i,\beta} + \beta^i_{3,\beta} u_{i,\alpha} + \beta^i_3 u_{i,\alpha\beta} \qquad (3.4.19)$$

In the numerical implementation, the above set of transformations are expanded and arranged in matrix form. The resultant expressions are documented in appendix A.

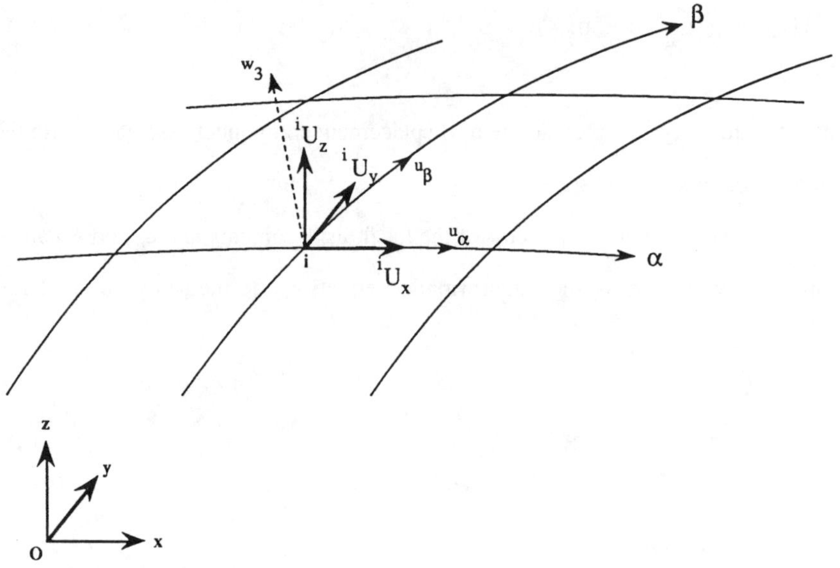


Figure 3.5: Nodal displacement components.

## 3.5 THE NUMERICAL IMPLEMENTATION OF THE GENERAL SURFACE STRESS, STRAIN AND DISPLACEMENT COMPONENTS

In section 3.4 and appendix A, the local displacements at any point within a curvilinear mesh were approximated by the equation:

$$\{u_i\} = \{ u_\alpha \; u_\beta \; w \}^t = [\Phi^u] \{U_n\} \qquad (3.5.1)$$

The matrix $[\Phi^u]$ contains the interpolation and transformation coefficients for a given $(\alpha,\beta)$ and $\{U_n\}$ contains the three unknown global displacement components assigned to each of the nodes within the mesh. By differentiating 3.5.1 with respect to the $\alpha$-$\beta$ coordinate system, similar expressions were obtained for the local partial derivatives of the middle surface displacements. i.e.,

$$\{u_{\alpha,\beta}\} = \{ u_{\alpha,\alpha} \; u_{\alpha,\beta} \; u_{\beta,\alpha} \; u_{\beta,\beta} \}^t = [\Phi^{u\alpha,\beta}] \{U_n\} \qquad (3.5.2a)$$

$$\{w_{,\alpha}\} = \{ w_{,\alpha} \; w_{,\beta} \}^t = [\Phi^{w,\alpha}] \{U_n\} \qquad (3.5.2b)$$

$$\{w_{,\alpha\beta}\} = \{ w_{,\alpha\alpha} \; w_{,\alpha\beta} \; w_{,\beta\alpha} \; w_{,\beta\beta} \}^t = [\Phi^{w,\alpha\beta}] \{U_n\} \qquad (3.5.2c)$$

In this section, the displacement approximations defined above are used extensively in the formulation of a series of matrix equations which represent the basic kinematic behaviour of a shell. Subsequently, these expressions are combined with equations 3.4.10 and 3.4.11 to enable the stiffness and mass matrices to be computed.

### 3.5.1 The General Surface Displacement Components

The covariant components of the general surface displacement, $v_\alpha$, are expressed in terms of the displacements of the middle surface by equation 2.6.67. This expression can be expanded and assembled in matrix form to yield

$$\{V_i\} = \{ v_\alpha \; v_\beta \; w \}^t = [D_{cv}] \{U_n\} \qquad (3.5.3a)$$

where

$$[D_{cv}] = [ [A][\Phi^u] - z[B][\Phi^{w,\alpha}] ] \qquad (3.5.3b)$$

$$[A] = \begin{bmatrix} \mu^1_\gamma \mu^\gamma_1 & \mu^2_\gamma \mu^\gamma_1 & 0 \\ \mu^1_\gamma \mu^\gamma_2 & \mu^2_\gamma \mu^\gamma_2 & 0 \\ 0 & 0 & 1 \end{bmatrix} \qquad (3.5.3c)$$

and

$$[B] = \begin{bmatrix} \mu^1_1 & \mu^2_1 \\ \mu^1_2 & \mu^2_2 \\ 0 & 0 \end{bmatrix} \qquad (3.5.3d)$$

A relationship defining the contravariant components of the displacement vector is obtained by applying the metric tensor $g^{\gamma\alpha}$. i.e.,

$$v^\alpha = v_\gamma \, g^{\gamma\alpha} \qquad (3.5.4a)$$

$$w^3 = w \qquad (3.5.4b)$$

Combining equation 3.5.3a with 2.6.9b and assembling it in matrix form gives

$$\{V^i\} = \{ v^\alpha \; v^\beta \; w \}^t = [D^{cn}] \{U_n\} \qquad (3.5.5a)$$

where $$[D^{cn}] = [C][D_{cv}] \qquad (3.5.5b)$$

and $$[C] = \begin{bmatrix} \lambda^1_\gamma\lambda^1_\delta a^{\gamma\delta} & \lambda^2_\gamma\lambda^1_\delta a^{\gamma\delta} & 0 \\ \lambda^1_\gamma\lambda^2_\delta a^{\gamma\delta} & \mu^2_\gamma\mu^2_\gamma a^{\gamma\delta} & 0 \\ 0 & 0 & 1 \end{bmatrix} \qquad (3.5.5c)$$

### 3.5.2 The General Strain Tensor

A similar approach to the one adopted for the formation of the general surface displacement components is employed in defining the strain tensor $\{\eta_{\alpha\beta}\}$. Here, each component of equation 2.6.69 is expanded and assembled in matrix form to yield

$$\{\eta_{\alpha\beta}\} = \{\ \eta_{\alpha\alpha}\ \eta_{\alpha\beta}\ \eta_{\beta\alpha}\ \eta_{\beta\beta}\ \}^t = [H_{\alpha\beta}]\{U_n\} \qquad (3.5.6a)$$

where $$[H_{\alpha\beta}] = [F]\{\varepsilon_{\alpha\beta}\} - z[G]\{\kappa_{\alpha\beta}\} \qquad (3.5.6b)$$

$$[F] = \begin{bmatrix} \mu^1_1\mu^1_1 & \mu^1_1\mu^2_1 & \mu^1_1\mu^2_1 & \mu^2_1\mu^2_1 \\ \mu^1_1\mu^1_2 & \mu^1_1\mu^2_2 & \mu^2_1\mu^1_2 & \mu^2_1\mu^2_2 \\ \mu^1_2\mu^1_1 & \mu^1_2\mu^2_1 & \mu^2_2\mu^1_1 & \mu^2_2\mu^2_1 \\ \mu^1_2\mu^1_2 & \mu^1_2\mu^2_2 & \mu^2_2\mu^1_2 & \mu^2_2\mu^2_2 \end{bmatrix} \qquad (3.5.6c)$$

$$\text{and} \qquad [G] = \begin{bmatrix} \mu_1^1 & \mu_1^2 & 0 & 0 \\ \frac{1}{2}\mu_2^1 & \frac{1}{2}\mu_2^2 & \frac{1}{2}\mu_1^1 & \frac{1}{2}\mu_1^2 \\ \frac{1}{2}\mu_2^1 & \frac{1}{2}\mu_2^2 & \frac{1}{2}\mu_1^1 & \frac{1}{2}\mu_1^2 \\ 0 & 0 & \mu_2^1 & \mu_2^2 \end{bmatrix} \qquad (3.5.6d)$$

The middle surface strain tensor, $\{\varepsilon_{\alpha\beta}\}$, and the change in curvature tensor, $\{\kappa_{\alpha\beta}\}$, defined by equations 2.6.50 and 2.6.68 respectively may be expanded to yield the following set of equations:

$$\{\varepsilon_{\alpha\beta}\} = [\,[N][\Phi^{u\alpha,\beta}] - [O][\Phi^{u}]\,]\,\{U_n\} \qquad (3.5.7)$$

$$\{\kappa_{\alpha\beta}\} = [\,[R][\Phi^{u}] + [S][\Phi^{u\alpha,\beta}] + [\Phi^{w,\alpha\beta}] - [T][\Phi^{w,\alpha}]\,]\,\{U_n\} \qquad (3.5.8)$$

In the above relationships, the coefficient matrices [N], [O], [R], [S] and [T] are defined as follows:

$$[N] = \begin{bmatrix} 1 & 0 & 0 & 0 \\ 0 & \frac{1}{2} & \frac{1}{2} & 0 \\ 0 & \frac{1}{2} & \frac{1}{2} & 0 \\ 0 & 0 & 0 & 1 \end{bmatrix} \qquad (3.5.9a)$$

$$[O] = \begin{bmatrix} \Gamma_{11}^1 & \Gamma_{11}^2 & b_{11} \\ \Gamma_{12}^1 & \Gamma_{12}^2 & b_{12} \\ \Gamma_{21}^1 & \Gamma_{21}^2 & b_{21} \\ \Gamma_{22}^1 & \Gamma_{22}^2 & b_{22} \end{bmatrix} \qquad (3.5.9b)$$

$$[R] = \begin{bmatrix} b^1_1\|_1 & b^2_1\|_1 & b^1_1 b_{11} + b^2_1 b_{21} \\ b^1_2\|_1 - b^1_2\Gamma^1_{11} - (b^2_2 - b^1_1)\Gamma^1_{21} + b^2_1\Gamma^1_{22} & b^2_2\|_1 - b^1_2\Gamma^2_{11} - (b^2_2 - b^1_1)\Gamma^2_{21} + b^2_1\Gamma^2_{22} & b^1_1 b_{12} + b^2_1 b_{22} \\ b^1_1\|_2 + b^1_2\Gamma^1_{11} - (b^1_1 - b^2_2)\Gamma^1_{12} - b^2_1\Gamma^1_{22} & b^2_1\|_2 + b^1_2\Gamma^2_{11} - (b^1_1 - b^2_2)\Gamma^2_{12} - b^2_1\Gamma^2_{22} & b^1_2 b_{11} + b^2_2 b_{21} \\ b^1_2\|_2 & b^2_2\|_2 & b^1_2 b_{12} + b^2_2 b_{22} \end{bmatrix}$$

(3.5.9c)

$$[S] = \begin{bmatrix} 0 & -b^2_1 & b^2_1 & 0 \\ b^1_2 & 0 & b^2_2 - b^1_1 & -b^2_1 \\ -b^1_2 & b^1_1 - b^2_2 & 0 & b^2_1 \\ 0 & b^1_2 & -b^1_2 & 0 \end{bmatrix} \qquad (3.5.9d)$$

$$[T] = \begin{bmatrix} \Gamma^1_{11} & \Gamma^2_{11} \\ \Gamma^1_{12} & \Gamma^2_{12} \\ \Gamma^1_{21} & \Gamma^2_{21} \\ \Gamma^1_{22} & \Gamma^2_{22} \end{bmatrix} \qquad (3.5.9e)$$

### 3.5.3 The General Stress Tensor

In section 2.5, the plane stress form of Hooke's law was derived for a material with isotropic properties. The formulation of the stiffness matrix using equation 3.4.10 requires the contravariant stresses, while Hooke's law (equation

2.5.9) has components of mixed variance. By applying the metric tensor $g^{\alpha\beta}$, the contravariant stress components may be written in the form

$$\sigma^{\alpha\delta} = \sigma^{\alpha}_{\gamma}\, g^{\gamma\delta} \tag{3.5.10}$$

Similarly, the pertinent strain components are expressed in terms of the covariant components of the strain tensor. i.e.,

$$\eta^{\alpha}_{\gamma} = \eta_{\gamma\delta}\, g^{\alpha\delta} \tag{3.5.11}$$

$$\eta^{\zeta}_{\zeta} = \eta_{\zeta\delta} g^{\zeta\delta} \tag{3.5.12}$$

Combining equations 3.5.10-3.5.12 and 2.5.9 yields

$$\sigma^{\alpha\beta} = \frac{E}{(1-\nu^2)}[(1-\nu)\, g^{\alpha\gamma}\, g^{\beta\delta} + \nu\, g^{\alpha\beta}\, g^{\gamma\delta}\,]\, \eta_{\gamma\delta} \tag{3.5.13}$$

which can be written in the form

$$\{\sigma^{\alpha\beta}\} = \{\, \sigma^{\alpha\alpha}\, \sigma^{\alpha\beta}\, \sigma^{\beta\alpha}\, \sigma^{\beta\beta}\, \}^t = [\Sigma^{\alpha\beta}]\, \{U_n\} \tag{3.5.14}$$

where

$$[\Sigma^{\alpha\beta}] = \frac{E}{(1-\nu^2)}\begin{bmatrix} g^{11}g^{11} & g^{11}g^{12} & g^{12}g^{11} & (1-\nu)g^{12}g^{12} + \nu\, g^{22}g^{11} \\ g^{11}g^{21} & (1-\nu)g^{11}g^{22} + \nu\, g^{12}g^{12} & g^{12}g^{21} & g^{12}g^{22} \\ g^{21}g^{11} & g^{21}g^{12} & (1-\nu)g^{22}g^{11} + \nu\, g^{21}g^{21} & g^{22}g^{12} \\ (1-\nu)g^{21}g^{21} + \nu\, g^{11}g^{22} & g^{21}g^{22} & g^{22}g^{21} & g^{22}g^{22} \end{bmatrix}$$

(3.5.15)

and

$$\begin{bmatrix} g^{11} & g^{12} \\ g^{21} & g^{22} \end{bmatrix} = \begin{bmatrix} \lambda^1_\mu \lambda^1_\nu a^{\mu\nu} & \lambda^1_\mu \lambda^2_\nu a^{\mu\nu} \\ \lambda^2_\mu \lambda^1_\nu a^{\mu\nu} & \lambda^2_\mu \lambda^2_\nu a^{\mu\nu} \end{bmatrix} \tag{3.5.16}$$

## 3.6 BOUNDARY CONDITIONS

It has been shown in the preceding sections that the curvilinear finite difference method can be combined with the principle of virtual displacements to yield a system of equations capable of representing the dynamic behaviour of shell structures. Clearly, the solution of this system of equations can only be completed when a minimum number of constraints have been applied to prevent rigid body displacement of the structure.

For shell structures, the boundary conditions most commonly referred to are: the fixed or clamped edge, the simply supported edge and the free edge. In addition, support conditions such as the diaphragm edge can be modelled by modifying some of the constraints associated with the three classical conditions. A mathematical treatment of the various edge conditions is detailed by Green and Zerna [85].

Generally, the constraints used to define the boundary conditions are classified as being either of force or displacement type. For example, a simply supported edge is restrained against deformation and is not capable of resisting normal bending stresses. In contrast, along a free edge the normal bending, membrane and shear forces must all vanish.

Previously, applications of the curvilinear finite difference method to shell and plate structures by Lau [73] and Kwok [71,72] have represented the constraints by using external 'fictitious' nodes to satisfy the derivative conditions which exist at the boundary.

The major difficulty encountered in using such an approach is the construction of the computational mesh along the boundary. The mesh, composed of fictitious and real nodes, must be aligned with the normals along the edge as illustrated in figure 3.6. In the implementations of Lau and Kwok, this was achieved manually by specifying the external node coordinates as part of the input data .

In the proposed method, an attempt has been made to avoid this difficulty by either introducing a lower order mesh along the edges and corners or applying the actual boundary conditions. Essentially, inplane force constraints are replaced by a 'backward differencing' scheme and normal derivative conditions are implemented by incorporating them into the displacement function.

Along a curved boundary, an orthogonal coordinate system normal to the edge is defined. In subsequent development, this frame of reference is known as the n-t-s coordinate system and is illustrated in figure 3.7. Its definition is dependent on which coordinate curve is used to define the boundary.

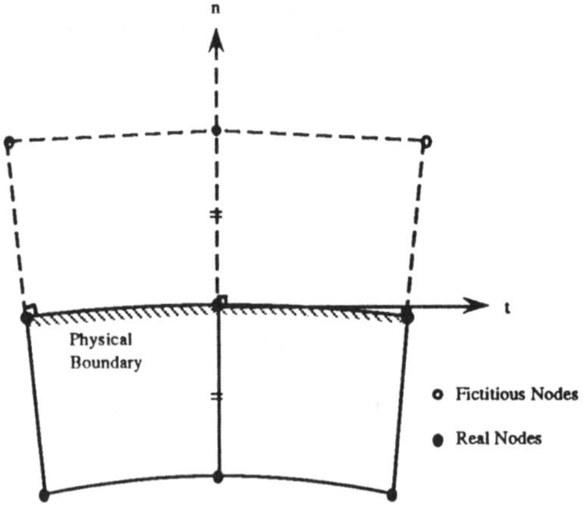


Figure 3.6: The approach adopted by Lau[73] and Kwok[72] for establishing a computational grid along a boundary.

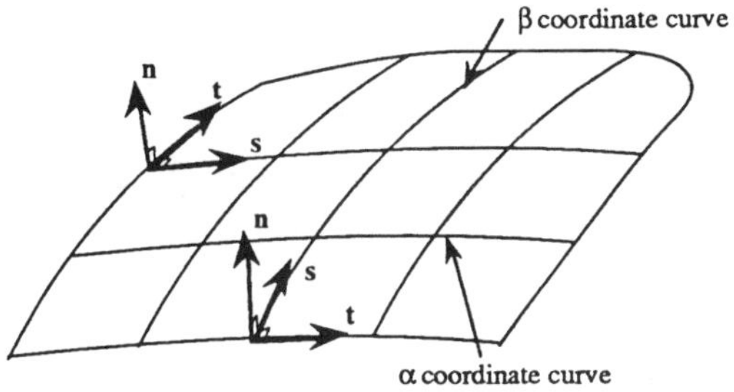


Figure 3.7: Definition of the n-t-s coordinate systems.

The normal vector $\mathbf{a}_n$ is a unit vector which is mutually perpendicular to the in-surface base vectors $\mathbf{a}_\alpha$ and $\mathbf{a}_\beta$. From equation 3.2.4 it may be concluded that

$$\mathbf{a}_n = \mathbf{a}_3 \tag{3.6.1}$$

Furthermore, if the boundary is defined by the β-coordinate curve then the base vector $\mathbf{a}_t$ becomes

$$\mathbf{a}_t = \mathbf{a}_\beta \tag{3.6.2}$$

Since the n-t-s coordinate system is an orthogonal one, then the base vector $\mathbf{a}_s$ must satisfy the relationship

$$\mathbf{a}_s = \mathbf{a}_t \times \mathbf{a}_n \tag{3.6.3}$$

and by using equation 3.2.8 the coordinate system may be represented by the following set of equations:

$$\begin{bmatrix} \mathbf{a}_n \\ \mathbf{a}_t \\ \mathbf{a}_s \end{bmatrix} = \begin{bmatrix} d_1/H & d_2/H & d_3/H \\ x,_\beta & y,_\beta & z,_\beta \\ (y,_\beta d_3 - z,_\beta d_2)/H & (z,_\beta d_1 - x,_\beta d_3)/H & (x,_\beta d_2 - y,_\beta d_1)/H \end{bmatrix} \begin{bmatrix} \mathbf{i} \\ \mathbf{j} \\ \mathbf{k} \end{bmatrix} \tag{3.6.4}$$

Similarly, if the $\alpha$-coordinate curve defines the boundary the following expressions may be used to represent the n-t-s coordinate system.

$$\begin{bmatrix} \mathbf{a}_n \\ \mathbf{a}_t \\ \mathbf{a}_s \end{bmatrix} = \begin{bmatrix} d_1/H & d_2/H & d_3/H \\ x,_\alpha & y,_\alpha & z,_\alpha \\ (y,_\alpha d_3 - z,_\alpha d_2)/H & (z,_\alpha d_1 - x,_\alpha d_3)/H & (x,_\alpha d_2 - y,_\alpha d_1)/H \end{bmatrix} \begin{bmatrix} \mathbf{i} \\ \mathbf{j} \\ \mathbf{k} \end{bmatrix} \tag{3.6.5}$$

A fixed edge is characterized by the partial differential equation

$$\frac{\partial w}{\partial s} = 0 \tag{3.6.6}$$

The derivative, $\partial w/\partial s$, is obtained by differentiating the normal displacement, w, with respect to $x^s$. i.e.,

$$\frac{\partial w}{\partial s} = \frac{\partial w}{\partial \alpha}\frac{\partial \alpha}{\partial s} + \frac{\partial w}{\partial \beta}\frac{\partial \beta}{\partial s} \tag{3.6.7}$$

In the above expression, approximations for the terms $\partial w/\partial \alpha$ and $\partial w/\partial \beta$ are obtained from equation 3.4.17. The partial derivatives $\partial \alpha/\partial s$ and $\partial \beta/\partial s$ are evaluated from the base vectors associated with the $\alpha$-$\beta$ and n-t-s coordinate systems. i.e.,

$$\frac{\partial \alpha}{\partial s} = \beta_s^\alpha = \mathbf{a}^\alpha \cdot \mathbf{a}_s \tag{3.6.8a}$$

$$\frac{\partial \beta}{\partial s} = \beta^{\beta}_{s} = \mathbf{a}^{\beta} \cdot \mathbf{a}_{s} \qquad (3.6.8b)$$

In addition to the above constraint, the displacement components $u_n$, $u_t$ and $u_s$ are zero. These constraints are applied by modifying the structure's stiffness and mass matrices in their fully assembled form. The techniques employed are well documented in many structural analysis texts (e.g. [46,47]) and will not be repeated here.

Similarly, the simply supported edge is represented by the conditions

$$M^{ss} = 0$$

and

$$u_n = u_t = u_s = 0 \qquad (3.6.9)$$

The bending moment $M^{ss}$ is a second order tensor and may be computed by transforming the inplane moment tensor $M^{\alpha\beta}$. i.e.,

$$M^{ss} = \beta^{s}_{\alpha}\, \beta^{s}_{\beta}\, M^{\alpha\beta} \qquad (3.6.10)$$

where

$$M^{\alpha\beta} = -K[\,(1-\nu)\,(\,b^{\alpha}_{\gamma}\,\varepsilon^{\gamma\beta} - b^{\lambda}_{\lambda}\,\varepsilon^{\alpha\beta}\,) + \nu\,(\,b^{\alpha\beta} - b^{\lambda}_{\lambda}\,a^{\alpha\beta}\,)\,\varepsilon^{\mu}_{\mu} - \tfrac{1}{2}(1-\nu)(\kappa^{\alpha\beta} + \kappa^{\beta\alpha}) - \nu\, a^{\alpha\beta}\,\kappa^{\lambda}_{\lambda}\,] \qquad (3.6.11)$$

and

$$K = \frac{E\,t^{3}}{12(1-\nu^{2})}$$

Unlike the previous boundary constraints, the definition of a free edge is substantially more complex. It not only requires a zero bending stress but in addition, inplane membrane stresses and the transverse shear stress along the boundary must also vanish. The complete application of these constraints can not be achieved in the present formulation due to a number of restrictions. Firstly, each 'fictitious' node adjacent to the boundary introduces only three degrees of freedom. The fictitious degrees of freedom are then expressed in terms of the internal displacements by incorporating the constraints into the displacement function. Therefore, only three of the four constraints can be incorporated into the formulation. Secondly, within the definition of the transverse shear stress (see Kwok[72]) third order displacement derivatives exist. Extending the transformation equations established in section 3.4 and appendix A, at least a fourth order polynomial would be required to successfully transform and compute approximations for the shear stress.

In the proposed method, only the normal bending moment constraint is imposed on the displacement function and the additional constraints are formulated by observing that the governing equations describing the inplane displacements, $u_\alpha$, do not require second order continuity. As a result, a first order finite difference equation can be used to approximate the partial derivatives along the boundary. This procedure is analogous to 'backward differencing' and can be achieved numerically, by the following procedure.

Using a displacement interpolation function which incorporates the fictitious degrees of freedom, the inplane displacements $u_t$ and $u_s$ are computed at the fictitious nodes and assembled in matrix form to yield

$$\{u_b\} = [\,\phi_{Hb}]\,\{U_n\} \tag{3.6.12a}$$

where $[\,\phi_{Hb}]$ is a matrix containing the transformed finite difference interpolation coefficients. A similarly set of equations is established by employing a lower order interpolation function which utilizes only the 'real' displacements within the mesh. i.e.,

$$\{u_b\} = [\,\phi_{Lb}]\,\{U_n\} \tag{3.6.12b}$$

Equations 3.6.12a and 3.6.12b are then combined to obtain the expression

$$\{0\} \quad = [\,[\,\phi_{Hb}] \; - \; [\,\phi_{Lb}]\,]\{U_n\} \qquad (3.6.13)$$

Generally, the constraint equations discussed so far can be written in the form

$$\{Bc\} \quad = [\phi_b^r]\,\{U_r\} + [\phi_b^f]\,\{U_f\} \qquad (3.6.14)$$

where $[\phi_b^r]$ and $[\phi_b^f]$ are matrices containing the constraint interpolation coefficients for the 'real' and 'fictitious' degrees of freedom respectively. Setting {Bc} = 0, the fictitious displacements can be expressed in terms of the real displacements by the equation

$$\{U_f\} \quad = -[\phi_b^f]^{-1}[\phi_b^r]\,\{U_r\} \qquad (3.6.15)$$

The fictitious degrees of freedom existing within the displacement interpolation functions can now be eliminated by firstly partitioning the coefficient matrix $[\phi_u]$ into its real and fictitious parts and secondly combining it with equation 3.6.15. i.e.,

$$\{u_i\} = [\phi_u]\,\{U_i\} = [\phi_u^r]\,\{U_r\} + [\phi_u^f]\,\{U_f\} = [\,[\phi_u^r] - [\phi_u^f]\,[\phi_b^f]^{-1}\,[\phi_b^r]\,]\,\{U_r\} \qquad (3.6.16)$$

where $[\phi_u^r]$ and $[\phi_u^f]$ contain the displacement coefficients for the 'real' and 'fictitious' degrees of freedom respectively. The resultant displacement field is independent of the 'fictitious' degrees of freedom and can be used for the generation of the stiffness and mass matrices associated with the mesh.

## 3.7 THE NUMERICAL SOLUTION OF THE EIGENVALUE PROBLEM

For a shell structure undergoing free vibration, the governing equations can be written as

$$[M]\{\ddot{U}\} + [K]\{U\} = \{0\} \qquad (3.7.1)$$

where [K] and [M] are the stiffness and mass matrices respectively. By assuming the displacement vector {U} to be of the form

$$\{U\} = e^{i\omega t}\{\phi\} \qquad (3.7.2)$$

equation 3.7.1 becomes

$$([K] - \lambda[M])\{\phi\} = \{0\}, \qquad \lambda = \omega^2 \qquad (3.7.3)$$

In the above expression, $\lambda$ is referred to as an eigenvalue and $\{\phi\}$ an eigenvector. Generally, if [K] is an $n \times n$ matrix, then the solution of equation 3.7.3 will yield n eigenvalues with corresponding eigenvectors.

In the proposed method, the numerical solution of equation 3.7.3 is achieved by a technique known as 'subspace iteration'. This method enables one to compute several of the lowest eigenvalues and corresponding vectors simultaneously. The theoretical basis of the algorithm is well documented in the literature [86, 89-93] and therefore only a brief overview will be presented.

If the p lowest eigenvalues and associated eigenvectors are required, then the first stage of the method is to select q starting iteration vectors, where $q>p$. The implementation used in this study follows the recommendations of Bathe and Wilson [86] where q is selected by the algorithm

$$q = \min(2p,p+8) \tag{3.7.4}$$

The first column of the starting vector $[X]_1$ is assumed to be a full unit vector and the remaining q-1 columns are unit coordinate vectors with the entries '+1' at the degrees of freedom with the smallest stiffness to mass ratios.

The next stage consists of performing the simultaneous iteration described by the relationship

$$[K]\,[\bar{X}]_{k+1} = [M]\,[X]_k \qquad k = 1,2,3.... \tag{3.7.5}$$

Using Gaussian elimination, equation 3.7.5 yields the n×q matrix $[\bar{X}]_{k+1}$ which is required to find the projections of the operators [K] and [M] onto the q dimensional subspace $\varepsilon_{k+1}$:

$$[K]_{k+1} = [\bar{X}]^t_{k+1}\,[K]\,[\bar{X}]_{k+1} \tag{3.7.6a}$$

$$[M]_{k+1} = [\bar{X}]^t_{k+1}\,[M]\,[\bar{X}]_{k+1} \tag{3.7.6b}$$

The eigensystem of the projected operators can then be expressed as

$$[K]_{k+1}\,[Q]_{k+1} = [M]_{k+1}\,[Q]_{k+1}\,[\Lambda]_{k+1} \tag{3.7.7}$$

where $[Q]_{k+1}$ contains the matrix of eigenvectors and $[\Lambda]_{k+1}$ is a diagonal matrix of eigenvalues. This equation is then converted to the standard eigenvalue problem (see appendix E) and solved using the Jacobian iteration method. The resultant eigenvectors are arranged in ascending order and used to compute $[X]_{k+1}$ for the next iteration. i.e.,

$$[X]_{k+1} = [\bar{X}]^t_{k+1} \, [Q]_{k+1} \qquad (3.7.8)$$

Provided the vectors in $[X]_1$ are not orthogonal to one of the required eigenvectors, then the ith diagonal entry in $[\Lambda]_{k+1}$ converges to $\lambda_i$ and the ith vector in $[X]_{k+1}$ converges to $\phi_i$. The iterative procedure is terminated when the convergence rate given by the following equation is below a predefined tolerance.

$$\frac{|\lambda_i^{(k+1)} - \lambda_i^{(k)}|}{\lambda_i^{(k+1)}} <= \text{tol} \qquad i = 1,\ldots,p \qquad (3.7.9)$$

In the numerical examples presented in this work a tolerance of $10^{-6}$ is assumed.

The final stage of the solution of equation 3.7.3 involves using the Sturm sequence to verify that the required eigenvalues and corresponding eigenvectors have been obtained.

# 4. NUMERICAL IMPLEMENTATION

In the previous chapter, the curvilinear finite difference method was derived in general form for two dimensional field problems and combined with the Principle of Virtual Displacements to formulate a technique capable of determining the free vibration characteristics of shell structures. Included within the following sections are details relating to the implementation of a second order nine node and a third order sixteen node finite difference approximation.

In section 4.1, discussion is confined to the second order implementation. Such an approach requires careful consideration in the selection of the displacement fields and in the integration scheme adopted for the evaluation of the shell's stiffness and mass characteristics. Like the finite element method, the use of the curvilinear finite difference energy method enables the Principle of Virtual Displacements to be written in discretized form. At each node a proportion of the shell, referred to as a 'subdomain', is assigned. Distorted mesh discretizations may result in 'subdomain overlap'. Theoretically, this effect violates the above principle by enabling adjacent subdomain limits to be non-coincident. In section 4.1.2, a procedure which enables integrals with non-fixed limits to be evaluated is outlined.

Section 4.1.3 presents a series of examples which substantiates the method and illustrate the need for adopting a higher order approximation.

In section 4.2, details referring to a third order approximation are presented. Using a sixteen node mesh as a basis, this approach eliminates the subdomain overlap problem encountered in the second order approximation. However, a number of difficulties are encountered in enforcing the conventional engineering boundary constraints. The scheme proposed in section 4.2.2 is based on a study examining the use of higher order finite difference approximations to model beam bending problems.

The approaches outlined above were employed in the construction of two general shell analysis computer programs. In addition, a pre and post processing program has been developed to facilitate the verification and interpretation of results.

These programs were developed using the AIX C [95] compiler on an 'IBM RT' (model 6151) workstation. The computer is configured with 12 megabytes of random access memory, a 720×512 monochrome graphics display, a twin button mouse and uses the 032 reduced instruction set central processing unit.

Both the analysis and processing programs were written entirely in the C programming language [96] to take advantage of the quality compiler available on the RT workstation and many of the software features inherent within the language. In particular, pointer arithmetic is used extensively in the iterative eigenvalue solution scheme in order to speed up the index manipulations of the sparse stiffness and consistent mass matrices. Furthermore, the ability to call functions that are indirectly referenced through a pointer variable improves software management and enables a greater level of flexibility in the program.

The pre and post processing program is based on the interactive mouse, menu, and window environment [97] commonly found on many computing systems. The current implementation uses the IBM RT 'Graphics Support Library' [98] as a basis. This set of routines only contains a primitive set of commands which enable line drawing, text writing, pixel block transfer and locator (e.g. a mouse) access. In order to provide the interactive environment, text editing, window and menu managing routines were developed.

The use of advanced software techniques coupled with the development and runtime environments currently available on workstations such as the IBM RT provides a means of improving the overall time taken for analysis and interpretation of results. These features have greatly contributed to the development of the numerical implementations discussed in the following chapter.

## 4.1 A SECOND ORDER IMPLEMENTATION

### 4.1.1 Second Order CFD Approximation

Like many shell analysis techniques, the application of the curvilinear finite difference method requires the geometry of a shell's middle surface to be defined. To achieve this, the structure is discretized to form a computational grid. At each node, the global coordinates are specified and a local computational mesh is formed.

For internal nodes, each mesh is based on the nine node configuration illustrated in figures 4.1 and 4.2(a). Similarly, at the edges and corners of the structure, the meshes are defined by six and four node configurations respectively.

As detailed in sections 3.1 and 3.2, the computational meshes provide a basis for computing the geometric properties of a shell's middle surface and also approximating its strain-displacement characteristics. In addition, representing the displacement field by a set of finite difference approximations enables the Principle of Virtual Displacements to be written in discretized form. To achieve this, a proportion of the shell, referred to as a 'subdomain', is assigned to each node. A system of equations defining the stiffness and mass properties of the structure may then be evaluated as outlined in sections 3.4 and 3.5.

The definitions of typical subdomains for internal, edge and corner nodes are illustrated in figure 4.2. If a structure is discretized in a 'regular' manner, the subdomain limits of $\alpha \pm \frac{1}{2}$ and $\beta \pm \frac{1}{2}$ may be assigned to internal meshes. However, assuming fixed limits for a 'distorted' mesh may result in adjacent subdomains overlapping. This effect is examined in detail in section 4.1.2.

It is important to emphasize that unlike the finite element method, the subdomain has no real physical significance. It serves as a convenient means of subdividing the shell's surface into regions which can be adequately represented by the assumed nodal displacement and geometric approximations. Continuity of displacements and derivatives is achieved by the finite differencing approach.

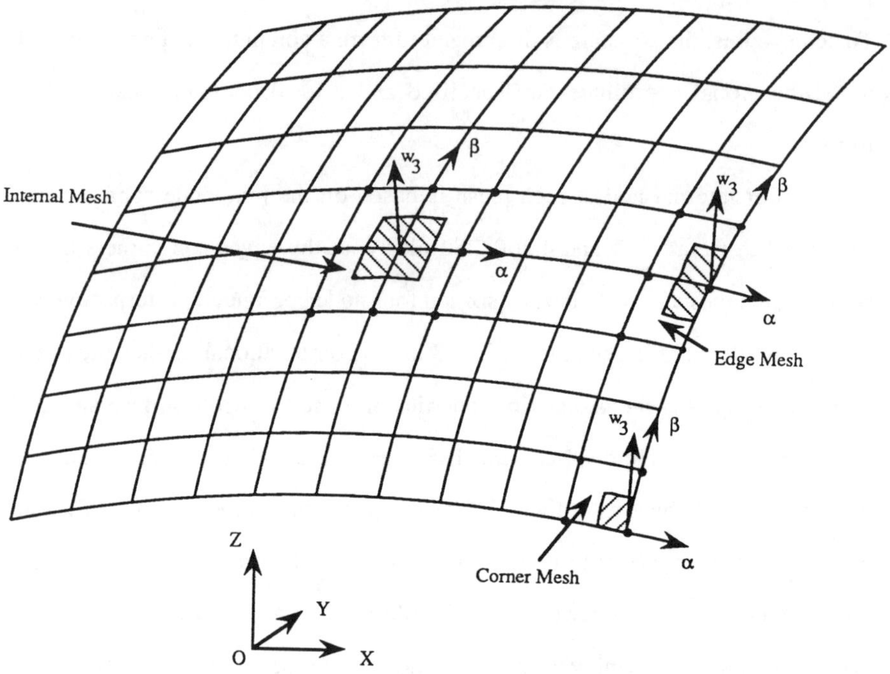


Figure 4.1: Mesh discretization.

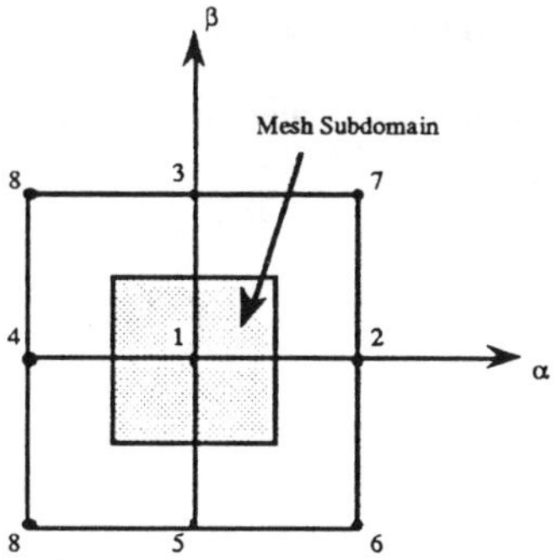


Figure 4.2 (a): Internal nine node mesh.

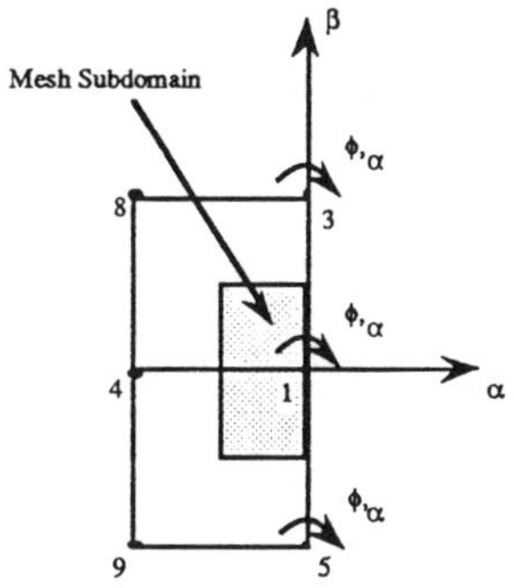


Figure 4.2 (b): Six node edge mesh.

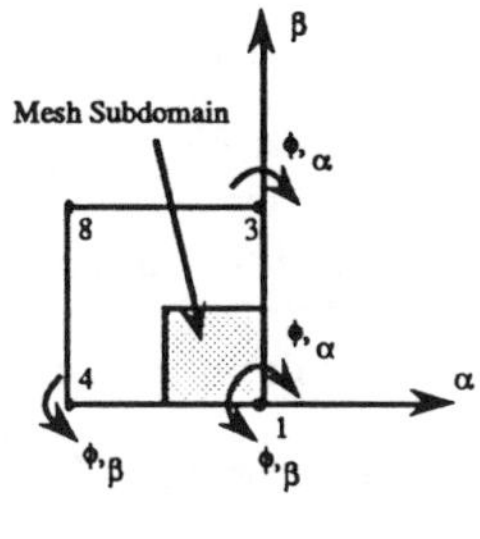


Figure 4.2 (c): Four node corner mesh.

At any point within the nine node mesh, the displacement components or geometric coordinates may be approximated by the expression

$$\psi = a_1 + a_2\alpha + a_3\beta + a_4\alpha^2 + a_5\beta^2 + a_6\alpha\beta + a_7\alpha^2\beta + a_8\alpha\beta^2 + a_9\alpha^2\beta^2 \tag{4.1.1}$$

where $a_1, a_2, a_3, \ldots, a_9$ are constant coefficients. The selection of a nine node mesh enables the displacement components to be described by a complete second order polynomial. This mesh establishes the minimum order of approximation which can be used to compute sufficiently the second order displacement derivatives appearing within the strain-displacement equations.

Following the approach detailed in section 3.1, equation 4.1.1 may be expressed in the form

$$\psi = [\,\phi\,]^t \{\Psi_p\} \tag{4.1.2}$$

where the vector $\{\Psi_p\}$ contains the nine nodal unknowns

$$\{\Psi_p\}^t = \{\,\Psi_1\ \Psi_2\,\Psi_3\ \Psi_4\,\Psi_5\ \Psi_6\,\Psi_7\ \Psi_8\,\Psi_9\,\} \tag{4.1.3}$$

and the matrix $[\phi]$ is a $9{\times}1$ matrix of interpolation functions. The algebraic expressions for $[\phi]$ and its associated derivatives are given by equation 3.1.4 and table 3.1 respectively.

A typical boundary mesh is formed by six nodes as shown in figure 4.2(b). Here the mesh is defined by a curvilinear coordinate system similar to the one used for an internal nine node mesh. However, nodes external to the continuum are excluded. Within an edge mesh, a scalar function may therefore be approximated by the expression

$$\psi = a_1 + a_2\alpha + a_3\beta + a_5\beta^2 + a_6\alpha\beta + a_8\alpha\beta^2 \qquad (4.1.4)$$

Where additional degrees of freedom are required for the boundary conditions, a complete second order polynomial identical to equation 4.1.1 is adopted. In this case, nodes external to the continuum are replaced by their corresponding derivatives at each of the boundary nodes. The vector $\{\Psi_p\}$ which contains the nine nodal unknowns becomes

$$\{\Psi_p\}^t = \{\Psi_1 \; \Psi_{1'\alpha} \; \Psi_3 \; \Psi_4 \; \Psi_5 \; \Psi_{5'\alpha} \; \Psi_{3'\alpha} \; \Psi_8 \; \Psi_9\} \qquad (4.1.5)$$

In appendix B, details of the interpolation functions associated with each boundary mesh are documented.

A similar scheme is adopted for a corner region where the mesh is formed by four nodes. In this case, the approximation function for $\psi$ becomes

$$\psi = a_1 + a_2\alpha + a_3\beta + a_6\alpha\beta \qquad (4.1.6)$$

Where fictitious degrees of freedom are required, the second order polynomial defined by equation 4.1.1 is modified by dropping the term $\alpha^2\beta^2$. i.e.,

$$\psi = a_1 + a_2\alpha + a_3\beta + a_4\alpha^2 + a_5\beta^2 + a_6\alpha\beta + a_7\alpha^2\beta + a_8\alpha\beta^2 \qquad (4.1.7)$$

Given the corner mesh illustrated in figure 4.2(c), the assumed nodal unknowns will be represented by the vector:

$$\{\Psi_p\}^t = \{\Psi_1 \; \Psi_{1'\alpha} \; \Psi_3 \; \Psi_4 \; \Psi_{1'\beta} \; 0 \; \Psi_{3'\alpha} \; \Psi_8 \; \Psi_{4'\beta}\} \qquad (4.1.8)$$

The corresponding matrix of interpolation functions which are required by equation 4.1.2 are documented in appendix B.

The above set of functions provide a basis for approximating the geometric properties and displacement components associated with the shell's middle surface. Together with the expressions developed in the previous chapter, the stress and strain characteristics of general thin shells may be established.

### 4.1.2 Numerical Integration Scheme

The formation of the stiffness and mass matrices defined by equations 3.4.10 and 3.4.11 require the evaluation of the following integrals:

$$\int_V [H_{\alpha\beta}]^t [\Sigma^{\alpha\beta}] \, dV, \qquad \int_V \rho \, [D^{cn}]^t [D_{cv}] \, dV$$

In previous work [71, 72], the integrand associated with the stiffness matrix was assumed to be constant over a subdomain. Based on this assumption, a one-point integration scheme was adopted and shown to produce adequate results when applied to a series of shell structures subjected to normal static loads.

However, in cases where direct in-plane forces are applied, it has been found that a one point integration scheme may, in some cases, produce inaccurate results. This is demonstrated in section 4.1.3 by considering the example of a cantilever plate subjected to a number of concentrated loads. The deflected shape and stress distributions are obviously not in agreement with those predicted by a quadrilateral plane stress finite element analysis [99]. Details highlighting the cause of this discrepancy are discussed in Appendix C. By selecting a uniform mesh and evaluating the stiffness matrix explicitly, it is shown that many terms vary as parabolic functions within a subdomain. In such cases, a one-point integration scheme would approximate these terms with zero stiffness. Numerical simulation has confirmed that in some instances this approximation is poor and the stiffness coefficients need to be evaluated using a higher order integration scheme.

In the finite element method, it is common to evaluate the local stiffness and mass matrices of isoparametric elements using a Gauss-Legendre numerical integration scheme [46,86,87]. Adopting this approach as a basis, the volume integral is firstly expressed in terms of the curvilinear coordinates $\alpha$ and $\beta$. i.e.,

$$\int_V [H_{\alpha\beta}]^t\,[\Sigma^{\alpha\beta}]\,dV = \int_V [H_{\alpha\beta}]^t\,[\Sigma^{\alpha\beta}]\,dA\cdot dz$$

$$= \int_\alpha \int_\beta \int_{z=-t/2}^{z=+t/2} [H_{\alpha\beta}]^t[\Sigma^{\alpha\beta}]\,\det[J]\,d\alpha\,d\beta\,dz \qquad (4.1.9)$$

where $\det[J]$ denotes the determinant of the Jacobian matrix and is evaluated by the expression

$$\det[J] = \begin{vmatrix} x,_\alpha & x,_\beta & d_1/H \\ y,_\alpha & y,_\beta & d_2/H \\ z,_\alpha & z,_\beta & d_3/H \end{vmatrix}$$

$$= H \qquad (4.1.10)$$

Secondly, the three dimensional integral given by equation 4.1.9 is evaluated numerically using the relationship

$$\int_V [H_{\alpha\beta}]^t\,[\Sigma^{\alpha\beta}]\,dV = \sum_{\alpha,\beta,z} w_\alpha\, w_\beta\, w_z\, [H_{\alpha\beta}]^t[\Sigma^{\alpha\beta}]\,\det[J] \qquad (4.1.11)$$

where the integrand $[H_{\alpha\beta}]^t[\Sigma^{\alpha\beta}]\,\det[J]$ is computed at the sampling points $r_\alpha$, $r_\beta$, $r_z$ and combined with the corresponding weighting factors $w_\alpha$, $w_\beta$, $w_z$.

The weights and sampling points given in table 4.1 are for the integration limits -1 to +1. These numerical values may be modified to accommodate the alternative limits *'a'* and *'b'* by using the following equations:

$$\frac{a+b}{2}+\frac{b-a}{2}r_i \quad \text{and} \quad \frac{b-a}{2}w_i$$

| n | $r_i$ | $w_i$ |
|---|---|---|
| 1 | 0. | 2. |
| 2 | ±0.577350269189626 | 1.000000000000000 |
| 3 | ±0.774596669241483 | 0.555555555555556 |
| | 0.000000000000000 | 0.888888888888889 |
| 4 | ±0.861136311594053 | 0.347854845137454 |
| | ±0.339981043584856 | 0.652145154862546 |

Table 4.1: Sampling points and weights in Gauss-Legendre numerical integration.

Having defined a subdomain for each node, it would appear that the stiffness and mass matrices may be computed directly from equation 4.1.11 by setting the integration limits $a = \pm\frac{1}{2}$ and $b = \pm\frac{1}{2}$. Provided the mesh discretization is 'regular' (see section 3.4), numerical tests have confirmed this approach. However, for a distorted mesh, the use of fixed integration limits may violate the Principle of Virtual Displacements by introducing an effect referred to as 'subdomain overlap'. This is illustrated in figure 4.3 where the global coordinates of the integration limits are determined and plotted to scale. Clearly, the integration limits associated with adjacent meshes do not coincide and a region of overlap exists.

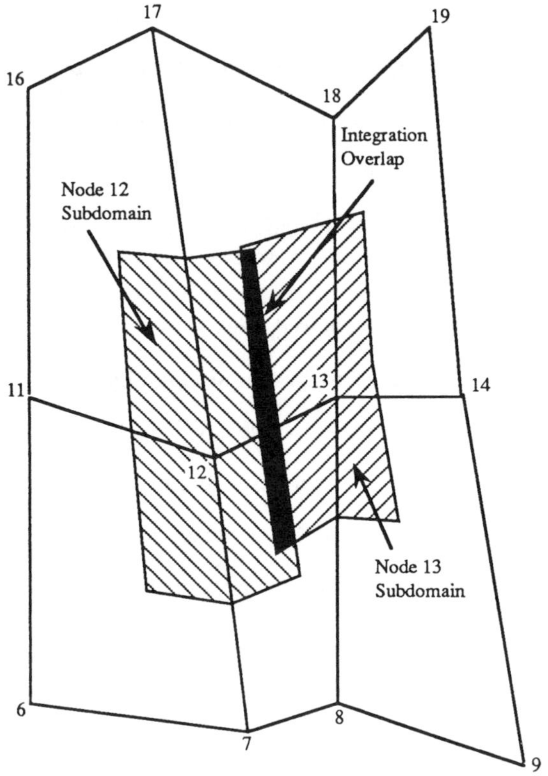


Node Data

| Node | X | Y |
|---|---|---|
| 6 | -0.5 | -0.5 |
| 7 | -0.15 | -0.55 |
| 8 | 0.05 | -0.5 |
| 9 | 0.3 | -0.6 |
| 11 | -0.5 | 0.0 |
| 12 | -0.2 | -0.1 |
| 13 | -0.05 | 0.05 |
| 14 | 0.2 | 0.05 |
| 16 | -0.5 | 0.5 |
| 17 | -0.3 | 0.6 |
| 18 | -0.05 | 0.45 |
| 19 | 0.15 | 0.6 |

NODE 12: Integration Limits

| Alpha | Beta | X | Y |
|---|---|---|---|
| 0.5 | 0.5 | -0.14375 | 0.23353975 |
| 0.0 | 0.5 | -0.24375 | 0.21875 |
| -0.5 | 0.5 | -0.3625 | 0.22421875 |
| -0.5 | 0.0 | -0.33125 | -0.08125 |
| -0.5 | -0.5 | -0.3125 | -0.33515625 |
| 0.0 | -0.5 | -0.16875 | -0.35625 |
| 0.5 | -0.5 | -0.06875 | -0.31328125 |
| 0.5 | 0.0 | -0.10625 | -0.05625 |

NODE 13: Integration Limits

| Alpha | Beta | X | Y |
|---|---|---|---|
| 0.5 | 0.5 | 0.046875 | 0.30078125 |
| 0.0 | 0.5 | -0.0625 | 0.26875 |
| -0.5 | 0.5 | -0.159375 | 0.24140625 |
| -0.5 | 0.0 | -0.1375 | -0.00625 |
| -0.5 | -0.5 | -0.103125 | -0.25546875 |
| 0.0 | -0.5 | -0.0125 | -0.20625 |
| 0.5 | -0.5 | 0.103125 | -0.20859375 |
| 0.5 | 0.0 | 0.0625 | 0.06875 |

Figure 4.3: Subdomain overlap.

To overcome this effect, the integration limits are redefined to avoid subdomain overlap between adjacent meshes. The new limits are mapped to a curvilinear coordinate system ($\eta$-$\xi$) and the required weights and sampling points for a Gauss-Legendre numerical integration scheme are computed:

The following procedure has been developed to enable the integration limits to be redefined:

(1) For each mesh, nine nodes are required to establish the curvilinear coordinate system $\eta$-$\xi$ (see figure 4.5). Adjacent meshes are initially checked to verify if any of the nine integration points have been previously defined.

(2) If an integration point exists, the geometric coordinates (x,y,z) are calculated using the adjacent mesh and the interpolation functions introduced in section 4.1.1 as a basis. The corresponding $\alpha$-$\beta$ coordinates of the current mesh are then computed using the procedure outlined below.

(3) If none of the integration points have been previously defined, then their positions are assumed to be located along the regular subdomain limits defined in figure 4.2.

Central to the above procedure is the curvilinear finite difference method. Based on the nine node mesh described by equations 4.1.1 and 3.1.4, the geometric coordinates (x,y,z) of any point may be expressed in the form

$$f_x(\alpha,\beta) = [\phi]^t\{X_p\} - x \qquad (4.1.12a)$$

$$f_y(\alpha,\beta) = [\phi]^t\{Y_p\} - y \qquad (4.1.12b)$$

$$f_z(\alpha,\beta) = [\phi]^t\{Z_p\} - z \qquad (4.1.12c)$$

Equations 4.1.12 (a) - (c) constitute a system of nonlinear equations and can be solved using the Newton-Raphson method [88]. Given the global coordinates of a point within a mesh, the corresponding $\alpha$-$\beta$ coordinates can be obtained from the relationship

$$\begin{bmatrix}\alpha\\ \beta\end{bmatrix}_{i+1} = \begin{bmatrix}\alpha\\ \beta\end{bmatrix}_{i} - [J]^{-1}\begin{bmatrix}f_x\\ f_y\\ f_z\end{bmatrix}_i \qquad (4.1.13)$$

where the matrix [J] and the components $f_x$, $f_y$ and $f_z$ are defined in equations 4.1.10 and 4.1.12 respectively.

By operating on the Jacobian matrix [J], it is possible to derive the first two rows of its inverse explicitly. Equation 4.1.13 can then be expanded to form

$$\begin{bmatrix}\alpha\\ \beta\end{bmatrix}_{i+1} = \begin{bmatrix}\alpha\\ \beta\end{bmatrix}_{i} - \frac{1}{H}\begin{bmatrix} y,_\beta d_3 - z,_\beta d_2 & z,_\beta d_1 - x,_\beta d_3 & x,_\beta d_2 - y,_\beta d_1 \\ z,_\alpha d_2 - y,_\alpha d_3 & x,_\alpha d_3 - z,_\alpha d_1 & y,_\alpha d_1 - x,_\alpha d_2 \end{bmatrix}\begin{bmatrix}f_x\\ f_y\\ f_z\end{bmatrix}_i$$

(4.1.14)

Even though $H = \det [J] \neq 0$, there is no guarantee that the above algorithm will converge to a solution. For example, figure 4.4 illustrates the mesh discretization for a region of contraflexure. Since the interpolation functions assumed in equation 4.1.12 are second order, the designated point P computed from neighbouring meshes is not unique. A solution to equation 4.1.14 will therefore not exist.

To overcome this problem, one of the geometric constraints given by equation 4.1.12 is released. The modified form of equation 4.1.14 can then be written as

$$\begin{bmatrix}\alpha\\ \beta\end{bmatrix}_{i+1} = \begin{bmatrix}\alpha\\ \beta\end{bmatrix}_{i} - \frac{1}{C}\begin{bmatrix} f_{k'\beta} & -f_{j'\beta} \\ -f_{k'\alpha} & f_{j'\alpha} \end{bmatrix}\begin{bmatrix}f_j\\ f_k\end{bmatrix}_i \qquad (4.1.15)$$

where $\quad C = f_{j'\alpha}\, f_{k'\beta} - f_{j'\beta}\, f_{k'\alpha}$

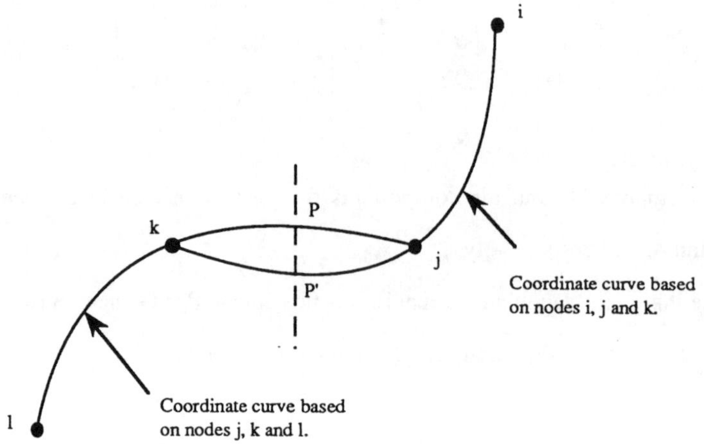


Figure 4.4: Mesh discretization for a region of contraflexure.

Using this form, the determinant of the Jacobian , C, can always be tested prior to evaluating equation 4.1.15. If C = 0, as in the case where the surface is perpendicular to the X-Y plane and the functions $f_x$, $f_y$ are selected, then an alternative set of functions ($f_x$ & $f_z$ or $f_y$ & $f_z$) may be chosen such that $C \neq 0$.

Having computed the integration limits, a relationship between the coordinate systems $\alpha$-$\beta$ and $\eta$-$\xi$ (see figure 4.5) is established using the expressions.

$$\alpha = [\,\phi(\eta,\xi)\,]^t \{A_p\} \tag{4.1.16a}$$

$$\beta = [\,\phi(\eta,\xi)\,]^t \{B_p\} \tag{4.1.16b}$$

The vectors $\{A_p\}$ and $\{B_p\}$ contain the nine nodal values of the integration limits $\alpha_i$ and $\beta_i$ respectively, and the matrix $[\,\phi(\eta,\xi)\,]^t$ is identical to equation 3.1.4 except it is defined with respect to the coordinates $\eta$ and $\xi$.

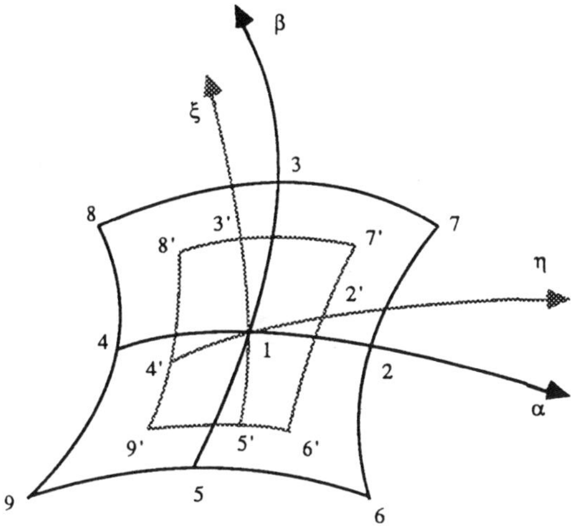


Figure 4.5: Integration limits.

Equations 4.1.16a and 4.1.16b enable the term $d\alpha\, d\beta$ to be written in the form

$$d\alpha\, d\beta \;=\; \det\,[J_{\eta\xi}]\; d\eta\, d\xi \tag{4.1.17}$$

where

$$\det\,[J_{\eta\xi}] = \begin{vmatrix} \alpha,_{\eta} & \alpha,_{\xi} \\ \beta,_{\eta} & \beta,_{\xi} \end{vmatrix}$$

$$= \alpha,_{\eta}\, \beta,_{\xi} \;-\; \alpha,_{\xi}\, \beta,_{\eta} \tag{4.1.18}$$

Evaluating the stiffness and mass matrices is achieved by selecting the appropriate sampling point locations with respect to the $\eta$-$\xi$ coordinate system, determining the corresponding $\alpha$-$\beta$ coordinates from the expressions 4.1.16a and 4.1.16b and evaluating the following integrands.

$$[K]_i = \int_V [H_{\alpha\beta}]^t [\Sigma^{\alpha\beta}] \, dV$$

$$= \sum_{\eta,\xi,z} w_\eta \, w_\xi \, w_z \, [H_{\alpha\beta}]^t [\Sigma^{\alpha\beta}] \det[J] \det [J_{\eta\xi}] \qquad (4.1.19)$$

and

$$[M]_i = \int_V \rho \, [D^{cn}]^t [D_{cv}] \, dV$$

$$= \sum_{\eta,\xi,z} w_\eta \, w_\xi \, w_z \, \rho \, [D^{cn}]^t [D_{cv}] \det[J] \det [J_{\eta\xi}] \qquad (4.1.20)$$

### 4.1.3 Numerical Examples

The purpose of presenting the following case studies is to verify numerically the theoretical concepts discussed in the previous chapters. Briefly, these concepts may be summarized in the following way:

(i) The need to evaluate the stiffness matrix using an integration order higher than the one point approximation adopted in previous research [71, 72].

(ii) The selection of a displacement field whose components are associated with one common global coordinate system.

(iii) The elimination of subdomain overlap for distorted mesh discretizations.

(iv) The technique's performance in evaluating the dynamic characteristics of shell structures of arbitrary geometry.

The first three concepts are illustrated by analysing either a cantilever or simply supported flat plate subjected to concentrated static loads. The fourth involves the dynamic analysis of a series of shell structures and comparing the generated results with those published within the literature.

(i) Numerical Integration:

To illustrate the effect of using a reduced one-point integration scheme, the flat cantilever plate defined in figure 4.6 was analysed. The plate is subjected to a number of static inplane concentrated loads located at the coordinate system origin and at the plates edge. Figures 4.7 - 4.12 summarize the displacement and stress resultants obtained from the analysis using a 16x16 mesh discretization. As illustrated, the problem was analysed using three approaches and the results compared. In cases 1 and 2 the curvilinear finite difference method was employed using a 3x3 Gauss-Legendre and a one-point integration scheme respectively. In case 3, the results of an analysis using a quadrilateral plane stress finite element [99] are presented and serve as a basis for comparison.

Clearly, a one point integration scheme cannot sufficiently approximate the stiffness characteristics of the plate. As shown in figures 4.7 - 4.12, the displacement and stress characteristics do not compare favourably with either the finite element approximation or the employment of a 3x3 integration scheme. The details of a theoretical investigation into the effect of using a low order integration scheme are discussed in Appendix C and section 4.1.2.

Further plane stress analysis has indicated that it is possible to obtain displacement characteristics similar to the ones illustrated in figures 4.8 - 4.10 for a variety of problems. This effect appears to be sensitive to both geometric and load configurations.

A higher order numerical integration scheme will therefore be adopted for the numerical examples presented in the following chapters.

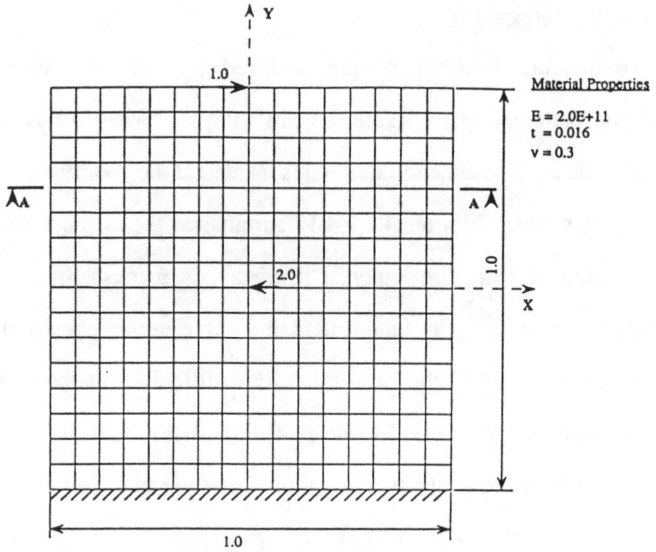


Figure 4.6: Geometric and Material Properties for the Cantilever Plate.

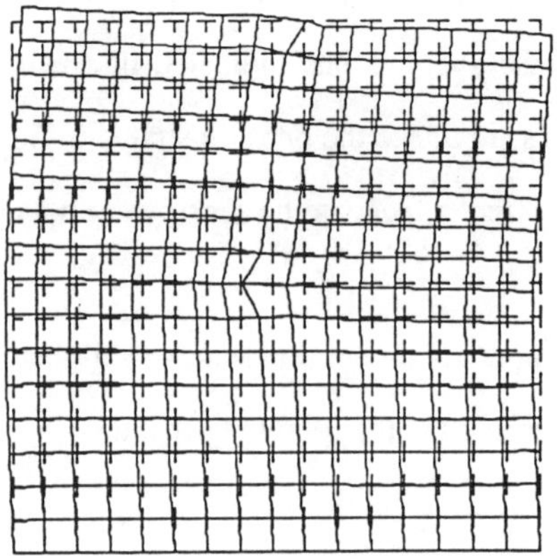

Figure 4.7: Deflected Shape (3x3 Gauss-Legendre Numerical Integration).

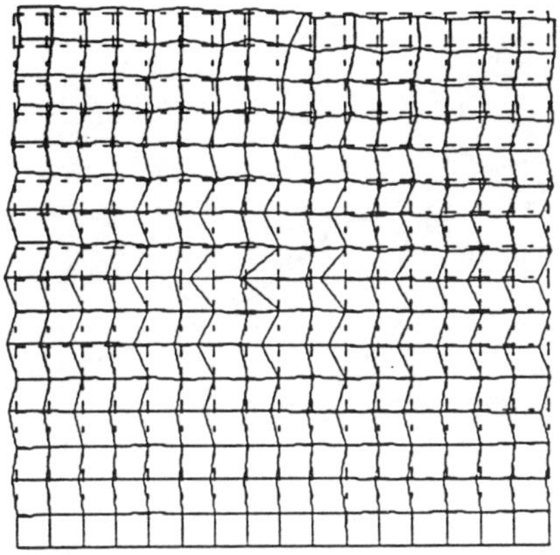

Figure 4.8: Deflected Shape (1-Point Numerical Integration Scheme).

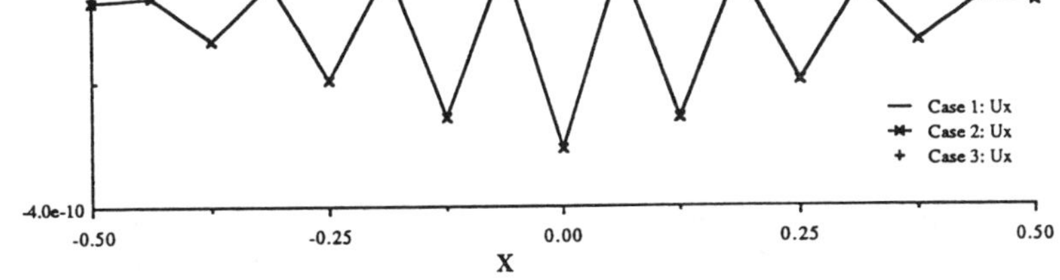


Figure 4.9: Displacement Component Ux along Section A-A.

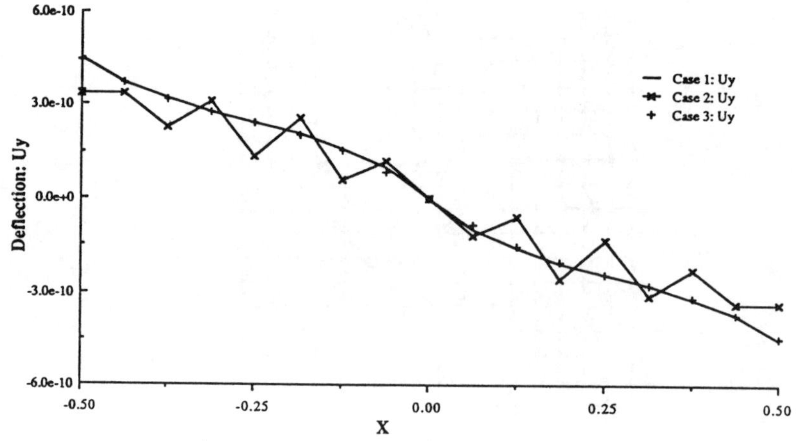


Figure 4.10: Displacement Component $U_y$ along Section A-A.

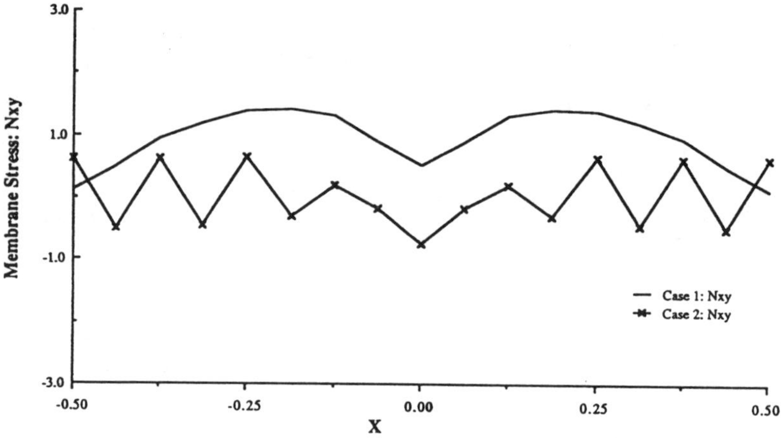


Figure 4.11: Membrane Stress $N_{xy}$ along Section A-A.

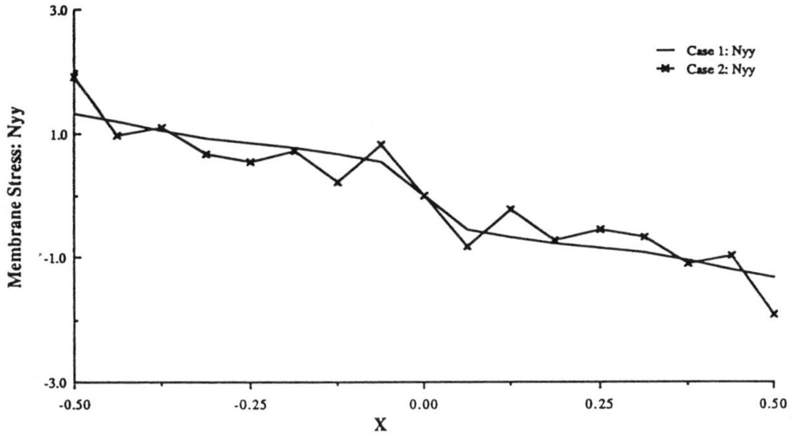


Figure 4.12: Membrane Stress Nyy along Section A-A.

(ii) Displacement Fields:

One of the most critical steps in applying the curvilinear finite difference method to the analysis of shell structures is the selection of the displacement field. In section 3.4 the effect of defining the displacement components $u_\alpha$, $u_\beta$ and w as nodal unknowns was examined with particular reference to distorted mesh discretizations. Figure 4.13 defines the cantilever plate example which will be studied in this section.

Table 4.2 provides a comparison of the displacement components $U_x$ and $U_y$ at two points for various orthogonal mesh discretizations and compares them to a finite element plane stress analysis. For the purposes of comparison, a 16×16 mesh discretization was employed in the following example. As with the previous example, three cases are investigated and the results were compared directly. In case 1, the plate was discretized using an orthogonal mesh arrangement (figure 4.13). For case 2, the mesh was distorted randomly as illustrated in figure 4.14.

Additionally, the mesh line along the x-axis was retained for comparative purposes and the problem analysed using the displacement components $u_x$, $u_y$ and $u_z$ as nodal unknowns. In case 3, the same distorted mesh was employed however the nodal unknowns were assumed to be the covariant displacement components $u_\alpha$, $u_\beta$ and w.

| Mesh | x | y | Displacements | | x | y | Displacements | |
|---|---|---|---|---|---|---|---|---|
| | | | Ux | Uy | | | Ux | Uy |
| 4x4 | -0.5 | 0.0 | 9.0113E-10 | 7.9021E-10 | 0.5 | 0.5 | 2.0991E-9 | -9.4176E-10 |
| 8x8 | -0.5 | 0.0 | 9.0648E-10 | 7.7085E-10 | 0.5 | 0.5 | 2.0691E-9 | -9.2402E-10 |
| 16x16 | -0.5 | 0.0 | 9.0360E-10 | 7.6057E-10 | 0.5 | 0.5 | 2.0537E-9 | -9.1553E-10 |
| 32x32 | -0.5 | 0.0 | 9.0176E-10 | 7.5753E-10 | 0.5 | 0.5 | 2.0494E-9 | -9.1573E-10 |
| 32x32[FEM] | -0.5 | 0.0 | 8.959E-10 | 7.551E-10 | 0.5 | 0.5 | 2.041E-9 | -9.146E-10 |

Table 4.2: Convergence Results

The distorted mesh discretization illustrated in figure 4.14 has been presented to highlight the numerical problems which may occur if the displacement components $u_\alpha$, $u_\beta$ and w are selected as nodal unknowns. This is clearly evident in figures 4.15 - 4.21 where the computed displacements and membrane stresses for case 3 are compared against the orthogonal mesh results. Further numerical analysis has shown that the deterioration in the results is dependant on the degree of mesh distortion. In contrast, the results computed in the second case duplicate the orthogonal mesh displacements and stresses to within acceptable limits.

It is important to emphasize that in a large number of cases, establishing a displacement field with the components $u_\alpha$, $u_\beta$ and w as nodal unknowns will yield results well within acceptable limits of accuracy . This has been shown in previous research by Kwok [71,72]. The use of the global components $u_x$, $u_y$ and $u_z$ enhances the technique's generality in analysing shells of arbitrary geometry.

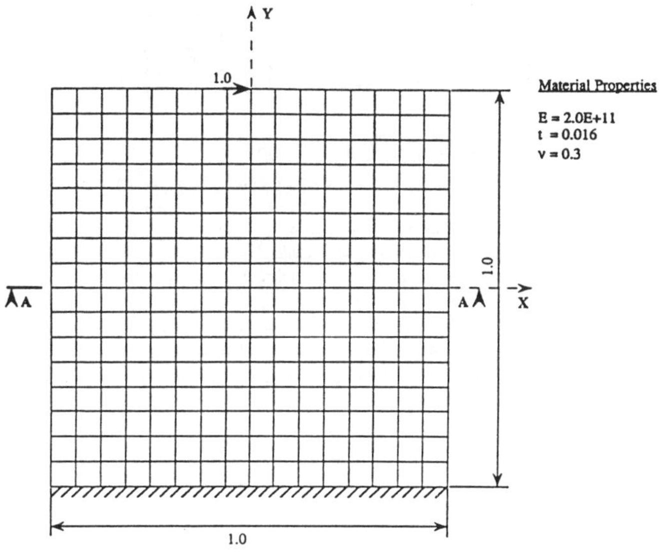


Figure 4.13: Cantilever Plate, Orthogonal Mesh Discretization.

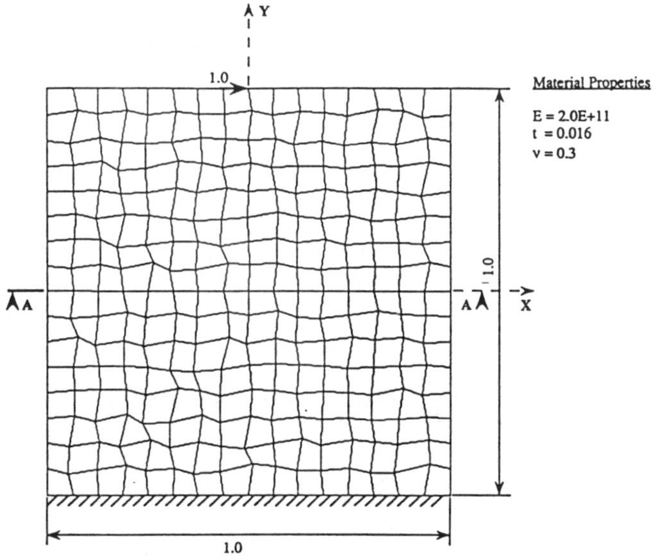


Figure 4.14: Cantilever Plate, Distorted Mesh Discretization.

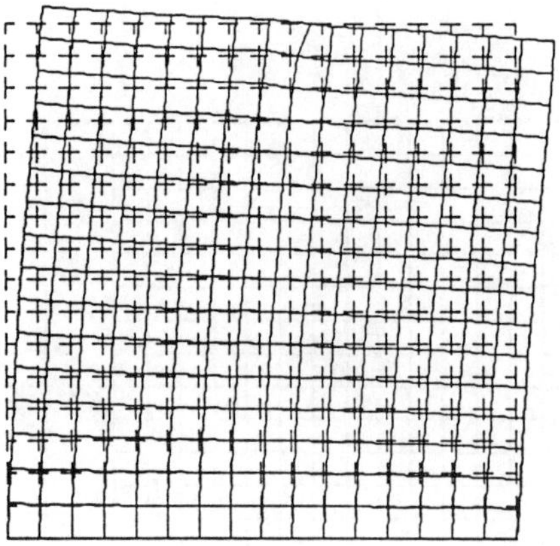

Figure 4.15: Deflected Shape, Orthogonal Mesh.

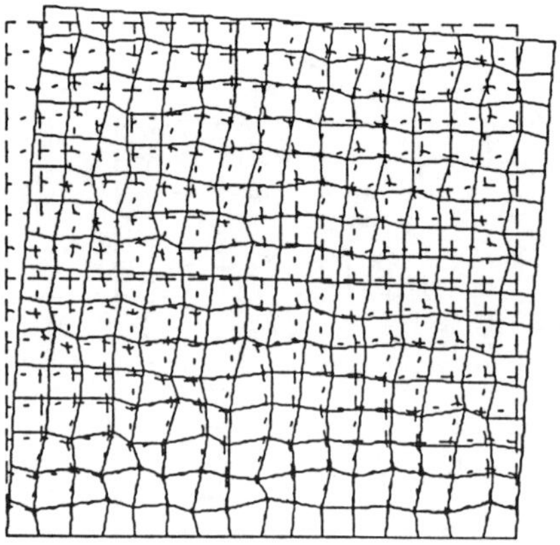

Figure 4.16: Deflected Shape, Distorted Mesh (case 2).

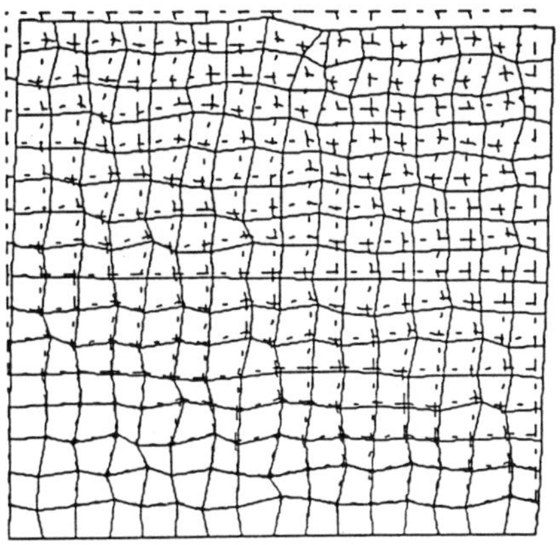

Figure 4.17: Deflected Shape, Distorted Mesh (case 3).

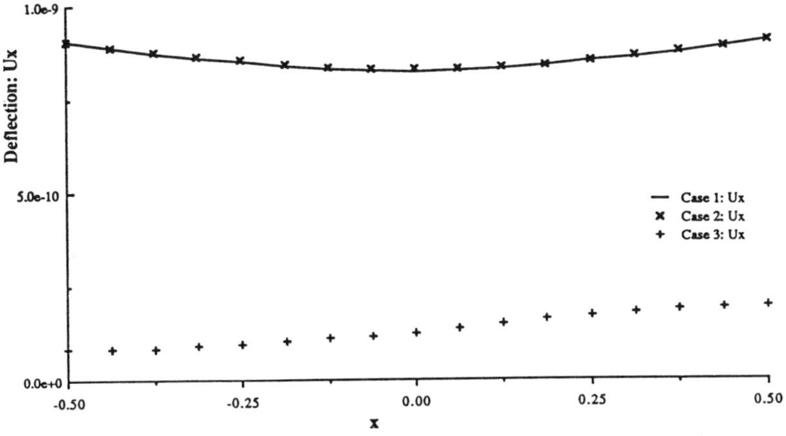


Figure 4.18: Displacement Component Ux along Section A-A.

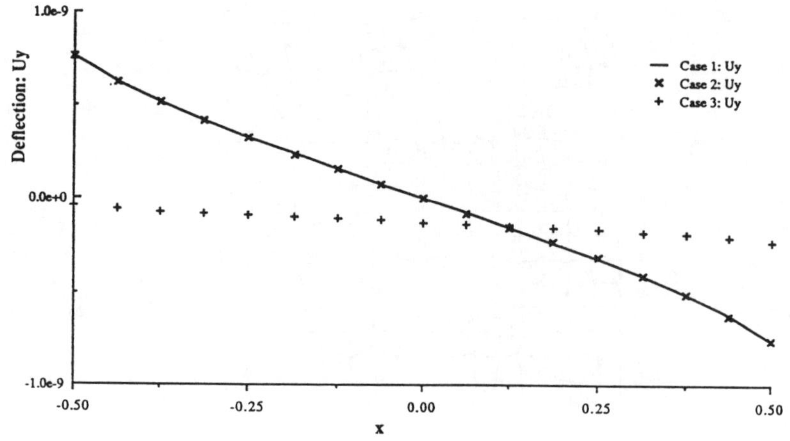


Figure 4.19: Displacement Component $U_y$ along Section A-A.

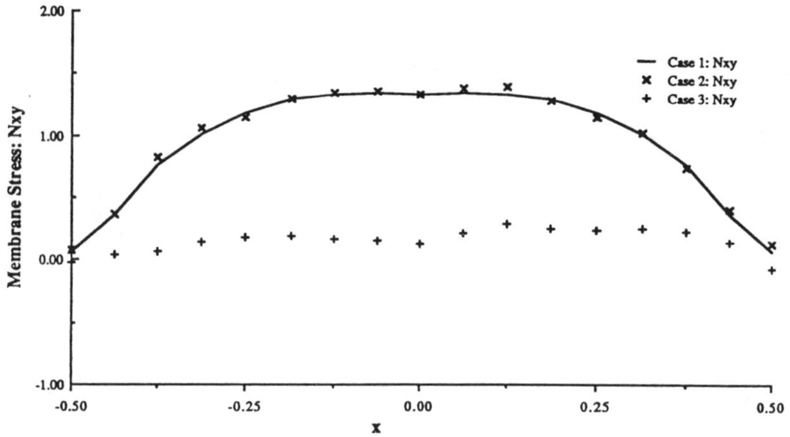


Figure 4.20: Membrane Stress $N_{xy}$ along Section A-A.

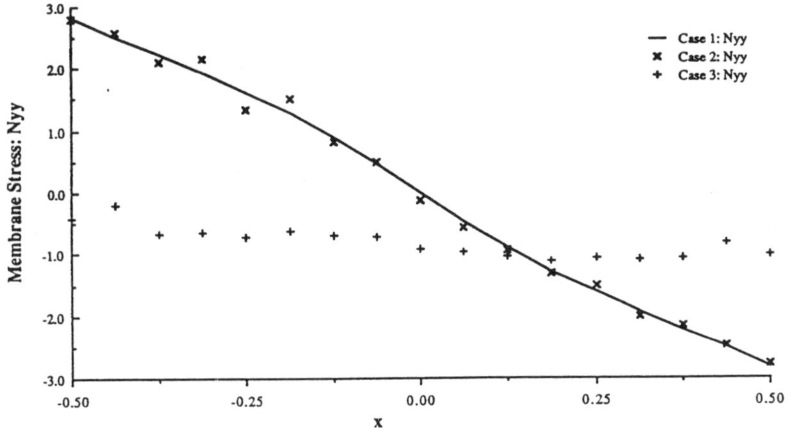


Figure 4.21: Membrane Stress $N_{yy}$ along Section A-A.

(iii) Subdomain Overlap:

Figure 4.22 defines the geometric and material properties of a simply supported flat plate. The plate is subjected to an out of plane concentrated load of unit magnitude at its origin. Table 4.3 summarizes the deflection $U_z$ at the origin for various orthogonal mesh discretizations and compares it to the exact solution [6].

For the purpose of illustrating the effect of subdomain overlap, an 8x8 mesh is selected and the nodal coordinates are randomly distributed as shown in figure 4.23. As with the previous example, the mesh line along the x-axis is retained as a basis for comparison.

Figures 4.24 and 4.25 compare the results from the three case studies. In case 1, the plate was discretized using an orthogonal 8x8 mesh while cases 2 and 3 are based on the distorted mesh arrangement illustrated in figure 4.23. A 3x3 Gauss-Legendre numerical integration scheme is used in all three cases for the evaluation of the plate's stiffness matrix. In addition, the modification described in section 4.1.2 which redefines each subdomain's limits to avoid overlap was in effect for case 2.

Although a slight deterioration is evident when the results of cases 2 and 3 are compared, it is important to note the loss of accuracy which results from truncating the displacement field and transforming from the global coordinate system to the local computational grid. This can gauged by comparing cases 1 and 2 directly. A qualitative look at the truncation errors associated with the curvilinear finite difference method is detailed by Kwok [72].

The above test has been applied to a number of different plate and shell structures with various discretization arrangements. A general conclusion on the effects of subdomain overlap is difficult as the loss of accuracy appears to be case dependent.

| Mesh | Displacement: Uz |
|---|---|
| 4x4 | -2.0176E-07 |
| 8x8 | -1.7013E-07 |
| 16x16 | -1.5942E-07 |
| 32x32 | -1.5606E-07 |
| Exact Solution | -1.5319E-07 |

Table 4.3: Convergence Results for the Simply Supported Plate.

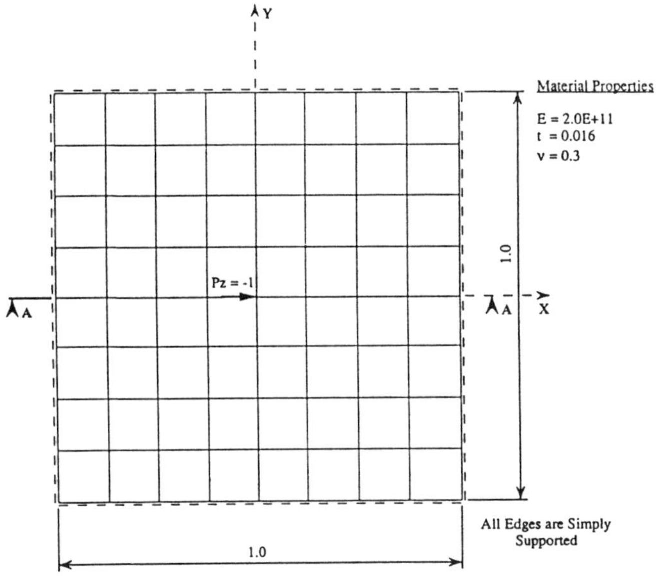


Figure 4.22: Simply Supported Plate, Orthogonal Mesh Discretization.

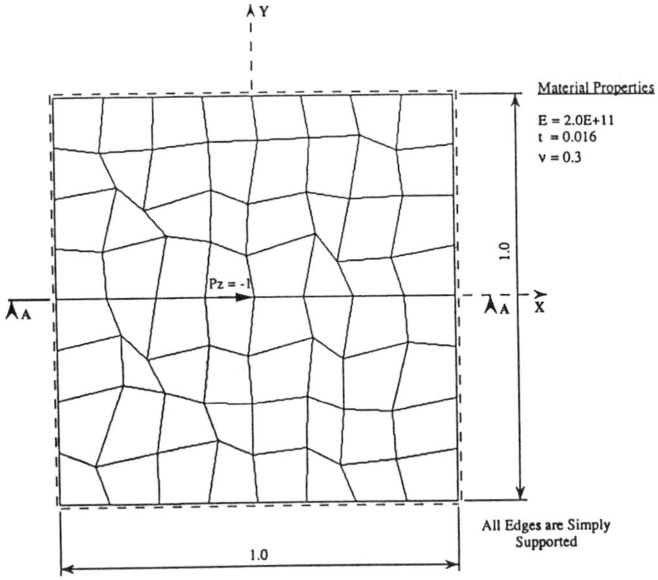


Figure 4.23: Simply Supported Plate, Distorted Mesh Discretization.

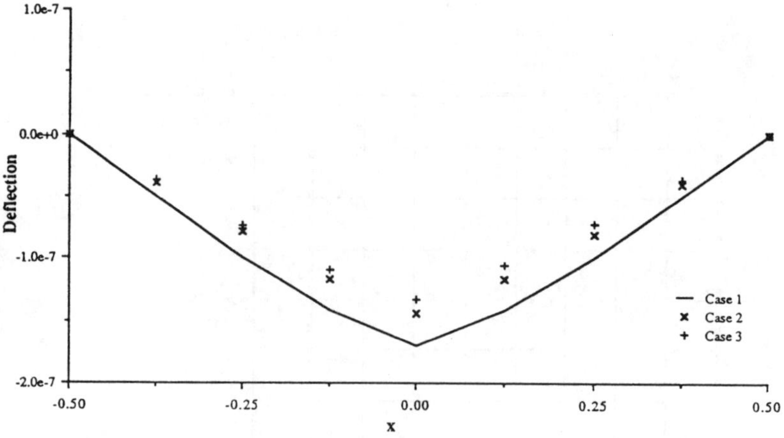


Figure 4.24: Displacement Component Uz along Section A-A.

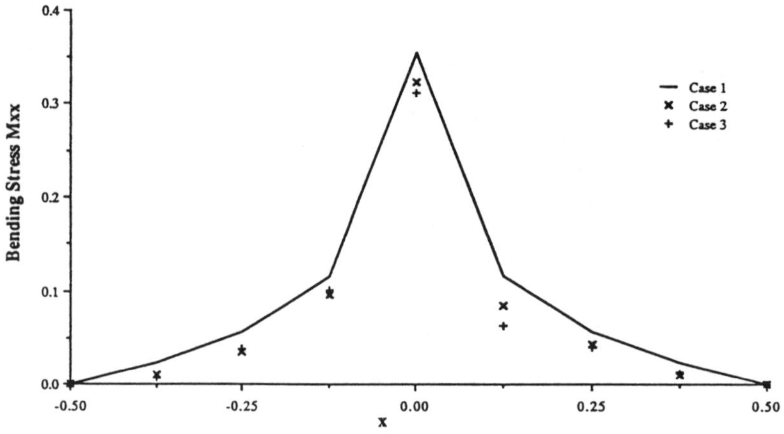


Figure 4.25: Bending Stress Mxx along Section A-A.

(iv) Dynamic Characteristics:

The examples presented so far have been confined to the static analysis of flat plates. These problems were selected to illustrate a number of theoretical concepts introduced in the previous chapters. In all cases, the results were compared with alternate numerical methods such as the finite element method or, where applicable, an exact solution. In this section, the technique is used to evaluate the dynamic characteristics of a number of shell structures.

*Simply Supported Plate*

The first example considered is a square plate simply supported at each boundary. The dimensions and material properties are given in figure 4.26. An equation defining the exact eigenvalues may be derived from plate theory [100,101]. It may be expressed in the form

$$\omega_{mn} = \frac{\pi^2}{a^2\sqrt{\frac{\rho t}{D}}}\left[m^2 + n^2\left[\frac{a}{b}\right]^2\right] \tag{4.1.21}$$

where m and n can be any combination of integers (m,n = 1,2,...). The mode shapes are given by the expression

$$\phi_{mn} = A_{mn}\sin(m\pi x/a)\sin(n\pi y/b) \tag{4.1.22}$$

For the square plate example considered here, the above equations clearly indicate that two or more entirely different mode shapes may occur at the same frequency. Theoretically, if multiple eigenvalues exist an infinite number of mode shapes are possible and an eigenvector satisfying the general equations of motion may be formed by a linear combination of the fundamental modes given by equation 4.1.22. Importantly, these modes which are referred to as superposition modes, are not necessarily orthogonal to each other. Thus, the identification and comparison of eigenvectors associated with multiple modes presents a number of difficulties.

Where multiple eigenvalues exist, the subspace iteration technique used in the present method computes an orthogonal set of eigenvectors. However, the vectors need not necessarily represent the fundamental modes. This observation is confirmed in numerous publications [81,82,102-105].

In order to recognize the fundamental mode shapes the numerical examples are discretized using a non-symmetric mesh. This approach forces multiple eigenvalues and eigenvectors to be unique. As the mesh discretization increases, the eigenvalues converge to the reference value.

For the simply supported plate defined in figure 4.26, the first six modes of vibration using symmetric and non-symmetric mesh discretizations are presented in figures 4.27 and 4.28 respectively. The corresponding eigenvalues for various mesh discretizations are given in Table 4.4 and compare favourably with the analytical solution.

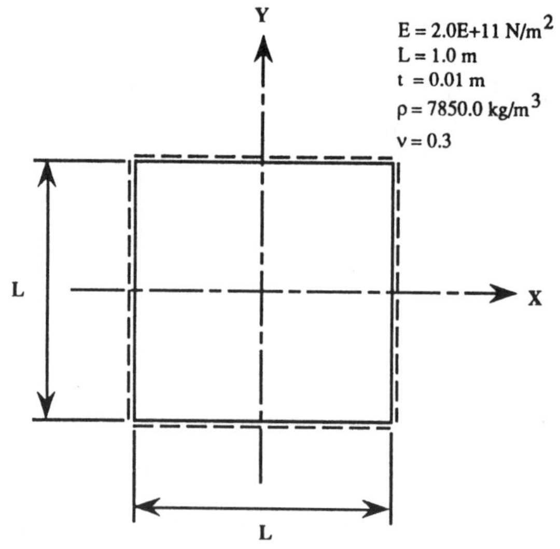


Figure 4.26: Geometric and Material Properties for the Simply Supported Plate.

Of particular interest in these examples are the shape of modes (1,2),(2,1), (1,3) and (3,1). The mode shapes generated using a non-symmetric mesh correspond to the fundamental modes given by equation 4.1.22. In contrast, the symmetric mesh yields mode shapes with nodal lines quite different from the analytical solution. These modes are superposition modes and may be duplicated by linearly combining the fundamental modes corresponding to the given eigenvalue.

| Mode | $\lambda = \omega a^2 \sqrt{\rho t/D}$ | | | |
|---|---|---|---|---|
| | 4x4 | 8x8 | 16x16 | 32x32 |
| (1,1) | 18.13 | 19.30 | 19.63 | 19.71 |
| (1,2) | 40.47 | 46.64 | 48.62 | 49.16 |
| (2,1) | 40.47 | 46.64 | 48.62 | 49.16 |
| (2,2) | 59.49 | 75.52 | 77.19 | 78.48 |
| (1,3) | 67.28 | 88.39 | 95.76 | 97.91 |
| (3,1) | 67.28 | 88.39 | 95.76 | 97.91 |

Table 4.4 a: Eigenvalues for the Simply-Supported Square Plate.

| Mode | $\lambda = \omega a^2 \sqrt{\rho t/D}$ | | | |
|---|---|---|---|---|
| | 8x6 | 16x14 | 32x30 | Analytical |
| (1,1) | 19.14 | 19.61 | 19.71 | 19.74 |
| (1,2) | 44.99 | 48.43 | 49.13 | 49.35 |
| (2,1) | 46.44 | 48.60 | 49.15 | 49.35 |
| (2,2) | 70.44 | 76.93 | 78.45 | 78.96 |
| (1,3) | 82.28 | 94.94 | 97.81 | 98.69 |
| (3,1) | 88.20 | 95.74 | 97.90 | 98.69 |

Table 4.4 b: Eigenvalues for the Simply-Supported Square Plate.

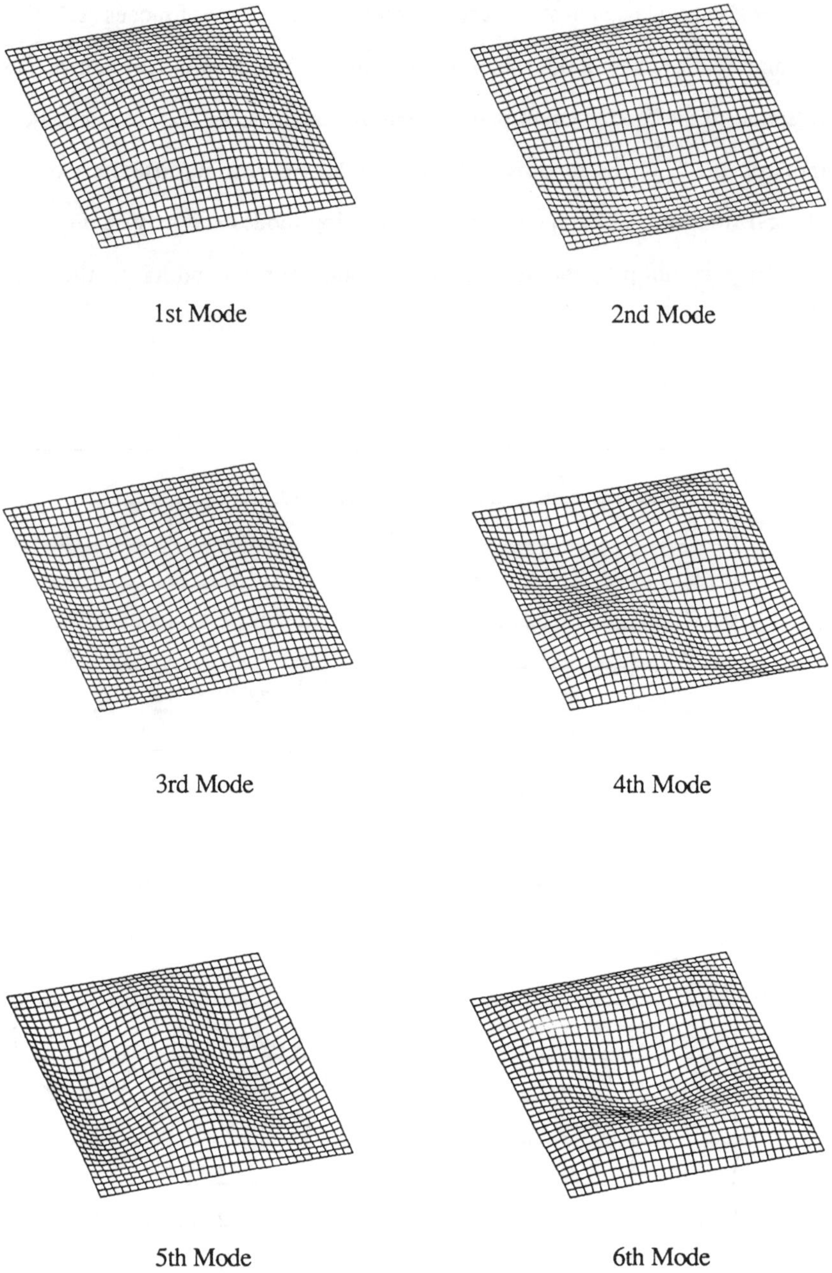


Figure 4.27: Mode Shapes for the Simply Supported Plate (Symmetric Mesh).

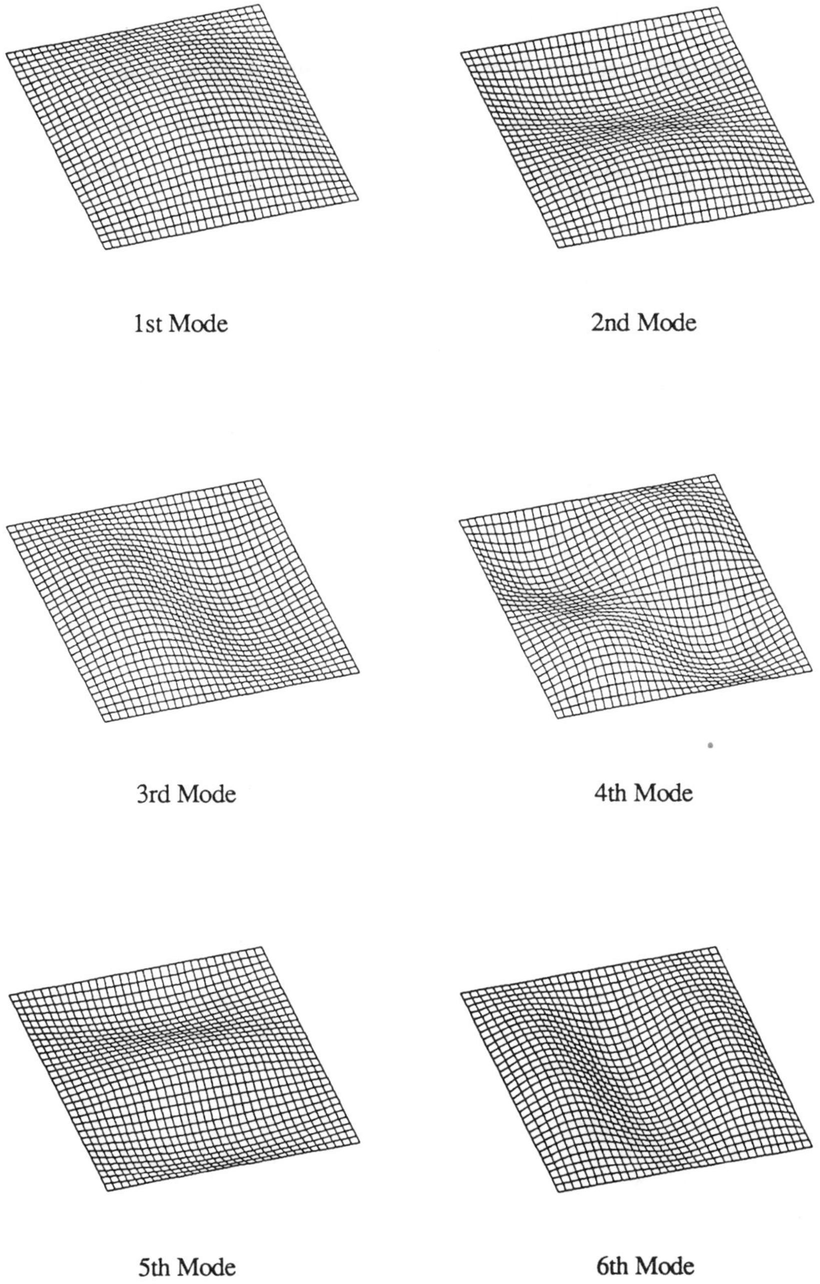


Figure 4.28: Mode Shapes for the Simply Supported Plate (Non-Symmetric Mesh).

*Spherical Cap on a Square Base*

The shallow spherical shell depicted in figure 4.29 was analysed using a number of different mesh discretizations. The shell is freely supported on a boundary of square planform. The displacements $u_n$ and $u_t$ are constrained along the shell's four edges. Shell structures with similar rise to radius ratios are generally considered to be shallow (see Kraus [10]) thereby enabling an approximate solution to be obtained from shell theory [54].

The numerical results for the first six modes of vibration are presented in table 4.5 and compare favourably with the analytical solution. Figure 4.30 illustrates the corresponding mode shapes.

| Mode | Eigenvalues (rad/sec) | | | |
|---|---|---|---|---|
| | 8x6 | 16x14 | 32x30 | Analytical |
| (1,1) | 23.70 | 24.96 | 25.24 | 25.42 |
| (1,2) | 23.58 | 25.62 | 26.09 | 26.34 |
| (2,1) | 23.76 | 25.63 | 26.09 | 26.34 |
| (2,2) | 22.54 | 26.34 | 27.44 | 27.97 |
| (1,3) | 26.05 | 28.33 | 28.96 | 29.40 |
| (3,1) | 26.39 | 28.39 | 28.99 | 29.40 |

Table 4.5: Eigenvalues for the Spherical Cap on a Square Base.

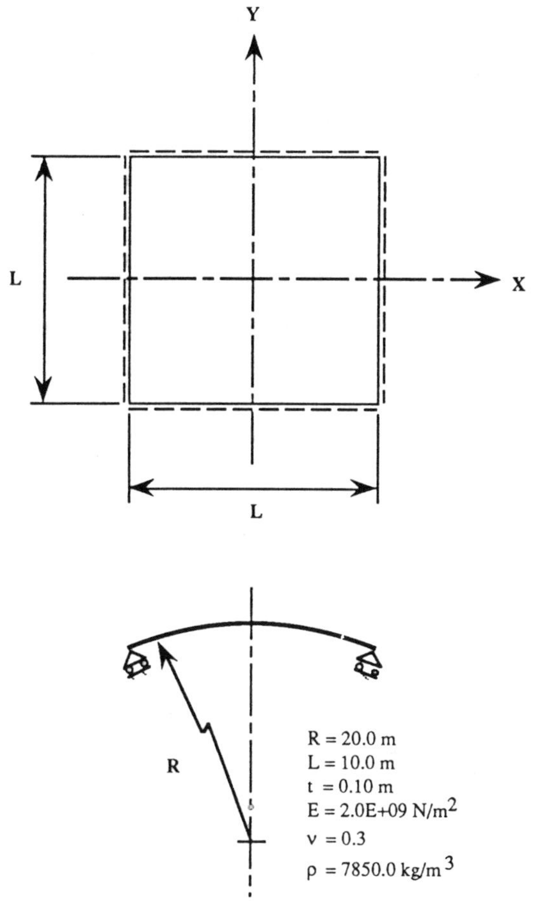


Figure 4.29: Geometric and Material Properties for the Spherical Shell.

In table 4.5, the modes are classified according to the numbers of half-waves parallel to the x and y axes. Importantly, this table highlights a problem which occurs when a small number of nodes are used to model the shell's geometric and displacement characteristics. It is apparent that when a 8×6 mesh is used to model the spherical cap, the resultant eigenvalues and eigenvectors are not in the same order as those predicted by the analytical solution. Yet, for higher mesh discretizations the mode shapes and eigenvalues are directly comparable. Briefly, this observation may be explained in the following manner. Firstly, the ability to model the shell's geometric properties and displacements improves as the mesh refinement increases. In addition, the above problem contains eigenvalues which are spaced relatively close together. Thus, any inaccuracy in representing the shell's geometry or displacement characteristics may not uniformly affect the numerical approximation of the eigenvalue problem. If this occurs, then it is possible that different eigenvalues are approximated with variable accuracy and results similar to the ones given in table 4.5 may arise. This situation is discussed in further detail later in this section where the results of a spherical shell with numerically close eigenvalues are presented.

Where repeated eigenvalues exist, the use of a non-symmetric mesh to model symmetric shells forces the values to differ slightly. These differences become smaller as the finite difference mesh is increasingly refined. It should be noted that not all the eigenvectors have nodal lines which run parallel to the shell's boundary. For example, the mode (1,3) has a circular nodal line. No rigorous explanation can be given for this result, although similar observations have been presented in the literature [54].

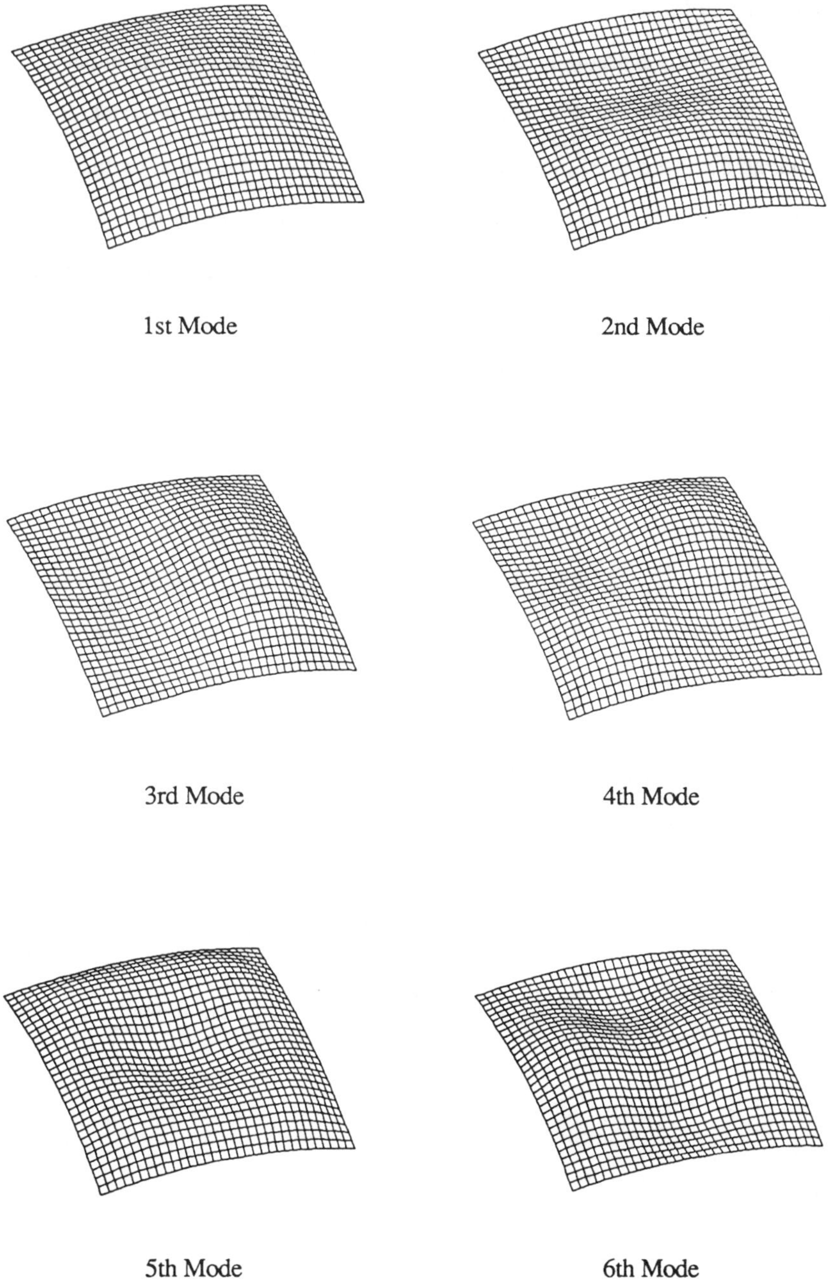


Figure 4.30: Mode Shapes for the Spherical Cap on a Square Base (32×30 Mesh).

*Clamped Cylindrical Shell Panel*

The third example is a cylindrical panel clamped at each boundary. The dimensions and material properties are given in figure 4.31 and are identical to the ones given by Olsen and Lindberg [ 54]. There is no exact solution to this problem although a number of different numerical techniques have been used to evaluate its dynamic characteristics [54-56,106]. The results from this analysis are presented in table 4.6 and are in good agreement with those given by the finite element analysis of Olsen and Lindberg. The corresponding eigenvectors are shown in figure 4.32.

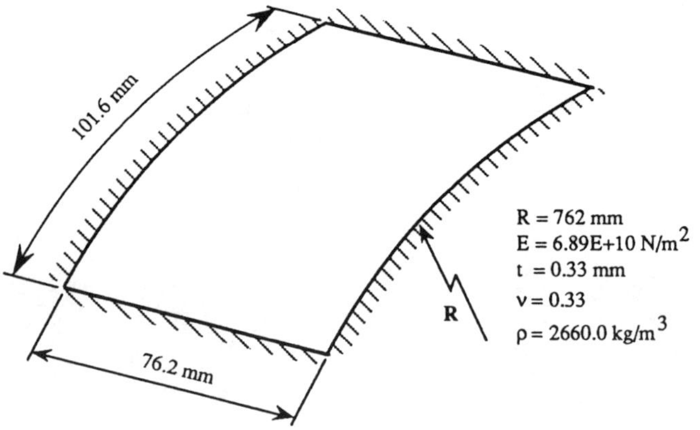


Figure 4.31: Geometric and Material Properties for the Clamped Cylindrical Shell Panel.

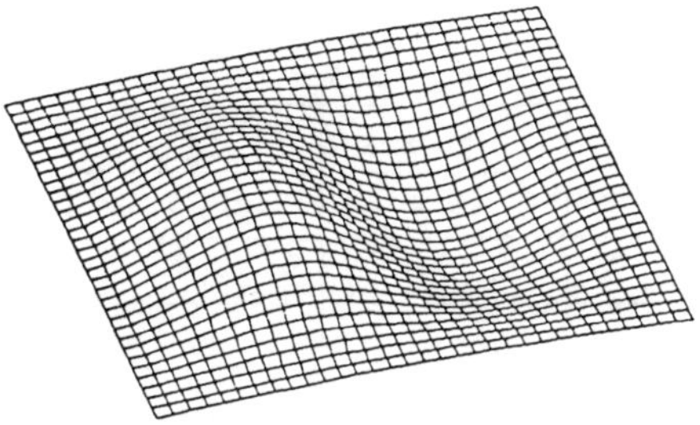

1st Mode

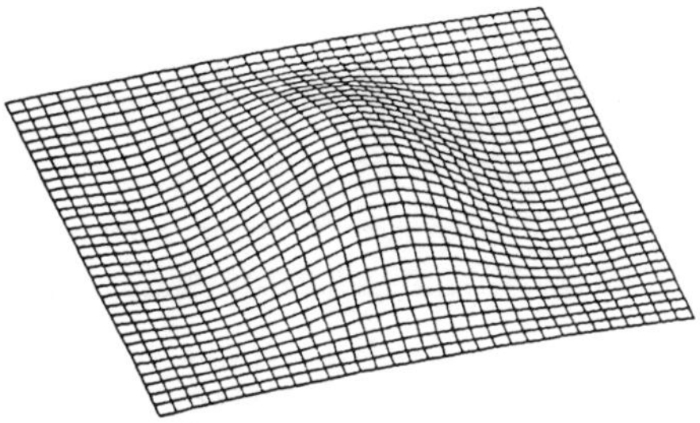

2nd Mode

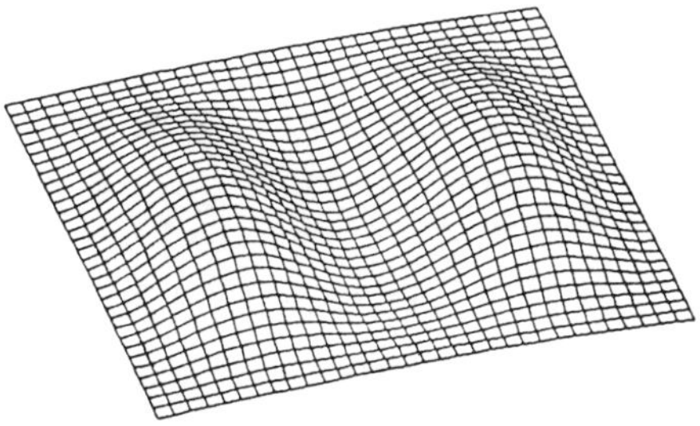

3rd Mode

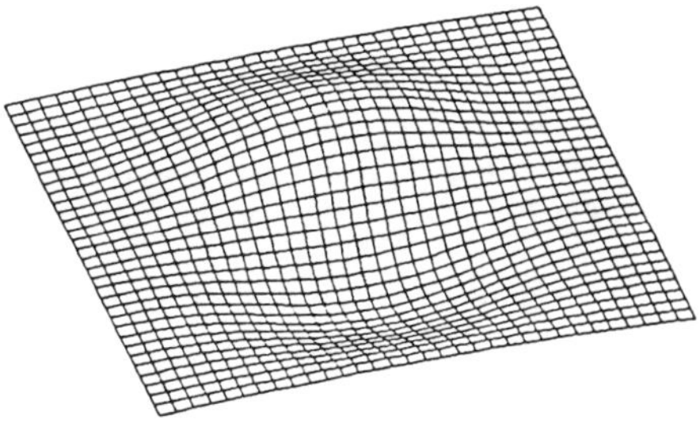

4th Mode

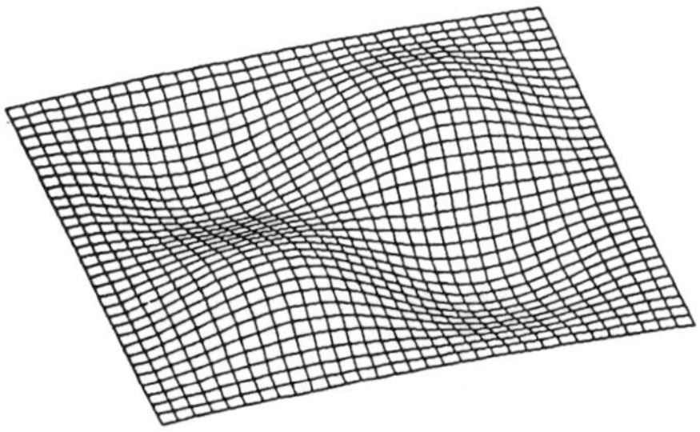

5th Mode

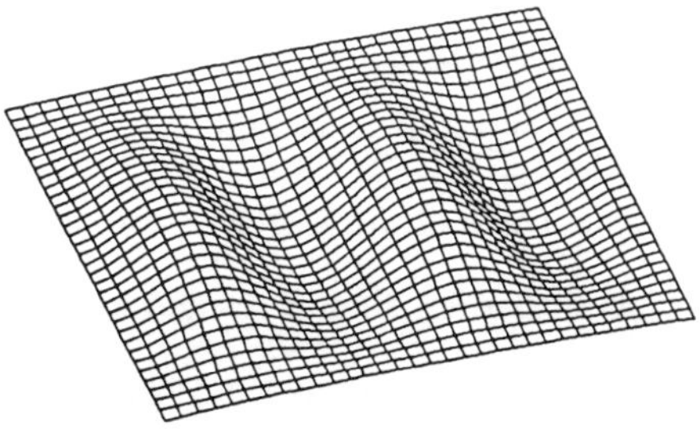

6th Mode

Figure 4.32: Mode Shapes for the Clamped Cylindrical Shell (32×32 Mesh).

| Mode | Natural Frequency (Hz) | | | |
|---|---|---|---|---|
| (x,θ) | 8x8 | 16x16 | 32x32 | Finite Element [54] |
| (1,2) | 785.2 | 842.5 | 861.6 | 869 |
| (1,3) | 861.8 | 929.9 | 950.0 | 957 |
| (1,3) | 1176.6 | 1242.6 | 1273.5 | 1287 |
| (2,1) | 1236.2 | 1328.6 | 1353.6 | 1362 |
| (2,2) | 1239.6 | 1375.8 | 1421.1 | 1437 |
| (1,4) | 1389.5 | 1630.9 | 1734.3 | 1752 |

Table 4.6: Eigenvalues for the Clamped Cylindrical Shell.

*Discussion.*

The case studies examined so far demonstrate the application of the curvilinear finite difference method to the linear analysis of flat plate, cylindrical and spherical shells. In all three cases, the computed dynamic characteristics compare favourably with results previously published within the literature. More importantly, the spherical shell example highlights the need for introducing a large number of nodes in order to reproduce the shell's eigenvalues and corresponding eigenvectors in correct order.

In previous work put forward by this author [107], the covariant components of the displacement vector were assumed as nodal unknowns and a spherical shell similar to the one illustrated in figure 4.29 was analysed. The material properties and geometric dimensions are summarized below.

$R = 2.5$ m

$L = 1.0$ m

$t = 0.0025$ m

$E = 2.0E{+}11$ N/m$^2$

$\nu = 0.3$

$\rho = 7850.0 \text{ kg/m}^3$

In this case, the problems physical characteristics result in a small difference between some of the adjacent eigenvalues. For example, if the first and second modes of vibration are considered and their corresponding eigenvalues computed from the analytical solution presented in reference [54], their percentage difference is approximately 0.4 percent. Despite this relatively small difference, the technique was capable of predicting the eigenvalues and their corresponding modes in the correct order.

In contrast, the same problem was analysed using the current implementation and the results for a 32×30 mesh are presented in table 4.7 and figure 4.33. Even at this level, it is evident that the technique has not converged sufficiently enough to be able to define the eigenvalues and their corresponding eigenvectors in their correct order.

Importantly, the only major difference between the methodology presented in reference [107] and the current implementation is the selection of the global displacement components as nodal unknowns (see section 3.4). As such, a series of transformation equations are required to compute the covariant displacement components which are used to evaluate the stiffness and mass matrices of the shell. These transformation equations are defined in chapter 3 and are expanded in matrix form in appendix A.

In a qualitative study of the truncation errors associated with a local nine node computational grid, Kwok [72] concluded that the curvilinear finite difference method could approximate a scalar function $\phi$ with 2nd-order accuracy in the local $\alpha$-$\beta$ plane. However, this accuracy decreased when the function was referenced in the global coordinate system due to the numerical approximations involved in computing the transformation coefficients.

In the current implementation, the global displacement components can be assumed to be of 2nd order accuracy in the local $\alpha$-$\beta$ plane. However, the governing strain-displacement equations require these components to be transformed to the local $\alpha$-$\beta$ coordinate system. Clearly, such a transformation lowers the order of accuracy of the displacement components. In particular, the second order derivatives of the normal displacement , w, are computed from the expression

$$w_{,\alpha\beta} = \beta^i_{3,\alpha\beta} u_i + \beta^i_{3,\alpha} u_{i,\beta} + \beta^i_{3,\beta} u_{i,\alpha} + \beta^i_3 u_{i,\alpha\beta} \qquad (4.1.23)$$

This equation requires terms such as $x_{,\alpha\alpha\alpha}$ and $z_{,\beta\beta\beta}$ to be evaluated (see appendix A). However, a nine node mesh can only provide at best a second order approximation.

Together with the numerical results presented in this chapter one may conclude that in some instances, (e.g. where eigenvalues are clustered together), a higher order approximation may be required to more accurately describe the surface geometry and the displacement components within a mesh. In the next section, a sixteen node, third order mesh is derived and applied to the analysis of shell structures.

| Mode | Natural Frequency (Hz) | | |
|---|---|---|---|
| (x,y) | 32x30 | 20x20 [107] | Analytical [54] |
| (1,1) | 2007. | 2016. | 2020. |
| (1,2) | 2014. | 2026. | 2028. |
| (2,1) | 2014. | 2026. | 2028. |
| (2,2) | 2010. | 2039. | 2041. |
| (1,3) | 2038. | 2051. | 2054. |
| (3,1) | 2041. | 2053. | 2054. |

Table 4.7: Eigenvalues for the Spherical Shell.

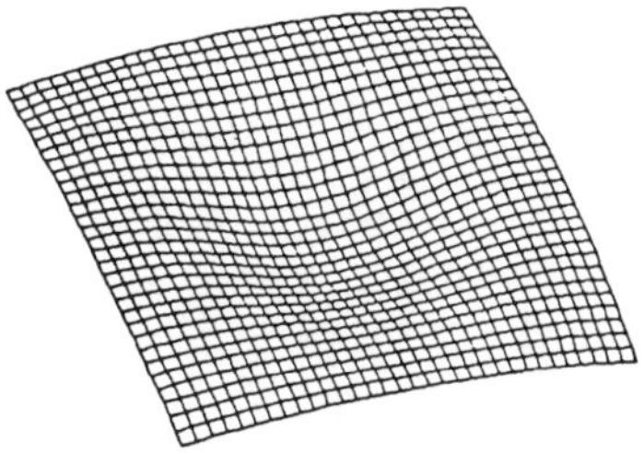

1st Mode

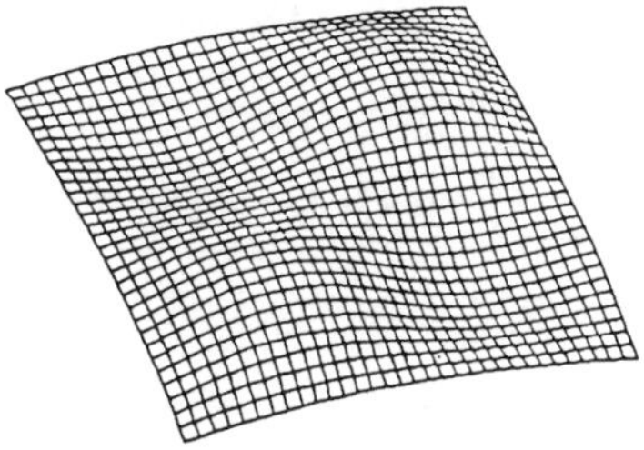

2nd Mode

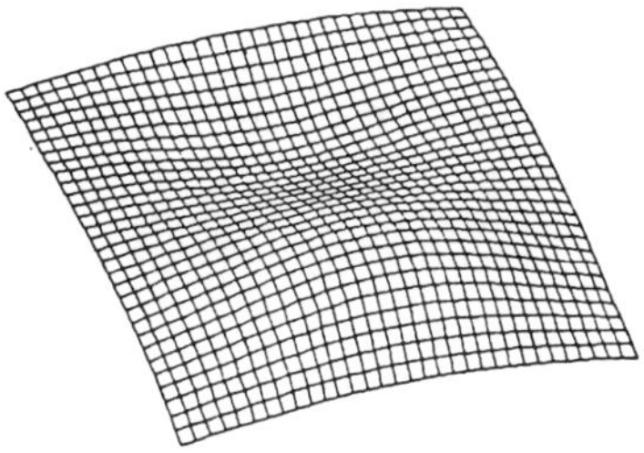

3rd Mode

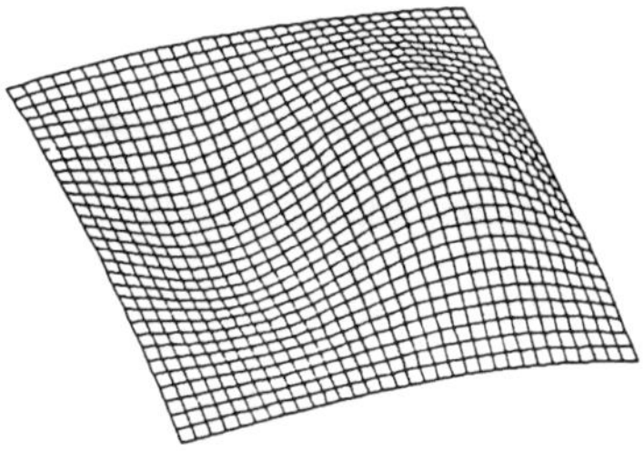

4th Mode

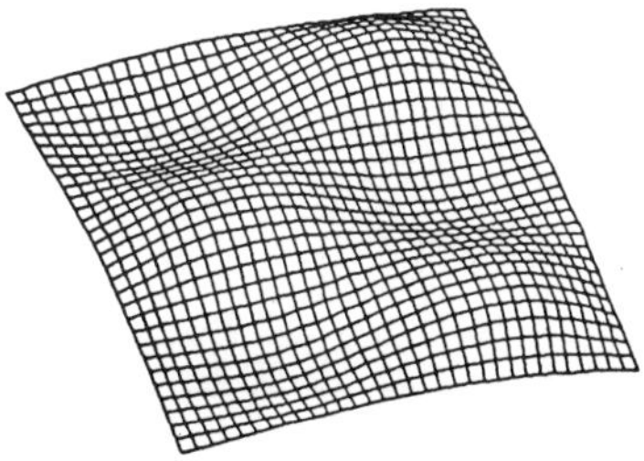

5th Mode

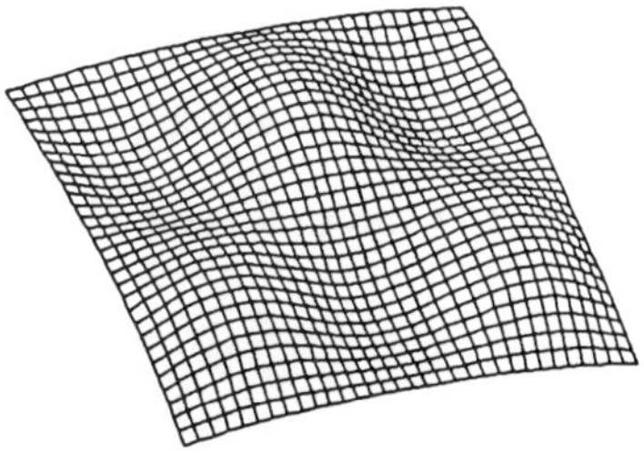

6th Mode

Figure 4.33: Mode Shapes for the Spherical Shell (32$\times$30 Mesh).

## 4.2 A THIRD ORDER IMPLEMENTATION

### 4.2.1 Third Order CFD Approximation

In this section, details relating to the use of a third order curvilinear finite difference mesh are presented. Essentially, the procedure is identical to the previous second order implementation with minor changes required in the application of the boundary conditions and the numerical integration scheme used in evaluating the stiffness and mass matrices.

A third order mesh may be constructed by assuming the sixteen node configuration illustrated in figure 4.34a. At any point within the mesh, a function $\psi$ may be approximated from the expression

$$\psi = a_1 + a_2\alpha + a_3\beta + a_4\alpha^2 + a_5\beta^2 + a_6\alpha\beta + a_7\alpha^2\beta + a_8\alpha\beta^2 + a_9\alpha^2\beta^2 + a_{10}\alpha^3 + a_{11}\beta^3 + a_{12}\alpha^3\beta + a_{13}\alpha\beta^3 + a_{14}\alpha^3\beta^2 + a_{15}\alpha^2\beta^3 + a_{16}\alpha^3\beta^3 \tag{4.2.1}$$

where $a_1, a_2, a_3, \ldots, a_{16}$ are constant coefficients. In equation 4.2.1, $\psi$ may represent the shell's displacement components or geometric coordinates.

Adopting a similar approach to the one detailed in section 3.1, equation 4.2.1 may be expressed in the form

$$\psi = [\phi]^t \{\Psi_p\} \tag{4.2.2}$$

where the vector $\{\Psi_p\}$ contains the sixteen nodal unknowns

$$\{\Psi_p\}^t = \{\Psi_1\ \Psi_2\ \Psi_3\ \Psi_4\ \Psi_5 \ldots\ \Psi_{13}\ \Psi_{14}\ \Psi_{15}\ \Psi_{16}\} \tag{4.2.3}$$

and the matrix $[\phi]$ is a 16×1 matrix of interpolation functions. i.e.,

$$[\phi] = \begin{bmatrix} \frac{1}{4}(2 - \alpha - 2\alpha^2 + \alpha^3)(2 - \beta - 2\beta^2 + \beta^3) \\ \frac{1}{4}(2\alpha + \alpha^2 - \alpha^3)(2 - \beta - 2\beta^2 + \beta^3) \\ \frac{1}{4}(2\alpha + \alpha^2 - \alpha^3)(2\beta + \beta^2 - \beta^3) \\ \frac{1}{4}(2 - \alpha - 2\alpha^2 + \alpha^3)(2\beta + \beta^2 - \beta^3) \\ \frac{1}{12}(2 - \alpha - 2\alpha^2 + \alpha^3)(-2\beta + 3\beta^2 - \beta^3) \\ \frac{1}{12}(2\alpha + \alpha^2 - \alpha^3)(-2\beta + 3\beta^2 - \beta^3) \\ \frac{1}{36}(-\alpha + \alpha^3)(-2\beta + 3\beta^2 - \beta^3) \\ \frac{1}{12}(-\alpha + \alpha^3)(2 - \beta - 2\beta^2 + \beta^3) \\ \frac{1}{12}(-\alpha + \alpha^3)(2\beta + \beta^2 - \beta^3) \\ \frac{1}{36}(-\alpha + \alpha^3)(-\beta + \beta^3) \\ \frac{1}{12}(2\alpha + \alpha^2 - \alpha^3)(-\beta + \beta^3) \\ \frac{1}{12}(2 - \alpha - 2\alpha^2 + \alpha^3)(-\beta + \beta^3) \\ \frac{1}{36}(-2\alpha + 3\alpha^2 - \alpha^3)(-\beta + \beta^3) \\ \frac{1}{12}(-2\alpha + 3\alpha^2 - \alpha^3)(2\beta + \beta^2 - \beta^3) \\ \frac{1}{12}(-2\alpha + 3\alpha^2 - \alpha^3)(2 - \beta - 2\beta^2 + \beta^3) \\ \frac{1}{36}(-2\alpha + 3\alpha^2 - \alpha^3)(-2\beta + 3\beta^2 - \beta^3) \end{bmatrix} \qquad (4.2.4)$$

The above expressions may be differentiated with respect to $\alpha$ and $\beta$ to yield a system of equations which can be used to approximate the displacement and geometric local partial derivatives.

Like the nine node implementation, the geometric properties at the boundary and corner meshes are evaluated using a mesh which excludes the nodes external to the continuum. A typical edge mesh is therefore constructed by using twelve nodes as shown in figure 4.34b. Within an edge mesh, any scalar function may then be approximated by the expression

$$\psi = a_1 + a_2\alpha + a_3\beta + a_4\alpha^2 + a_5\beta^2 + a_6\alpha\beta + a_7\alpha^2\beta + a_8\alpha\beta^2 + a_9\alpha^2\beta^2 + a_{11}\beta^3 + a_{13}\alpha\beta^3 + a_{15}\alpha^2\beta^3 \qquad (4.2.5)$$

Where additional degrees of freedom are required for the boundary conditions the complete third order polynomial described by equation 4.2.1 is adopted. Unlike the previous nine node implementation, the 'fictitious' nodes external to the continuum are retained and not replaced by their corresponding derivatives at each of the boundary nodes.

A similar approach is adopted at a corner where the mesh is formed by nine nodes (see figure 4.34c). In this case, the approximation function for $\psi$ becomes

$$\psi = a_1 + a_2\alpha + a_3\beta + a_4\alpha^2 + a_5\beta^2 + a_6\alpha\beta + a_7\alpha^2\beta + a_8\alpha\beta^2 + a_9\alpha^2\beta^2 \qquad (4.2.6)$$

Where the external degrees of freedom are required, the third order polynomial defined by equation 4.2.1 is modified by dropping the term $\alpha^3\beta^3$. i.e.,

$$\psi = a_1 + a_2\alpha + a_3\beta + a_4\alpha^2 + a_5\beta^2 + a_6\alpha\beta + a_7\alpha^2\beta + a_8\alpha\beta^2 + a_9\alpha^2\beta^2 + a_{10}\alpha^3 + a_{11}\beta^3 + a_{12}\alpha^3\beta + a_{13}\alpha\beta^3 + a_{14}\alpha^3\beta^2 + a_{15}\alpha^2\beta^3 \qquad (4.2.7)$$

Each of the above interpolation functions can be written in the form described by equation 4.2.2 and the terms differentiated with respect to the local $\alpha,\beta$ coordinate system to arrive at a system of equations which approximate both the geometric and displacement partial derivatives. These quantities are essential in the implementation of the general strain-displacement equations derived in chapter 2.

In addition to defining the basic mesh configurations, figure 4.34 highlights the subdomain associated with each mesh. Unlike the previous nine node scheme, the limits are well defined along the coordinates $\alpha = 0, +1$ and $\beta = 0, +1$. These limits are independent of the mesh discretization allowing the stiffness and mass matrices to be computed using the conventional Gauss-Legendre numerical integration scheme. i.e.,

$$[K]_i = \int_V [H_{\alpha\beta}]^t [\Sigma^{\alpha\beta}]\, dV$$

$$= \sum_{\alpha,\beta,z} w_\alpha\, w_\beta\, w_z\, [H_{\alpha\beta}]^t [\Sigma^{\alpha\beta}]\, \det[J] \qquad (4.2.8)$$

and $$[M]_i = \int_V \rho\, [D^{cn}]^t\, [D_{cv}]\, dV$$

$$= \sum_{\alpha,\beta,z} w_\alpha\, w_\beta\, w_z\, \rho\, [D^{cn}]^t [D_{cv}]\, \det[J] \qquad (4.2.9)$$

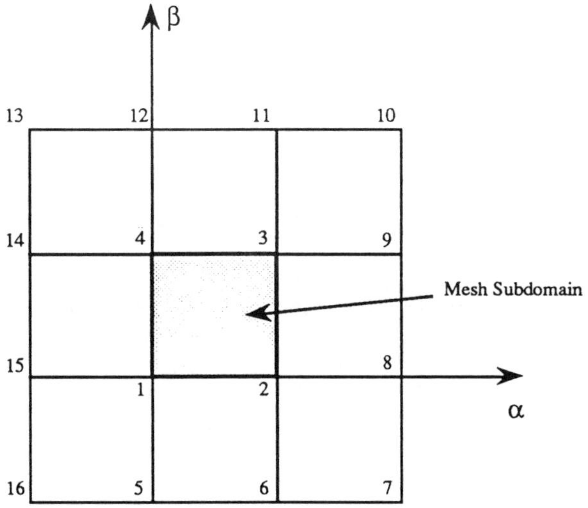


Figure 4.34a: Internal 16-node Mesh.

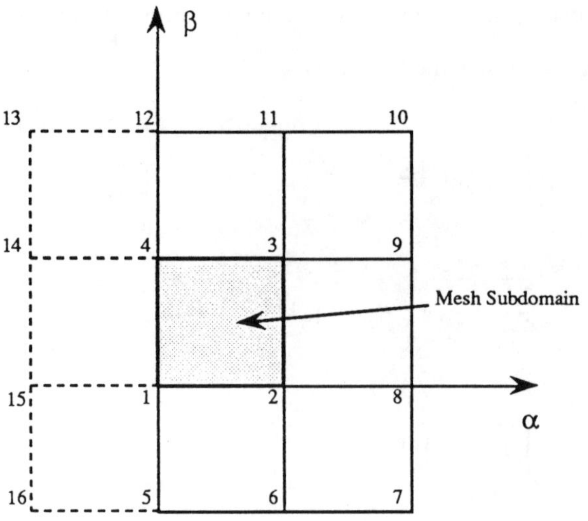


Figure 4.34b: A Typical Edge Mesh .

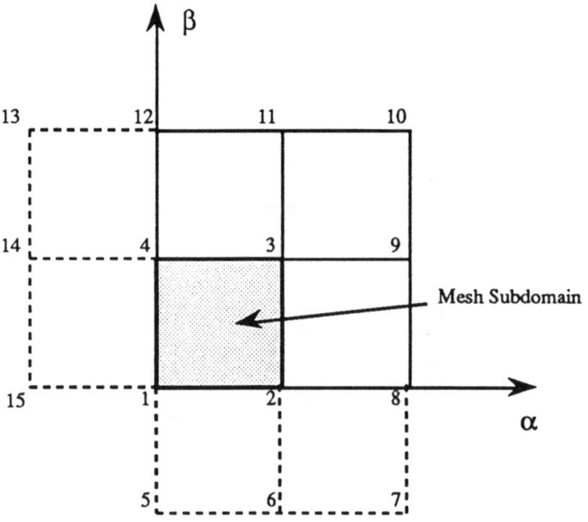


Figure 4.34c: A Typical Corner Mesh .

### 4.2.2 Boundary Conditions

In the proposed method, the boundary constraints pertaining to fixed, simply supported and free edges are implemented. These conditions are defined mathematically by Green and Zerna [85] and their application to the curvilinear finite difference (CFD) method has been discussed in general terms in section 3.6 .

As a first approach, the boundary conditions were incorporated as presented in section 3.6. Thus, in the case of a fixed edge where the constraint is characterized by the condition

$$\frac{\partial w}{\partial s} = 0 \qquad (4.2.10)$$

the displacement interpolation function is modified in accordance with equation 3.6.16 enabling the external fictitious degrees of freedom to be eliminated. Although the above expression theoretically defines the zero normal slope condition which exists along a fixed edge, its direct application when using a third order CFD approximation is not straight forward. Numerous case studies involving flat, spherical and cylindrical shell structures subjected to uniformly distributed loads could not reproduce acceptable results. With increasing mesh refinement, a deterioration in accuracy was evident particularly adjacent to the boundary where an apparent discontinuity in stress existed.

In order to study the problem in greater detail, the general shell equations were simplified to describe the strain-displacement behaviour of a beam. At a boundary, two separate schemes were used to model a fixed edge. In the first scheme, the constraint equation $\partial w/\partial \alpha = 0$ was derived using a third order displacement function as a basis. In contrast, the second scheme computed the constraint equation from a second order displacement function. The resultant load vectors and stiffness matrices are described in detail within Appendix D.

Figure 4.35 defines the geometric properties and dimensions of a beam fixed at both ends and subjected to a uniformly distributed load. The deflections and bending moments at the centre and edge of the beam are presented in Appendix D for various mesh discretizations. Figures 4.36 and 4.37 illustrate the deflected shape and bending moments adjacent to the beams edge .

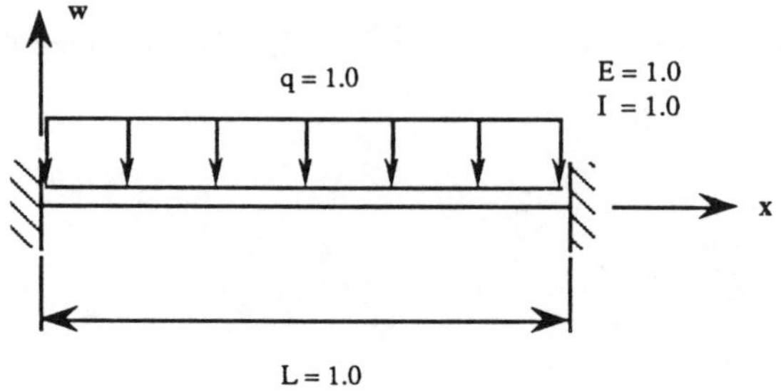


Figure 4.35: Beam with Fixed Boundary Conditions at Both Ends.

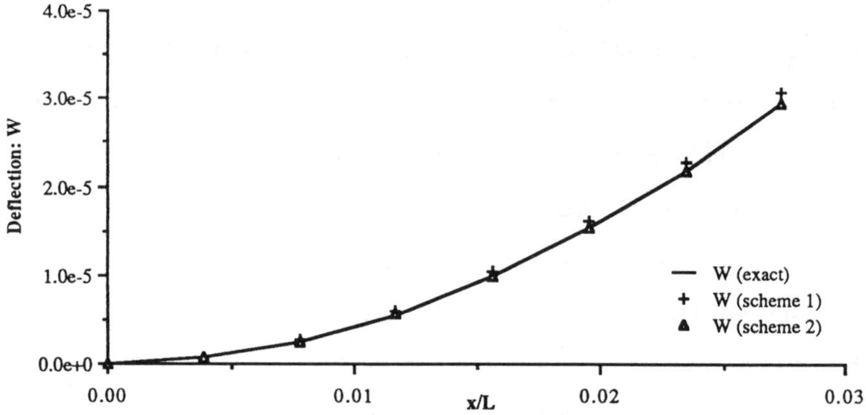


Figure 4.36: Deflected Shape Adjacent to the Beams Edge (257 nodes).

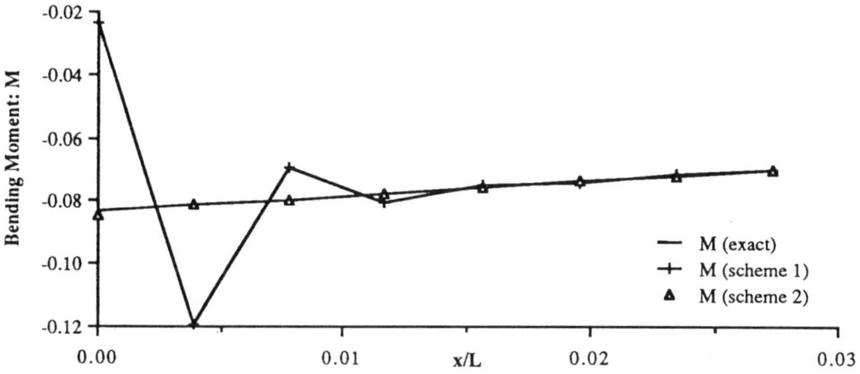


Figure 4.37: Bending Moments Adjacent to the Beams Edge (257 nodes).

Clearly, figure 4.37 illustrates the problems associated with approximating a fixed edge with scheme 1 and confirms the observations made previously when plate and shell examples were studied. The apparent discontinuities in the bending moment diagram are local and only marginally effect the behaviour at mid-span. In contrast, the second approach (scheme 2) produces results which compare favourably with the analytical solution.

Figure 4.38 compares the variation in slope within a boundary mesh. Notably, two zero slope conditions exist within the mesh when the constraint equation defined using scheme 1 is applied. In comparison, the analytical solution has only one zero slope condition within the same domain. Numerically, this difference has the effect of introducing a 'wave' into the displacement function between the external fictitious node and the boundary node. As a consequence, a local disturbance will be evident as the function fails to sufficiently approximate the beams deformation.

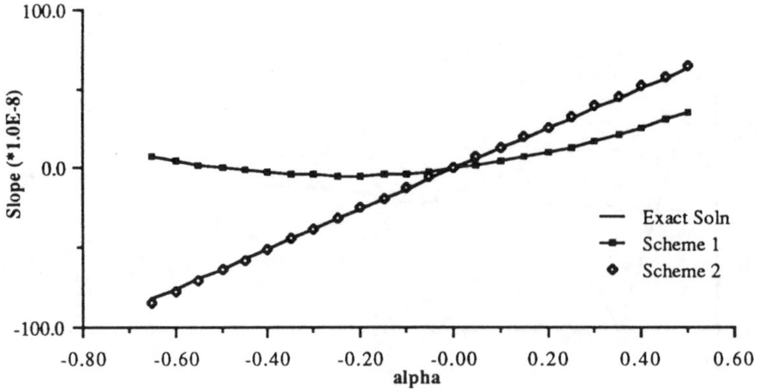


Figure 4.38: Variation in Slope Within a Boundary Mesh.

The above study confirms the need to approximate the constraint equations using a second order displacement interpolation function. This can be achieved within the framework of the existing method by using a number of different mesh configurations to approximate the shell's displacement characteristics. Therefore, at an edge similar to the one illustrated in figure 4.39 the constraint equations associated with the local nodes 4 and 12 are based on a different displacement function than the equations associated with the constraints at nodes 1 and 5. Similarly the boundary equations existing at the corner mesh shown in figure 4.40 would be based on three separate displacement functions.

Having computed the constraint equations, the external fictitious nodes are eliminated using an identical procedure to the one discussed in section 3.6.

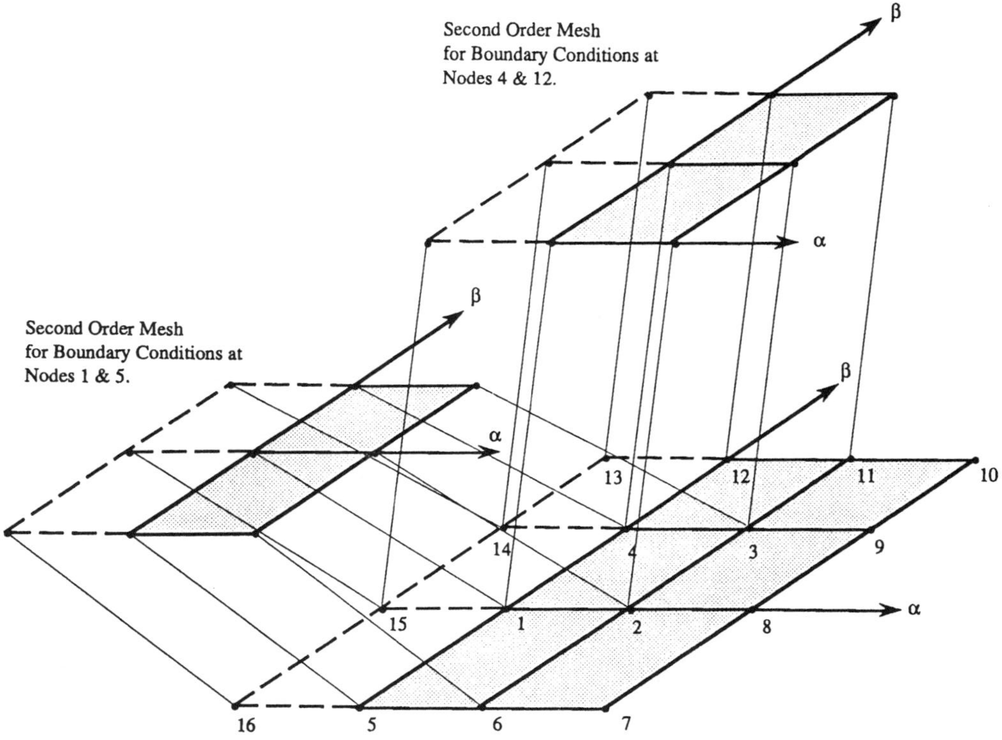


Figure 4.39: Displacement interpolation functions for the constraint equations along a typical edge mesh.

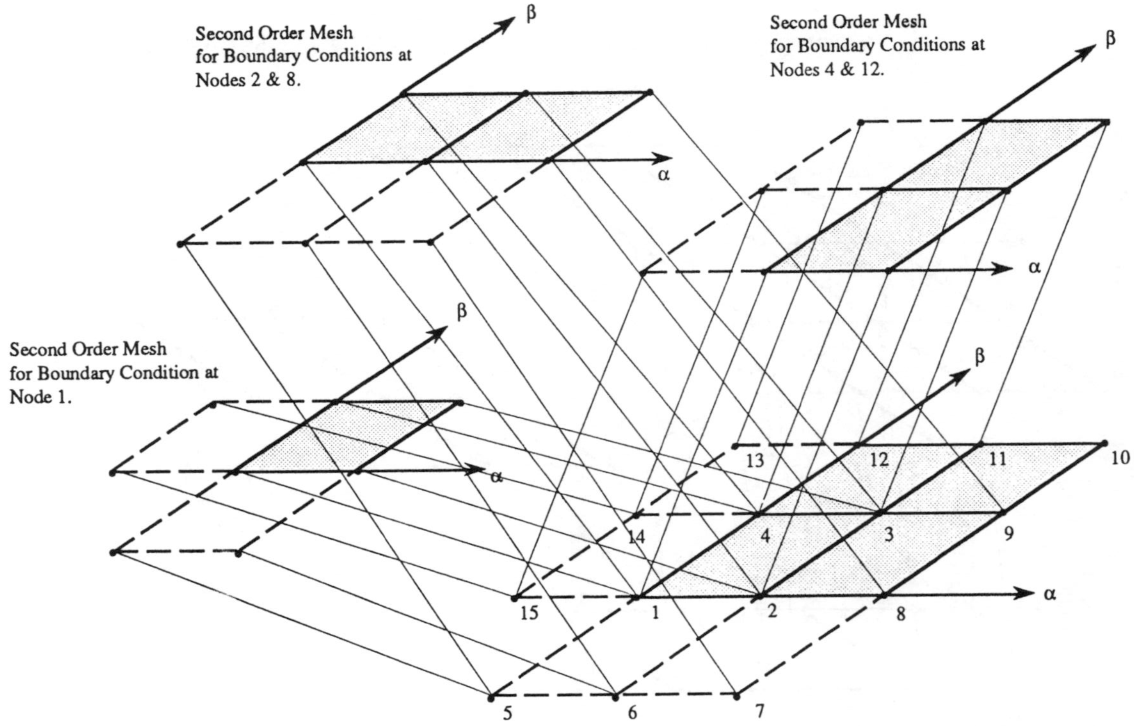


Figure 4.40: Displacement interpolation functions for the constraint equations at a typical corner mesh.

# 5. NUMERICAL APPLICATIONS

Details relating to a numerical method capable of determining the dynamic characteristics of shell structures have been presented in the preceding chapters. The methodology is based on the strain-displacement equations of Flugge [1] and uses the curvilinear finite difference method in order to approximate the geometric properties and displacement behaviour of the shell. Such an approach enables shells of arbitrary geometry to be analysed.

In chapter 4 (see section 4.1.3), the technique was applied successfully to a number of shallow shell and flat plate problems. The results were computed using a second order approximation as a basis. As demonstrated by the spherical shell example, this level of approximation has difficulties in correctly isolating eigenvalues within a clustered group.

A closer investigation into the shell's governing equations revealed that its geometric properties, in particular those required in transforming the global displacements and their associated derivatives, needed a higher order approximation. To achieve this, the basic nine-node curvilinear mesh was extended to sixteen nodes. This enables a complete third-order polynomial to be used.

As noted by Leissa [9], the literature relating to the dynamic characteristics of shells is extensive. In many publications it has been common practice by authors to summarize the performance of a technique by comparing its computed eigenvalues with those of alternate numerical or experimental methods. Numerical details relating to the eigenvectors are seldom published.

In presenting the numerical results in this chapter, it is the authors' intention to firstly, validate the overall methodology by demonstrating its ability to compute the dynamic characteristics of both shallow and non-shallow shells. Secondly, to provide an insight into the convergence characteristics of the technique and finally document the numerical results for future reference. Only the eigenvalues and a perspective view of the accompanying modes will be given in this chapter. Numerical details relating to the eigenvectors at selected cross-sections are summarized in appendix F.

## 5.1 SIMPLY SUPPORTED PLATE

Figure 5.1 defines the geometric and material properties associated with a square plate simply supported at each boundary. The dynamic characteristics of the plate were computed using a series of non-symmetric mesh discretizations. This problem has identical properties to the plate example analysed in the previous chapter. It therefore provides an indication of the convergence characteristics of a 3rd-order 16 node approximation in comparison to a second-order 9 node scheme.

The numerical results for the first six modes of vibration are presented in table 5.1 and compare favourably with the analytical solution. Figure 5.2 illustrates the corresponding mode shapes and in addition, Appendix F summarizes the numerical results generated for each eigenvector at a selected number of cross-sections. The 28×26 mesh defines the finest discretization which could be analysed successfully on the IBM RT workstation.

Comparing directly the results presented in tables 4.4b and 5.1 shows that in the case of non-repeating eigenvalues a third order approximation will yield only a minor improvement in the methods convergence characteristics. This improvement is more pronounced at coarse discretizations. For example, using an 8×6 mesh, the 1st and 4th modes improve by approximately 0.8% and 3.2% respectively. This difference decreases at higher mesh discretizations.

In contrast, the second-order approximation yields marginally better results where repeated eigenvalues exist. No rigorous explanation can be given for this result.

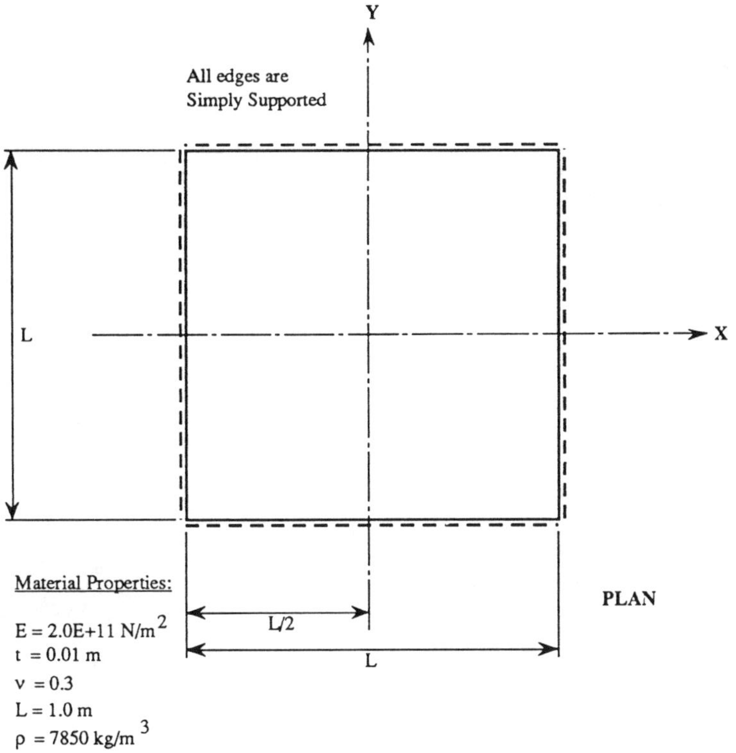


Figure 5.1: Geometric and Material Properties for the Simply Supported Plate.

| Mode | $\lambda = \omega a^2 \sqrt{\rho t / D}$ | | | |
|---|---|---|---|---|
| | 8x6 | 16x14 | 28x26 | Analytical |
| (1,1) | 19.30 | 19.64 | 19.71 | 19.74 |
| (1,2) | 44.14 | 48.21 | 49.00 | 49.35 |
| (2,1) | 46.03 | 48.45 | 49.04 | 49.35 |
| (2,2) | 72.80 | 77.46 | 78.47 | 78.96 |
| (1,3) | 75.64 | 92.87 | 96.88 | 98.69 |
| (3,1) | 83.35 | 94.13 | 97.12 | 98.69 |

Table 5.1: Eigenvalues for the Simply-Supported Square Plate.

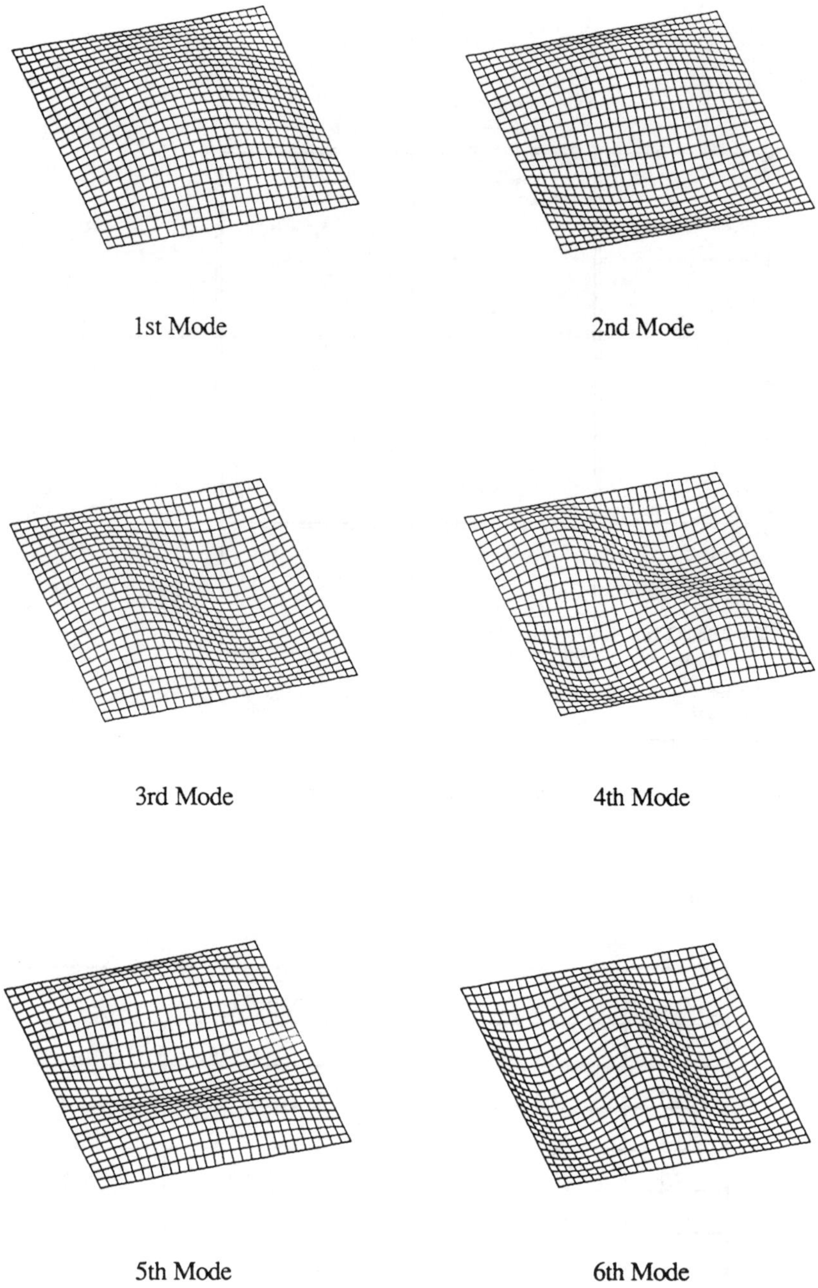


Figure 5.2: Mode Shapes for the Simply Supported Plate.

## 5.2 CANTILEVER PLATE

Figure 5.3 details the geometric characteristics and material properties of a square plate fixed along one edge and allowed to deform freely along the remaining three edges. Unlike the simply supported plate example where the displacement behaviour was one dimensional, the dynamic behaviour of the cantilever plate is essentially three dimensional. In addition, since the plate has no curvature, its inplane and out of plane displacements are not coupled and the resultant eigenvectors will not be composed of a combination of these deformation modes.

For cantilever plates, a considerable number of results have been published within the literature. In particular, an extensive study detailed by Klieb, Leissa and MacBain [108] presents and compares results derived from beam theory, shell theory and the finite element method. However, a more complete set of numerical values are presented in an early publication by Barton [109], where the Ritz method was applied to the problem of determining the vibrational characteristics of rectangular and skew cantilever plates.

The eigenvalues associated with the first six modes of vibration are presented in table 5.2 and compare favourably with the results of Barton[109] and the finite element method. The element adopted in this comparison is based on the non-conforming rectangular flat shell element described by Zienkiewicz [46].

The corresponding eigenvectors are illustrated in figure 5.4 and numerical values at selected cross sections are provided in Appendix F. Importantly, the method has computed the 6th mode of vibration as being composed entirely of inplane deformations. Its shape is comparable to the 1st bending mode associated with a cantilevered beam model.

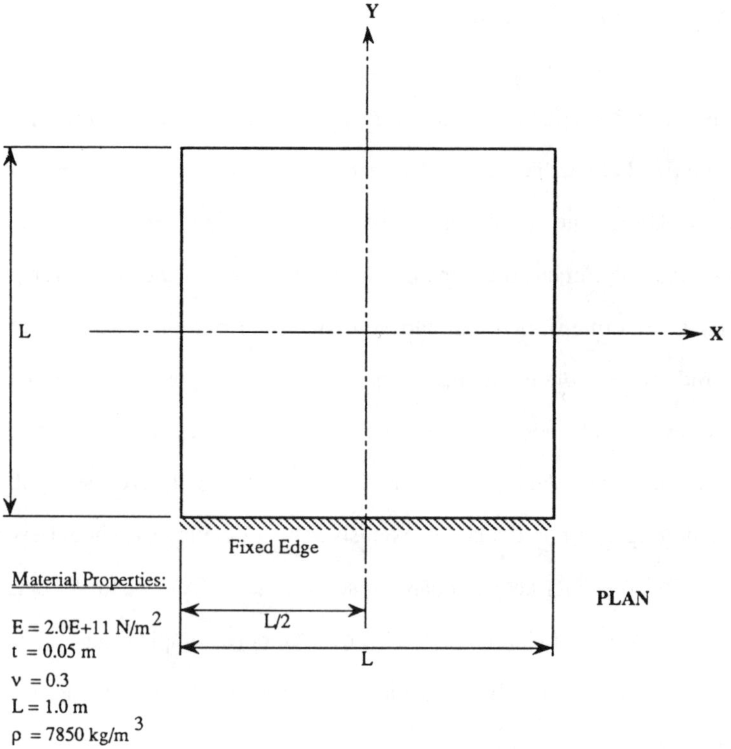


Figure 5.3: Geometric and Material Properties for the Cantilever Plate.

| Mode | Eigenvalues (rad/sec) | | | | | |
|---|---|---|---|---|---|---|
| | 4x4 | 8x8 | 16x16 | 28x28 | Barton [109] | FEM [16x16] |
| 1 | 250.4 | 260.9 | 263.9 | 264.6 | 266.9 | 265.1 |
| 2 | 603.5 | 633.0 | 644.2 | 647.1 | 652.8 | 649.7 |
| 3 | 1211.1 | 1493.7 | 1586.4 | 1608.6 | 1637.4 | 1626.9 |
| 4 | 1846.2 | 1986.8 | 2044.3 | 2059.1 | 2097.2 | 2076.2 |
| 5 | 1953.2 | 2202.8 | 2311.5 | 2339.6 | 2380.5 | 2364.4 |
| 6 | 3374.1 | 3335.7 | 3326.1 | 3323.6 | - | 3340.3 |

Table 5.2: Eigenvalues for the Square Cantilever Plate.

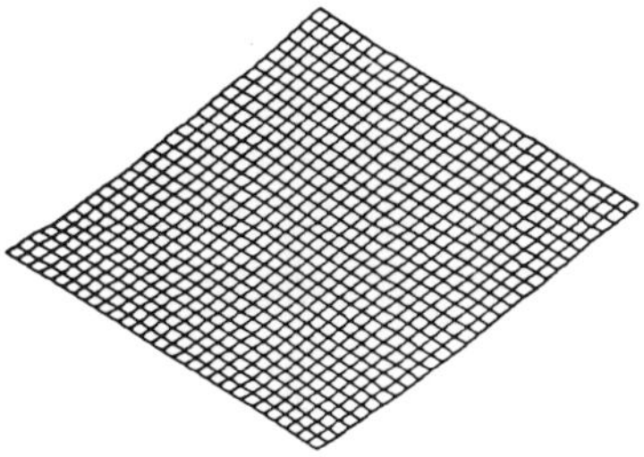

1st Mode

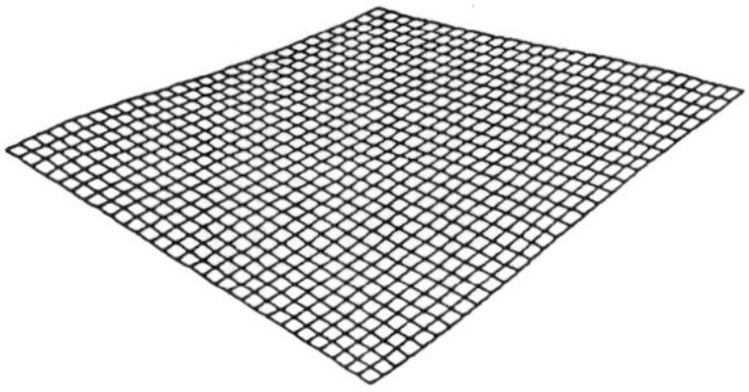

2nd Mode

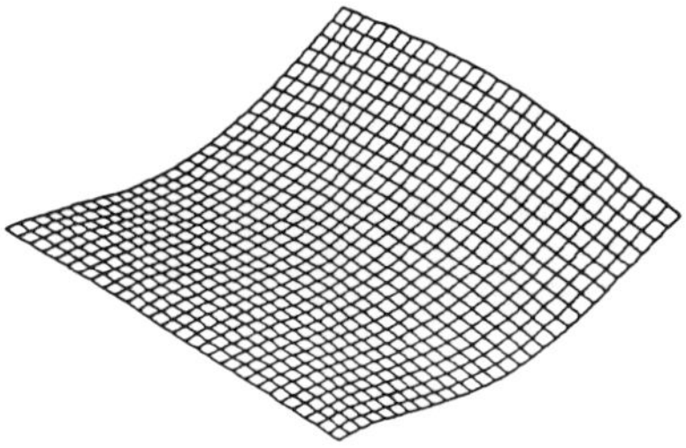

3rd Mode

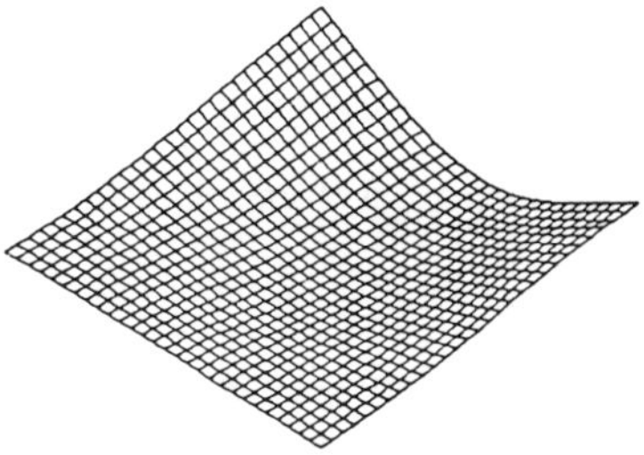

4th Mode

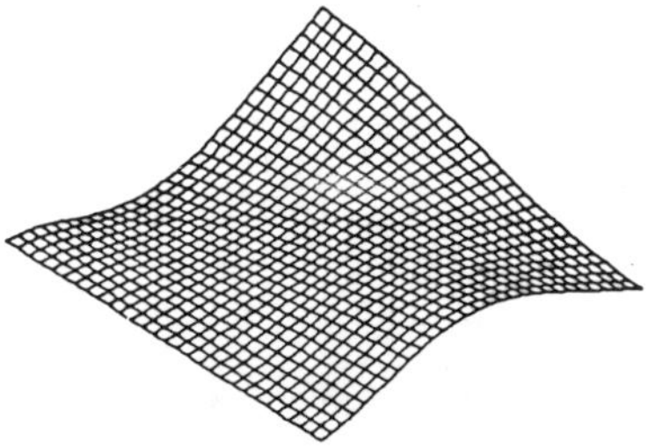

5th Mode

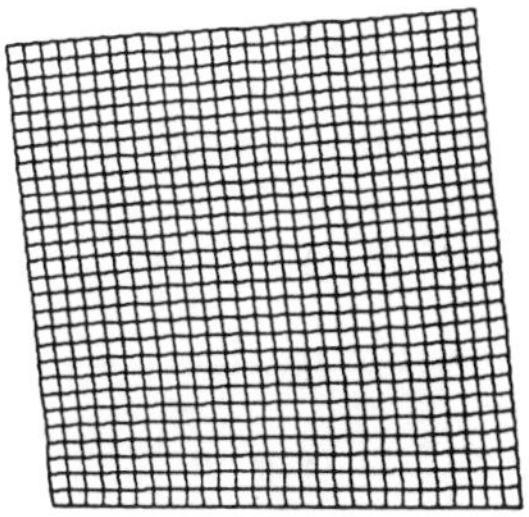

6th Mode

Figure 5.4: Mode Shapes for the Cantilever Plate.

## 5.3 SPHERICAL CAP ON A SQUARE BASE

In the previous chapter, the dynamic characteristics of a number of spherical shell examples were presented and discussed. The results were generated using a second order finite difference approximation the details of which were developed in chapters 3 and 4. In particular, the examples were selected to confirm the procedure and to highlight a deficiency which exists when a shell structure is characterized by a series of clustered eigenvalues.

Briefly, it was concluded that where eigenvalues were closely spaced together, a higher order approximation was required to determine more accurately the surface geometric properties and the displacement components within a mesh. Subsequently, a third order finite difference approximation was developed and utilized in the generation of results presented in this chapter.

For completeness, the examples analysed in the previous chapter are re-analysed using the third order approximation as a basis. The spherical shell defined in figure 5.5 is freely supported on a boundary of square plan form. The displacements $u_n$ and $u_t$ are constrained along the edges.

The numerical results for the first six modes of vibration are presented in table 5.3 and compare well with the analytical solution. Figure 5.6 illustrates the respective mode shapes and numerical details at selected cross-sections are summarized in Appendix F.

In comparison with the results generated using a second order approximation (see table 4.5), the eigenvalues given in table 5.3 illustrate a noticeable improvement in the methods convergence characteristics. Once again, despite the use of a non-symmetric mesh, the mode (1,3) has a circular nodal line similar to the one discussed previously.

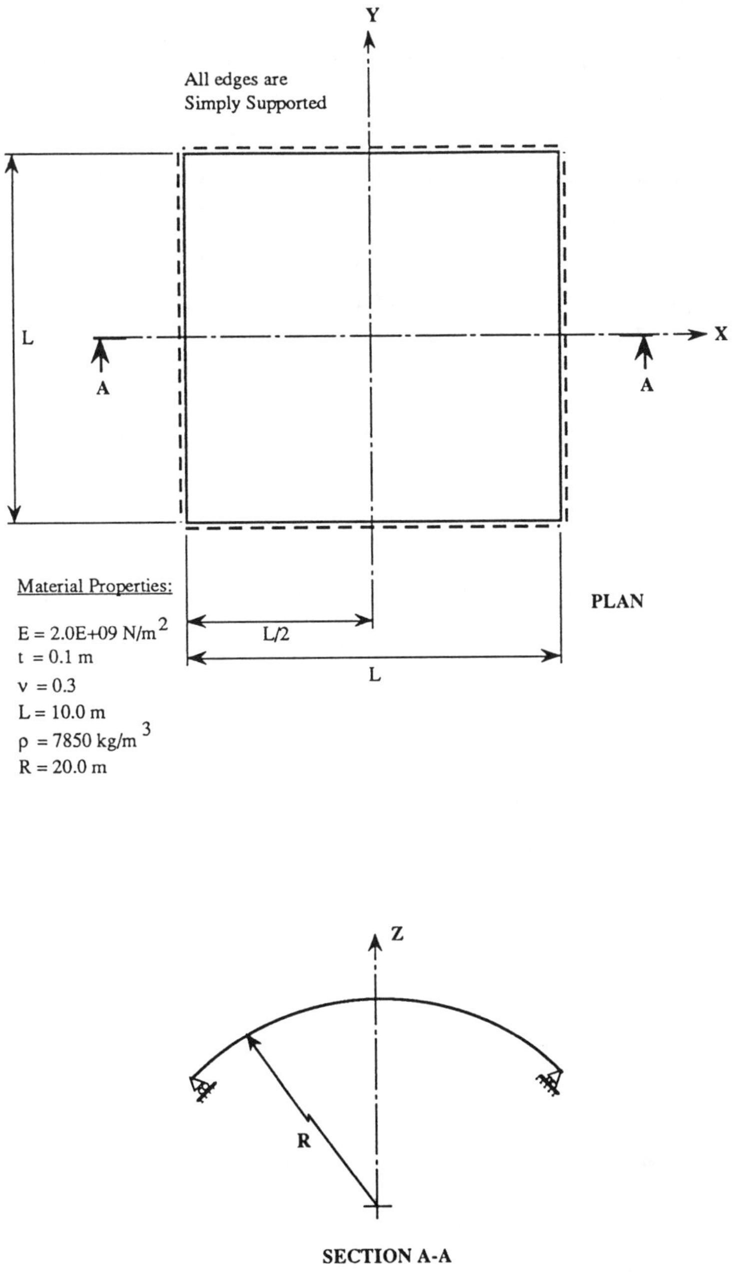


Figure 5.5: Geometric and Material Properties for the Spherical Shell.

| Mode | Eigenvalues (rad/sec) | | | |
|---|---|---|---|---|
| | 8x6 | 16x14 | 28x26 | Analytical |
| (1,1) | 25.44 | 25.34 | 25.34 | 25.42 |
| (1,2) | 26.44 | 26.22 | 26.22 | 26.34 |
| (2,1) | 26.69 | 26.22 | 26.22 | 26.34 |
| (2,2) | 28.65 | 27.77 | 27.78 | 27.97 |
| (1,3) | 28.97 | 28.81 | 29.04 | 29.40 |
| (3,1) | 30.14 | 28.89 | 29.07 | 29.40 |

Table 5.3: Eigenvalues for the Spherical Cap on a Square Base.

The second problem under consideration is similar to the one illustrated in figure 5.5 but has the following geometric and material properties:

$R = 2.5$ m

$L = 1.0$ m

$t = 0.0025$ m

$E = 2.0E{+}11$ N/m$^2$

$\nu = 0.3$

$\rho = 7850.0$ kg/m$^3$

The results obtained by using a second order approximation were given in table 4.7 and figure 4.33. Even with a 32x30 mesh discretization it was found that the technique had not converged sufficiently to be able to define the eigenvalues and their corresponding eigenvectors in the correct order.

In contrast, there is a substantial improvement in the results generated by the third order implementation (see table 5.4 and figure 5.7). Despite a discrepancy existing when the shell is modeled using a coarse 8x6 mesh discretization, the technique is capable of predicting the eigenvalues and eigenvectors in their correct

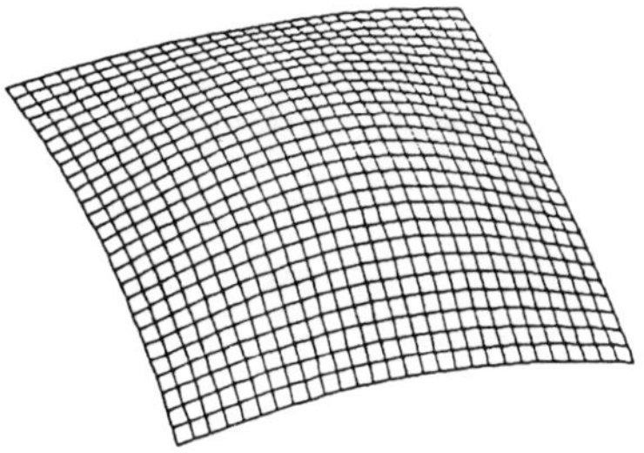

1st Mode

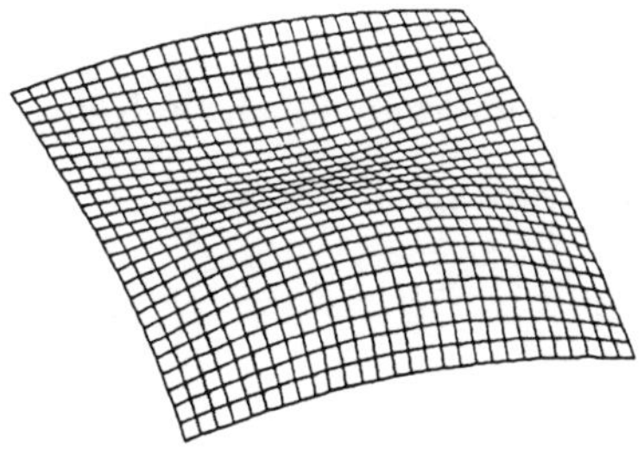

2nd Mode

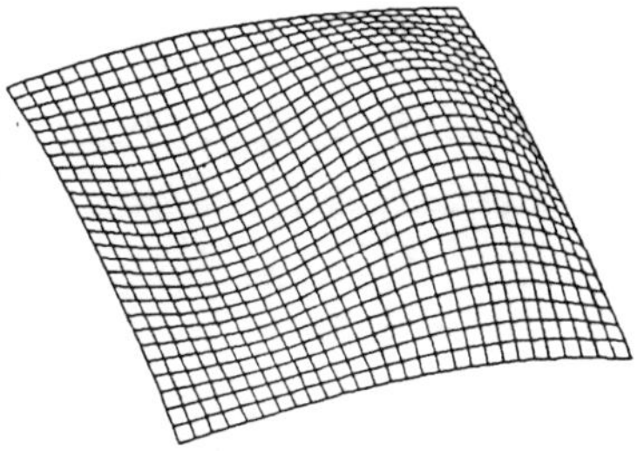

3rd Mode

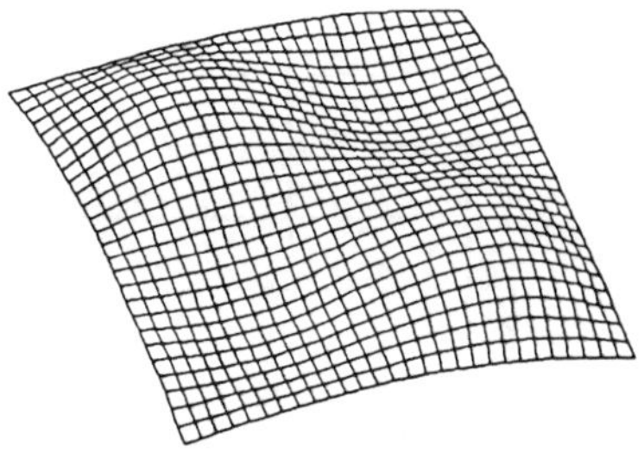

4th Mode

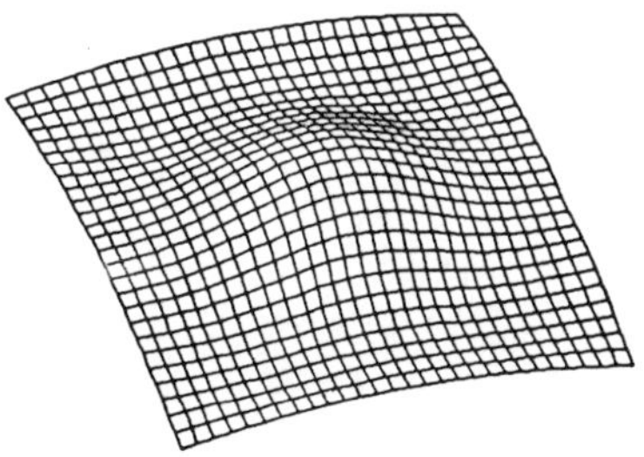

5th Mode

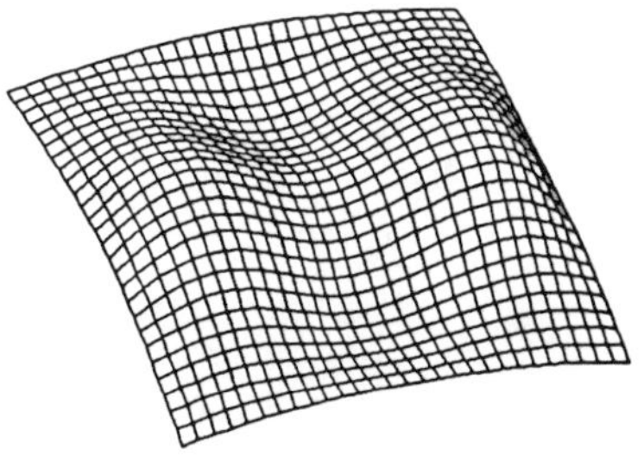

6th Mode

Figure 5.6: Mode Shapes for the Spherical Shell on a Square Base.

order. Further investigation has shown that the above anomaly is a result of using a coarse unsymmetrical mesh discretization and can be overcome by using a symmetric mesh.

| Mode | Eigenvalues (rad/sec) | | | |
|---|---|---|---|---|
| | 8x6 | 16x14 | 28x26 | Analytical |
| (1,1) | 2023. | 2015. | 2015. | 2020. |
| (1,2) | 2048. | 2027. | 2026. | 2028. |
| (2,1) | 2075. | 2027. | 2026. | 2028. |
| (2,2) | 2136. | 2043. | 2040. | 2041. |
| (1,3) | 2116. | 2053. | 2051. | 2054. |
| (3,1) | 2258. | 2056. | 2054. | 2054. |

Table 5.4: Eigenvalues for the Spherical Cap on a Square Base.

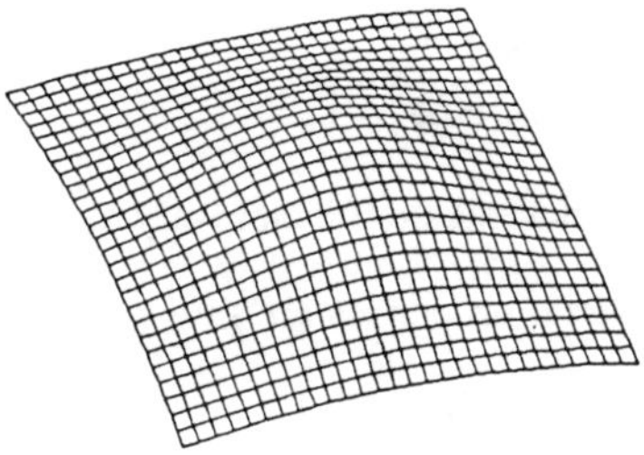

1st Mode

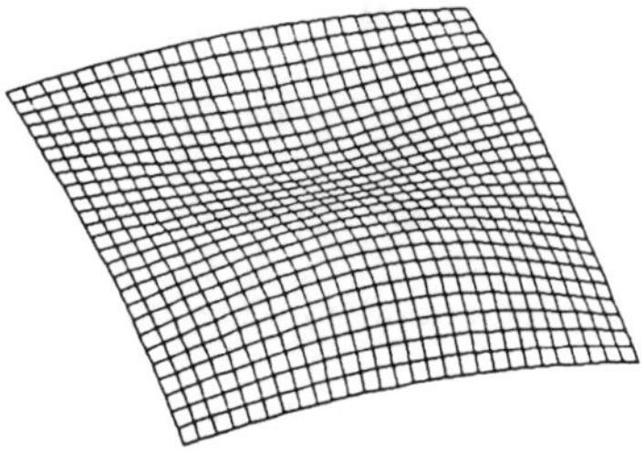

2nd Mode

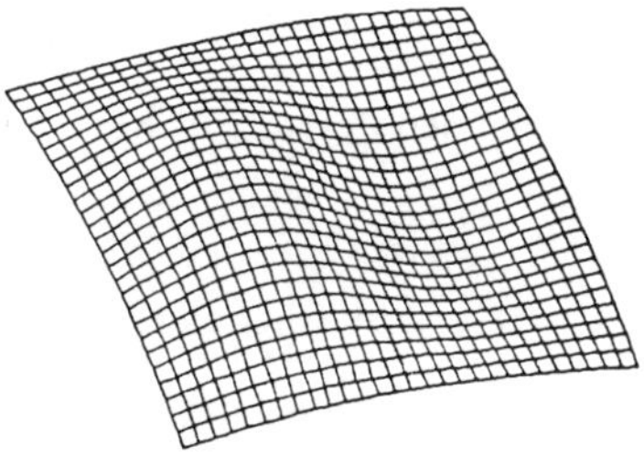

3rd Mode

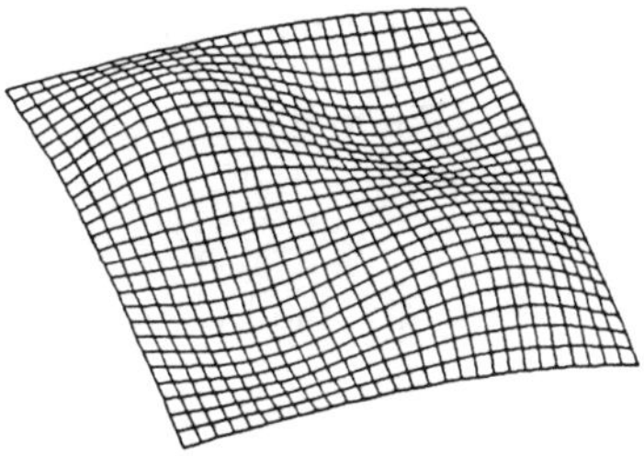

4th Mode

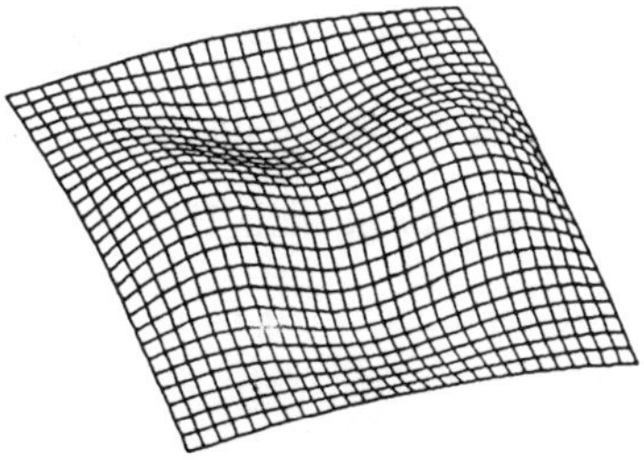

5th Mode

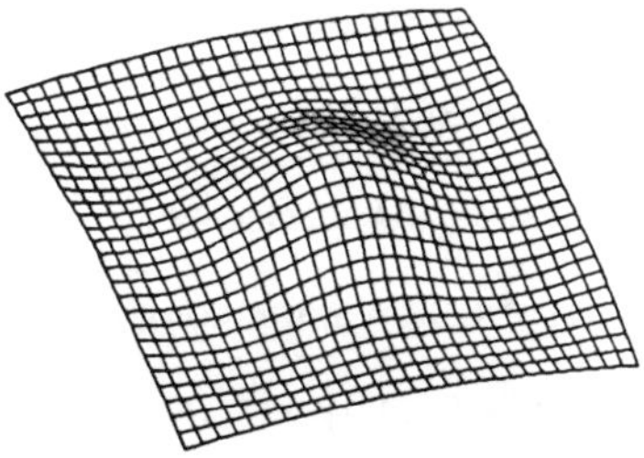

6th Mode

Figure 5.7: Mode Shapes for the Spherical Shell on a Square Base.

## 5.4 CYLINDRICAL PANEL

The next application considered is a cylindrical shell panel fully clamped along each boundary. The dimensions and material properties are given in figure 5.8 and are identical to the ones used by Olson and Lindberg [54].

This problem, referred to in the literature as the Lockheed shell panel, has been investigated extensively by a number of authors [54,55,56,106]. There is no exact solution to this problem, however detailed comparisons between the experimental natural frequencies and those obtained by a number of different numerical approaches (Extended Rayleigh-Ritz, Triangular Finite Element, Rectangular Finite Element and Kantorovich Method, see Petyt [55]) have been made.

Recently, the Lockheed shell panel was analysed by Cheung et al [106] using the Spline Finite Strip Method. The published values compared favourably with those already existing within the literature.

Using the proposed method, the panel was analysed using a variety of mesh discretizations. The numerical results for the first six modes of vibration are presented in table 5.5 and compared with the experimental and theoretical values provided by Olson and Lindberg [54] and the numerical values published by Cheung et al [106]. Figure 5.9 illustrates the respective mode shapes and numerical details at selected cross-sections are presented in Appendix F.

As shown, the computed eigenvalues and there associated mode shapes compare favourably with the published results.

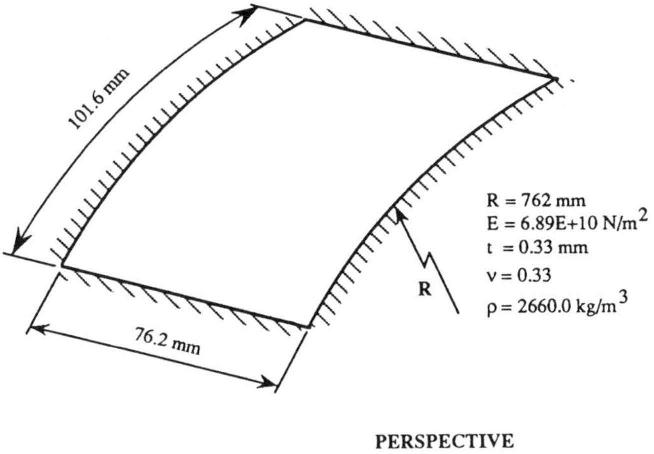


Figure 5.8: Geometric and Material Properties for the Cylindrical Shell Panel.

| Mode | Natural Frequency (Hz) | | | | | |
|---|---|---|---|---|---|---|
| | 8x8 | 16x16 | 28x28 | Experimental | FEM [54] | FSM [106] |
| 1 | 804. | 843. | 860. | 814. | 870. | 874. |
| 2 | 860. | 918. | 944. | 940. | 958. | 963. |
| 3 | 1170. | 1225. | 1262. | 1260. | 1288. | 1298. |
| 4 | 1226. | 1318. | 1347. | 1306. | 1363. | 1369. |
| 5 | 1267. | 1376. | 1416. | 1452. | 1440. | - |
| 6 | 1339. | 1535. | 1685. | 1735. | 1780. | - |

Table 5.5: Natural Frequencies for the Cylindrical Shell Panel.

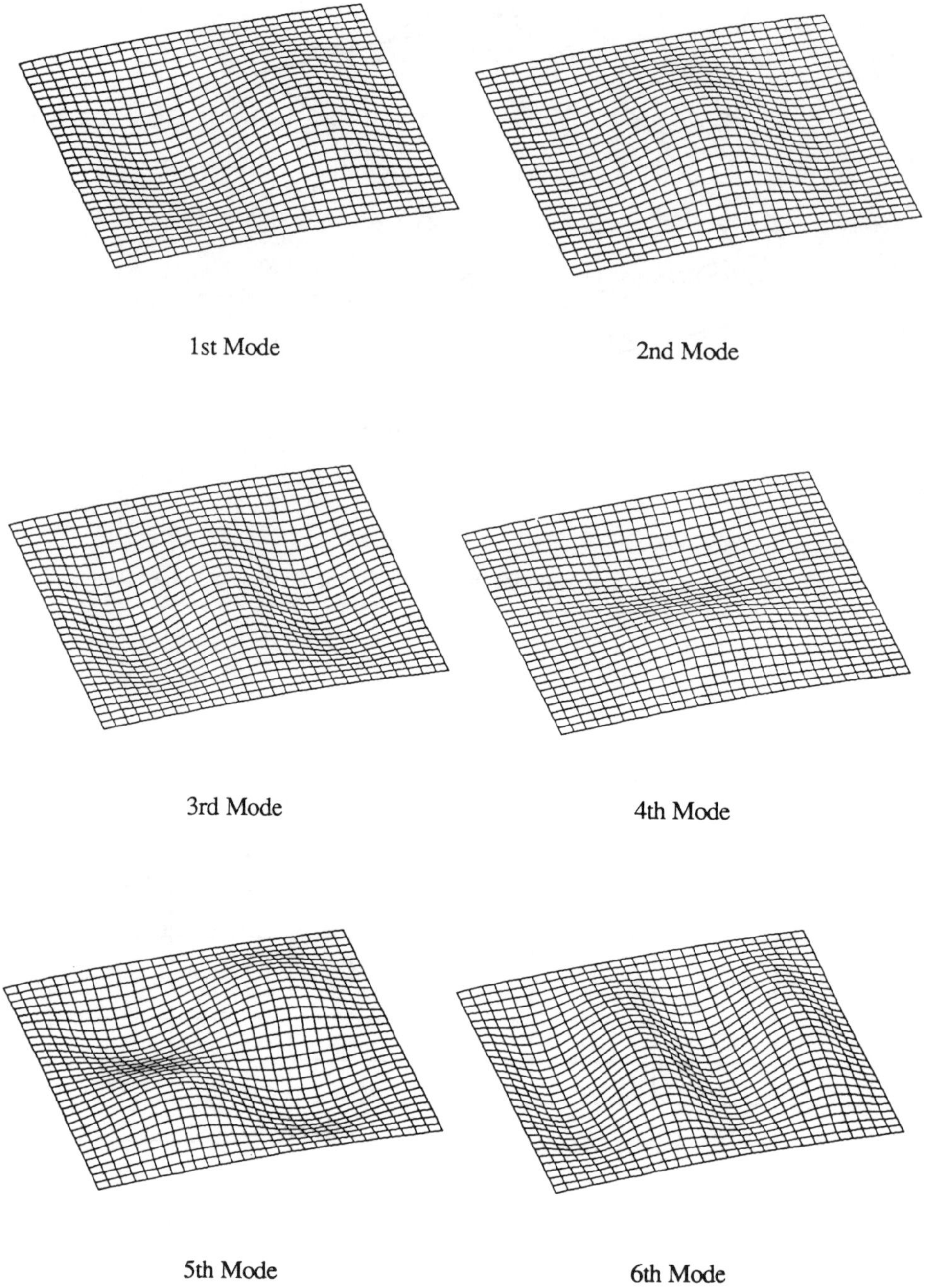


Figure 5.9: Mode Shapes for the Cylindrical Shell Panel with Fully Fixed Edges.

## 5.5 CURVED FAN BLADE

The curve fan blade illustrated in figure 5.10 was first studied experimentally by Olson and Lindberg [52,54]. The experimental model was constructed by rolling a piece of sheet steel to its required radius and then welding it to a steel block (4.0 in × 4.0 in × 16.0 in) to simulate the clamped boundary condition indicated in figure 5.10.

Numerically, this problem has been studied by many authors. Originally, Olsen and Lindberg [52] analysed the fan blade using a cylindrical shell finite element and later a curved shallow shell element [54]. The same problem has subsequently been analysed by Bossak and Zienkiewicz [60] using isoparametric solid elements, Hofmeister and Evensen using isoparametric shell elements and Martins and Owen [105] using the semiloof element. These are only a representative sample of the techniques adopted for the analysis of this shell problem (see also references [57-59,110-111]) and highlight the wide range of results which appear within the literature to base a comparison on.

Table 5.6 and figure 5.11 contain the first six modes of vibration generated using the curvilinear finite difference approach implemented in this study. Clearly, these results compare favourably with the experimental and numerical values provided by Olsen and Lindberg [54]. Further details pertaining to the mode shapes are given in Appendix F.

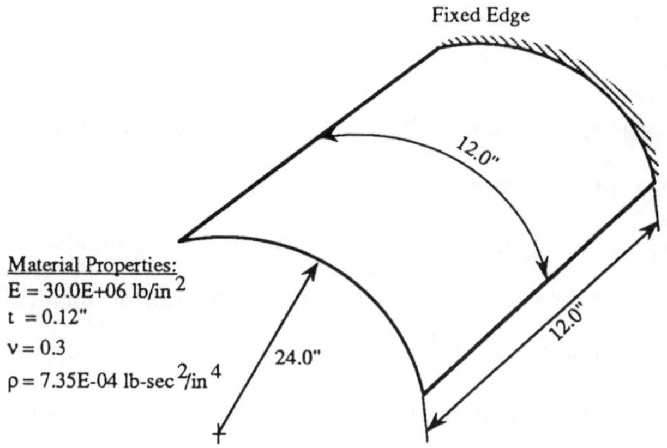


Figure 5.10: Geometric and Material Properties for the Curved Fan Blade.

| Mode | Natural Frequency (Hz) | | | | |
|---|---|---|---|---|---|
| | 8x8 | 16x16 | 28x28 | Experimental | FEM [54] |
| 1 | 95. | 87. | 86. | 86. | 87. |
| 2 | 144. | 139. | 139. | 134. | 139. |
| 3 | 261. | 248. | 247. | 259. | 251. |
| 4 | 348. | 341. | 342. | 351. | 349. |
| 5 | 391. | 385. | 385. | 395. | 393. |
| 6 | 541. | 520. | 525. | 531. | 533. |

Table 5.6: Natural Frequencies for the Curved Fan Blade.

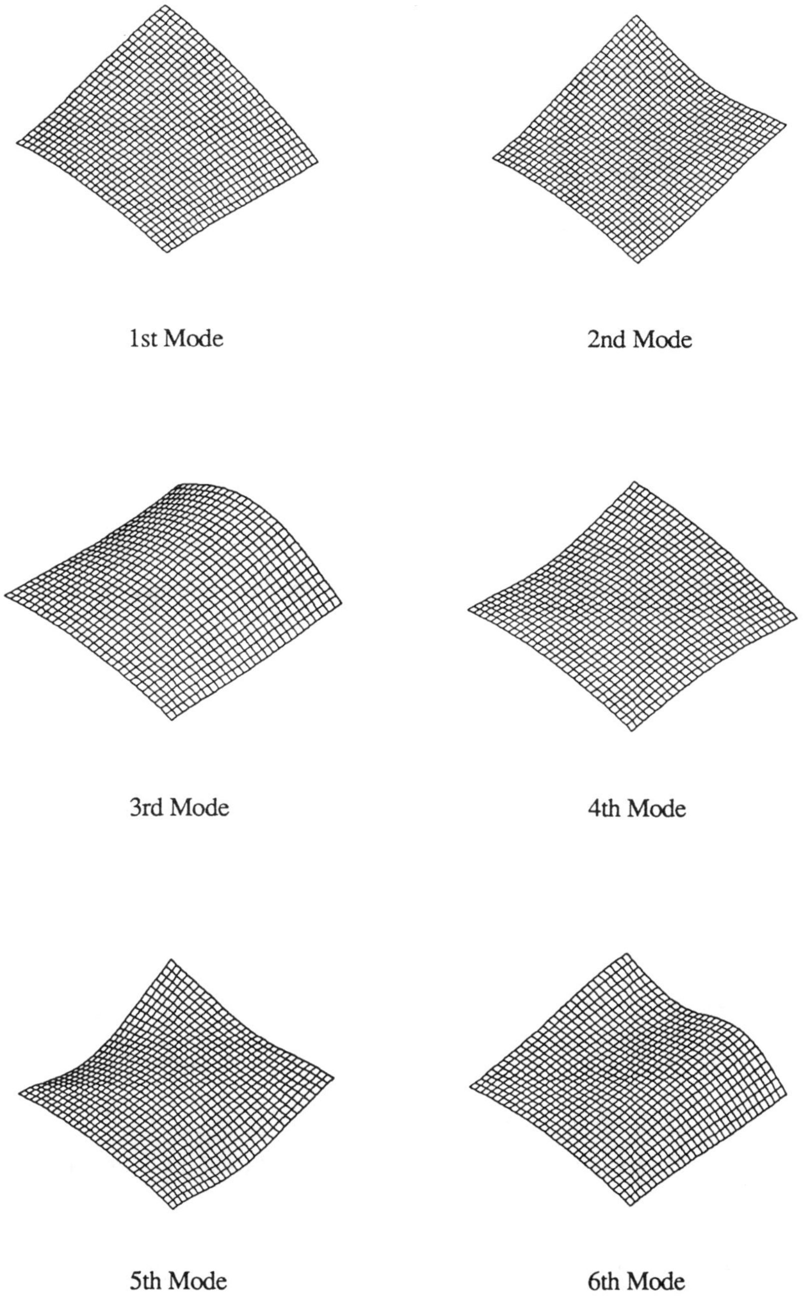


Figure 5.11: Mode Shapes for the Curved Fan Blade.

## 5.6 CONICAL SHELL PANEL

Conical shells are widely used in a variety of engineering applications ranging from structural components in aircrafts to the use of loudspeaker cones in the music industry. As such, knowledge of their dynamic characteristics becomes of prime importance in the design process.

Theoretically, the conical shell panel analysed in this section (see figure 5.12) can not be classified as a shallow shell. That is, the square of the geometric partial derivative $z,_x$ is comparable to unity. In such cases, it is important to compute the geometric properties accurately and represent the shell's behaviour using the general form of the strain-displacement equations. This was demonstrated in the work of Kwok [72] where a fully fixed conoidal shell subjected to a uniformly distributed static load was analysed. A comparison was then made between the results generated using a model based on shallow shell theory and general shell theory. Clearly significant discrepancies existed between the results obtained from the two models.

Based on the above study, the conical shell panel provides a good test for substantiating the methodology. However, as noted by Srinivasan and Krishnan [27], the free vibration analysis of conical shells has been confined mostly to structures which may be defined as a shell of revolution (see Leissa [9]).

Recently the panel discussed in the work of Srinivasan and Krishnan was considered by Cheung et al [106]. In contrast to the integral equation technique (IET) used by Srinivasan and Krishnan , Cheung et al used the spline finite strip method (FSM).

The numerical results for the first six modes of vibration are presented in table 5.7 and compared with the published values of Srinivasan & Krishnan, and Cheung et al. Figure 5.13 illustrates the respective mode shapes and numerical details at selected cross-sections are presented in Appendix F. As shown in table 5.7, the computed eigenvalues compare favourably with the published results.

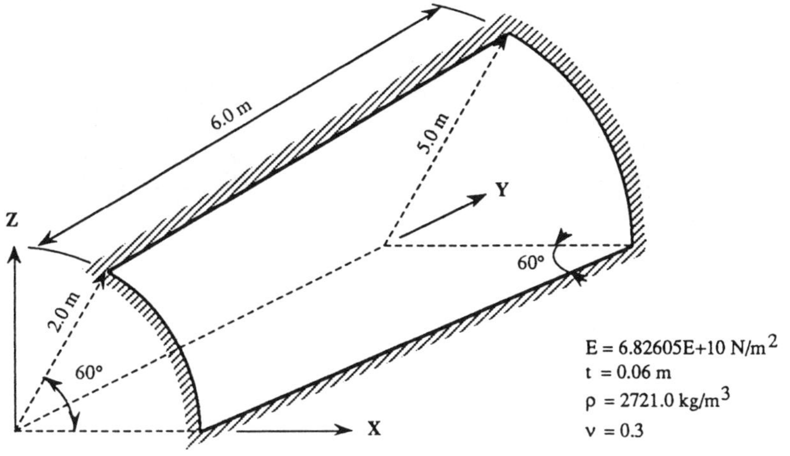


Figure 5.12: Geometric Details for the Conical Shell Panel.

| Mode | Natural Frequency (Hz) | | |
|---|---|---|---|
| | 28x28 | FSM [106] | IET [27] |
| 1 | 523. | 539. | 512. |
| 2 | 628. | 663. | 657. |
| 3 | 765. | 795. | 772. |
| 4 | 863. | 906. | 897. |
| 5 | 986. | - | 1018. |
| 6 | 1000. | - | - |

Table 5.7: Natural Frequencies for the Conical Shell Panel.

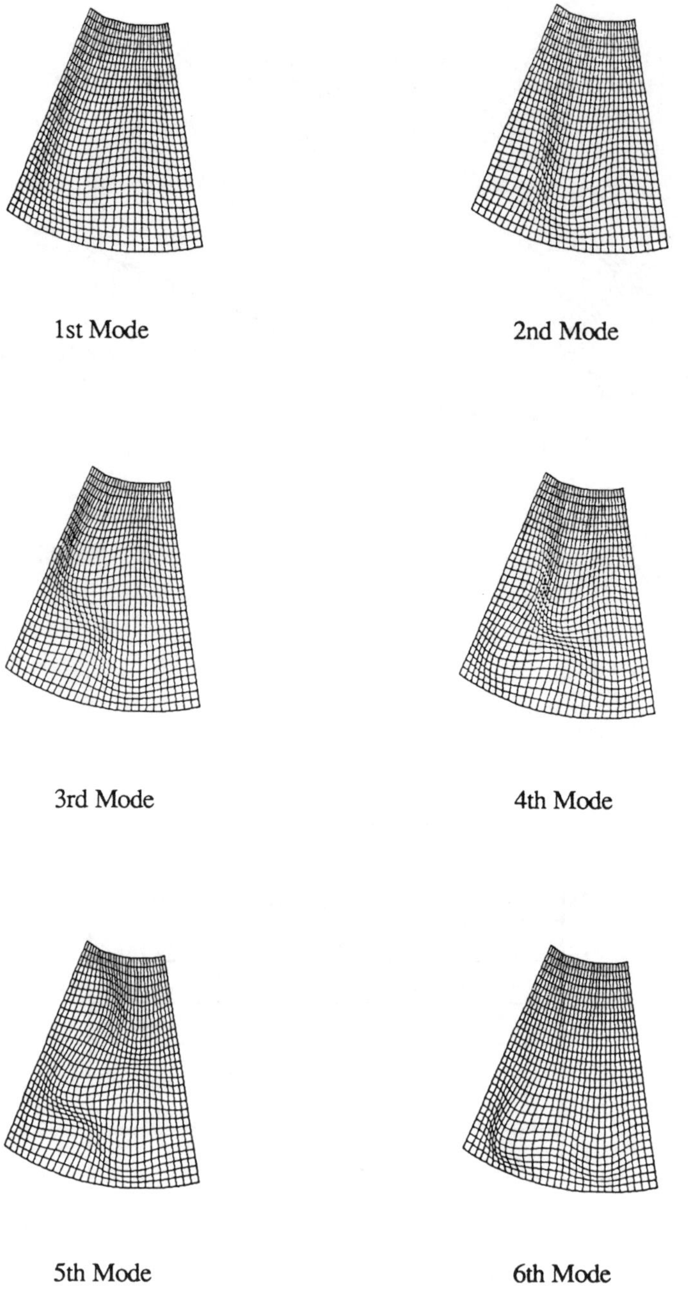


Figure 5.13: Mode Shapes for the Conoidal Shell.

## 5.7 CYLINDRICAL TANK

The final example presented in this chapter is that of a circular cylindrical shell which is clamped at one end and free at the other. The geometric and physical properties (see figure 5.14) are identical to the ones given by Leissa [9] where both experimental and theoretical frequencies are tabulated. The theoretical predictions are obtained using an energy method based on Sanders shell theory and the Rayleigh-Ritz method.

Due to the nature of the resultant mode shapes, this problem was modeled by attempting to make use of the existing lines of symmetry. It must be emphasized, that the assumed boundary conditions are at best only an approximation to the conditions which would exist if the whole problem was analysed.

The numerical results for the first six modes of vibration are presented in table 5.8 and compared with the experimental and theoretical values provided by Leissa [9]. Within this table the indices n and m refer to the circumferential and axial nodal patterns respectively. The computed mode shapes are illustrated in figure 5.15 and numerical values at selected cross-sections are documented in Appendix F.

Table 5.8 shows good agreement between the published theoretical results and the numerical values computed using the present method. Clearly, discrepancies exist when compared to the experimental values. This is noted by Leissa and attributed to insufficient axial constraint at the boundaries during the experimental process.

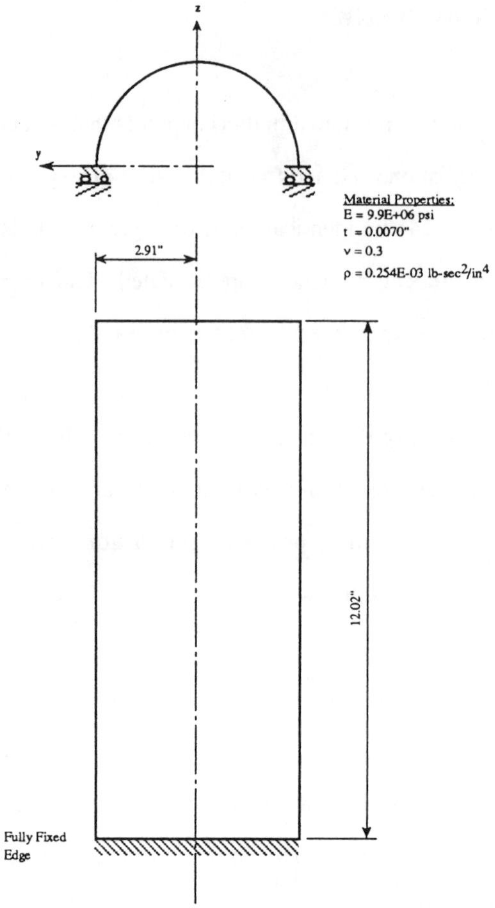


Figure 5.14: Geometric and Physical Properties for the Cylindrical Tank.

| Mode | Natural Frequency (Hz) | | |
|---|---|---|---|
| (m,n) | 21x53 | Leissa [9] | Experimental |
| (0,4) | 177. | 181. | 158. |
| (0,5) | 208. | 207. | 200. |
| (0,3) | 234. | 246. | 165. |
| (0,6) | 285. | 280. | 276. |
| (0,7) | 388. | 378. | 374. |
| (0,2) | 456. | 476. | 149. |

Table 5.8: Natural Frequencies for the Cylindrical Tank.

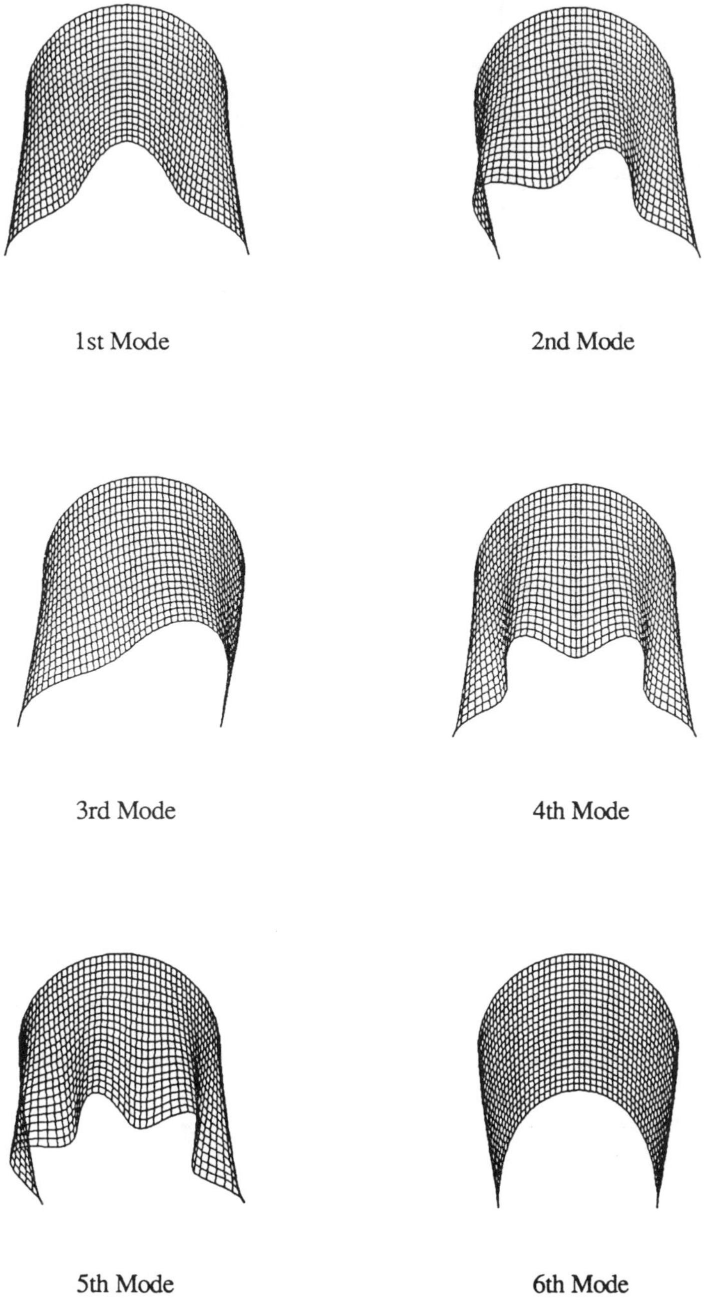


Figure 5.15: Mode Shapes for the Cylindrical Tank.

# 6. SUMMARY

A technique capable of computing the natural frequencies and mode shapes of thin shells has been developed and applied successfully to a number of shallow and non-shallow shell structures.

The methodology is based on the strain-displacement equations of Flugge. These relationships are expressed in general curvilinear coordinates and have been formulated entirely within the framework of tensor calculus. Provided the geometry of the shell remains 'smooth', the underlying theory does not restrict the application of the technique to any particular geometric form.

The shell's governing strain-displacement equations are approximated using the curvilinear finite difference method. This approach permits the use of irregular mesh discretizations and enables both the displacement characteristics and the geometric properties of the shell to be approximated numerically both at a local and global level.

In developing and verifying the technique, a number of fundamental observations have been made:

### Coordinate system

To enable the method to remain relatively insensitive to mesh distortions, the displacement fields assigned to each local computational grid must be expressed in terms of a common global coordinate system. This ensures that the displacement components and their associated base vectors are continuous and single valued.

### Integration

The evaluation of a structure's stiffness matrix using a single-point numerical integration scheme may in some cases produce inaccurate results. This was illustrated in chapter 4 by considering a cantilevered plate subjected to a set of inplane concentrated loads. Subsequent numerical and analytical studies showed that the acceptance of a reduced integration scheme was case dependent and greater reliability is achieved by adopting an integration scheme which can numerically evaluate the integral 'exactly'.

**Geometric properties**

The ability to accurately compute the geometric properties of a shell is an important component in the dynamic analysis of shell structures. In particular, the process of transforming the displacement components and their associated partial derivatives from the global reference frame to the local curvilinear coordinate system requires the evaluation of third-order geometric partial derivatives. In the spherical shell example analysed in chapters 4 and 5, the use of a lower second-order approximation was shown to have difficulties in correctly isolating eigenvalues within a clustered group. In contrast, a significant improvement in the accuracy of the computed eigenvalues and eigenvectors was obtained when the spherical shell was re-analysed using third-order curvilinear meshes.The third-order finite difference approximation is therefore recommended as a basis for numerically modelling curved shell structures of arbitrary geometry.

**Approximation**

In adopting a third-order approximation, special attention must be given to the implementation of the boundary conditions. The approximation of the constraint equations using a third-order displacement interpolation introduces a local disturbance in the displacements and bending stresses adjacent to the boundary. This effect becomes more pronounced as the mesh is refined. A method of overcoming this problem was achieved by deriving the constraint equations using a lower second-order displacement approximation.

Finally the method developed has been applied to a number of flat plate, cylindrical, spherical and conical shell examples. In all cases, the results have compared favourably with those existing within the literature. In addition, numerical details related to the computed eigenvectors are documented for future reference.

## REFERENCES


[1] Flugge, W. '*Tensor Analysis and Continuum Mechanics*', Springer Verlag, Berlin, (1972).

[2] Donnell, L. H. 'A discussion of thin shell theory', *Proc. 5th Int. Congr. Appl. Mech.*, (1938).

[3] Love, A. E. H. 'A treatise on the mathematical theory of elasticity', 1st Edn. Cambridge University Press (1892); 4th Edn. Dover Pub. Inc. New York, (1944).

[4] Mushtari, Kh. M. 'Certain generalizations of the theory of thin shells', *Izv. Fiz. Mat. ob-va. Pri Kaz.*, un-te **11**, in Russian (1938).

[5] Naghdi P. M. 'The theory of shells and plates', in: C. Truesdell (ed.), *Flugge's Handbuch der Physik* , VIa/2, 425-640, (1972).

[6] Timoshenko S. and Woinowsky-Krieger S. '*Theory of Plates and Shells*', McGraw-Hill, New York (1959).

[7] Reissner E. 'A new derivation of the equations of the deformation of elastic shells', *Am. J. Math.* **63**, 177-184, (1941).

[8] Vlasov V. Z. 'Osnovnye differentsialnye Uravnenie obshche teorii uprugikh obolochek', *Prikl. Mat. Metkh.* **8** (1944). (English translation: NACA TM 1241, 'Basic Differential Equations in the General Theory of Elastic Shells', Feb. 1951.)

[9] Leissa A. W. 'Vibration of Shells', NASA SP-288 (1973).

[10] Kraus H. '*Thin Elastic Shells: An Introduction to the Theoretical Foundations and the Analysis of their Static and Dynamic Behaviour*', Wiley, New York, (1967).

[11] Bushnell D. 'Computerized analysis of shells - governing equations', *Computers & Structures*, Vol. **18**, No. 3, 471-536, (1984).

[12] Naghdi P. M. and Berry J. G. 'On the equations of motion of cylindrical shells', *J. Appl. Mech.,* Vol. **21**, No. 2, 160-166, June (1964).

[13] Epstein P. S. 'On the theory of elastic vibrations in plates and shells', *J.Math. Phys.,* Vol. **21**, 198-209, (1942).

[14] Yu Y. Y. 'Free vibrations of thin cylindrical shells having finite lengths with freely supported and clamped edges', *J. Appl. Mech.,* Vol. **22**, No. 4, 547-552, Dec. (1955).

[15] Forsberg K. 'Influence of boundary conditions on the modal characteristics of thin cylindrical shells', *AIAA Journal,* Vol. **2**, No.12, 2150-2157, Dec. (1964).

[16] Warburton G. B. 'Vibration of thin cylindrical shells', *J. Mech. Engng. Sci.,* Vol. **7**, 399-407, Dec. (1965).

[17] Kurt C. E. and Boyd D. E. 'Free vibrations of noncircular cylindrical shell segments', *AIAA Journal,* Vol. **9**, No. 2, 239-244, Feb. (1971).

[18] Reissner E. 'On vibrations of shallow spherical shells', *Journal of Applied Physics*, Vol. **17**, Dec., 1038-1042, (1946).

[19] Reissner E. 'On axi-symmetrical vibrations of shallow spherical shells', *Quart. Appl. Math.*, Vol **13**, No. 3., Oct., 279-290, (1955).

[20] Arnold R. N. and Warburton G. B. 'Flexural vibrations of the walls of thin cylindrical shells having freely supported ends', *Proc. Roy. Soc.* (London), ser A, **197**, 238-256, June (1949).

[21] Webster J. J. 'Free vibration of rectangular curved panels', *International Journal of Mechanical Sciences,* Vol . **10**, 571-582, (1968).

[22] Szechenyi E. 'Approximate methods for the determination of the natural frequencies of stiffened and curved plates', *Journal of Sound and Vibration,* Vol. **14**, No. 3, 401-418, (1971).

[23] Leissa A. W. and Kadi A. S. 'Curvature effects on shallow shell vibrations', *Journal of Sound and Vibration,* Vol. **16**, No. 2, 173-187, (1971).

[24] Srinivasan R. S. and Bobby W. 'Free vibration of non-circular cylindrical shell panels', *Journal of Sound and Vibration,* Vol. **46**, No. 1, 43-49, (1976).

[25] Leissa A. W., Lee J. K. and Wang A. J. 'Vibrations of cantilevered doubly-curved shallow shells', *Int. J. Solids Structures,* Vol. **19**, No. 5, 411-424, (1983).

[26] Srinivasan R. S. and Krishnan P. A. 'Vibration of cylindrical shell panels', *Journal of Sound and Vibration,* Vol. **114**, No. 3, 583-587, (1987).

[27] Srinivasan R. S. and Krishnan P. A. 'Free vibration of conical shell panels', *Journal of Sound and Vibration,* Vol. **117**, No. 1, 153-160, (1987).

[28] Norrie D. H. and deVries G. 'A finite element bibliography', parts I,II & III, Report No. **57-59**, Dept of Mech. Eng., University of Calgary, June (1974).

[29] Klein S. and Sylvester R. J. 'The linear elastic dynamic analysis of shells of revolution by the matrix displacement method', *Proc. of Conf.on Matrix Methods in Structural Mechanics,* AFFDL-TR-66-80, Air Force Inst. Tech., Oct., (1966).

[30] Grafton P. E. and Strome D. R. 'Analysis of axisymmetrical shells by the direct stiffness method', *AIAA Journal,* Vol **1**, No. 10, (1963).

[31] Jones R. E. and Strome D. R. 'Direct stiffness method analysis of shells of revolution utilizing curved elements', *AIAA Journal,* Vol. **4**, No. 9, 1519-1525, Sept., (1966).

[32] Webster J. J. 'Free vibrations of shells of revolution using ring finite elements', *Int. J. Mech. Sci.*, Vol **9**, No 8, 559-570, (1967).

[33] Carter R. L., Robinson A. R. and Schnobrich W. C. 'Free vibrations of hyperboloidal shells of revolution', *Proc. Am. Soc. Civil Engrs. J., Eng. Mech. Div.*, Vol **95**, EM 5, 1033-1052,Oct. (1969).

[34] Webster J. J. 'The accuracy of finite-element solutions for the modal characteristics of shells of revolution', *Int. J. Mech. Sci.*, Vol. **12**, No. 2, 157-168, (1970).

[35] Gladwell G.M. and Tahbildar U.C. 'Finite elements analysis of the axisymmetric vibrations of cylinders', *Journal of Sound and Vibration*, Vol **22**, No. 2, 143-157, (1972).

[36] Sen S. K. and Gould P.L. 'Free vibration of shells of revolution using the finite element method', *ASCE J. Eng. Mech. Div.*, Vol. **100**, No. EM 2, 283-303, April (1974).

[37] Wilson J. 'Vibration of Axisymmetric shells', in *Vibrations of Engineering Structures*, chapter 11, ( Brebbia C. A., Tottenham H., Warburton G. B., Wilson J. and Wilson R.), Computational Mechanics Ltd., (1976).

[38] Deb Nath J. M. 'Free vibration, stability and "non-classical modes" of cooling tower shells', *Journal of Sound and Vibration*, Vol. **33**, No.1, 79-101, (1974).

[39] Tou S. K. and Wong K. K. 'High-Precision finite element analysis of cylindrical shells', *Computers and Structures*, Vol. **26**, No. 5, 847-854, (1987).

[40] Tedesco J. W., Kostem C. N. and Kalnins A. 'Free vibration analysis of circular cylindrical shells', *Computers and Structures*, Vol. **25**, No. 5, 677-685, (1987).

[41] Altman W. and Neto E. L. 'Vibration of thin shells of revolution based on a mixed finite element formulation', *Computers and Structures,* Vol. **23**, No. 3, 291-303, (1986).

[42] Dungar R., Severn R. T. and Taylor P. R. 'Vibration of plate and shell structures using triangular finite elements', *Journal of Strain Analysis,* Vol. **2**, No. 1, 73-83, (1967).

[43] Clough R. W. and Johnson C. P. 'A finite element approximation for the analysis of thin shells', *Int. J. Solids and Structures,* Vol. **4**, (1968).

[44] Argyris J. H., Fried I. and Scharpf D. W. 'The TUBA family of plate elements for the matrix displacement method', *The Aeronaut. J. Royal Aeronaut. Soc.,* Vol. **72**, 701-709, (1968).

[45] Clough R. W. and Wilson E. L. 'Dynamic finite element analysis of arbitrary thin shells', *Computers & Structures*, Vol. **1**, 33-56, (1971).

[46] Zienkiewicz O. C. *'The Finite Element Method'*, Third edition, McGraw-Hill Book Company (UK) Limited, (1977).

[47] Bathe K. J. *'Finite Element Procedures in Engineering Analysis'*, Prentice-Hall Inc., Englewood Cliffs, New Jersey, (1981).

[48] Hrabok M. M. and Hrudey T. M. 'A review and catalogue of plate bending finite elements', *Computers & Structures*, Vol. **19**, No. 3, 479-495, (1984).

[49] Argyris J. H. and Scharpf D. W. 'The SHEBA family of shell elements for the matrix displacement method', *The Aeronautical Journal of the Royal Aeronautical Society,* Vol. 72, 873-883, Oct. (1968).

[50] Gallagher R. H. 'Problems and progress in thin shell finite element analysis', *Finite Elements for Thin Shells and Curved Members,* (Eds. Ashwell D. G. and Gallagher R. H.), John Wiley & Sons Ltd., chapter 1, London, (1976).

[51] Greene B. E., Jones R. E., McLay R. W. and Strome D. R. 'Dynamic analysis of shells using doubly-curved finite elements', *Proceedings of the Second Conference on Matrix Methods in Structural Mechanics,* AFFDL-TR-68-150, Wright-Patterson Air Force Base, Ohio, 185-212, 15-17 Oct. (1968).

[52] Olson M. D. and Lindberg G. M. 'Vibration analysis of cantilevered curved plates using a new cylindrical shell finite element', *Proceedings of the Second Conference on Matrix Methods in Structural Mechanics,* AFFDL-TR-68-150, Wright-Patterson Air Force Base, Ohio, 247-269, 15-17 Oct. (1968).

[53] Tahbildar U., Tottenham H. and Brebbia C. 'Dynamic analysis of shells using curved finite elements', *Bul. Int. Ass. Shell Structures,* No. **40**, 15-34, Dec. (1969).

[54] Olson M. D. and Lindberg G. M. 'Dynamic analysis of shallow shells with a doubly-curved triangular finite element', *Journal of Sound and Vibration,* Vol. **19**, No. 3, 299-318, (1971).

[55] Petyt M. 'Vibration of curved plates', *Journal of Sound and Vibration,* Vol. **15**, No. 3, 381-395, (1971).

[56] Tabarrok B. and Gass N. 'Vibration of cylindrical shells by hybrid finite element method', *AIAA Journal,* Vol. **10**, No. 12, 1553-1554, Dec. (1972).

[57] Ahmad S., Anderson R. G. and Zienkiewicz O. C. 'Vibration of thick curved shells, with particular reference to turbine blades', *Journal of Strain Analysis,* Vol. **5**, No. 3, 200-206, (1970).

[58] Gill P. A. T. and Ucmaklioglu M. 'Isoparametric finite elements for free vibration analysis of shell segments and non-axisymmetric shells', *Journal of Sound and Vibration,* Vol. **65**, No. 2, 259-273, (1979).

[59] Hofmeister L. D. and Evensen D. A. 'Vibration problems using isoparametric shell elements', *Int. J. Num. Meth. Engn.*, Vol. **5**, No. 1, 142-145, Sep-Oct (1972).

[60] Bossak M. A. J. and Zienkiewicz O. C. 'Free vibration of initially stressed solids, with particular reference to centrifugal-force effects in rotating machinery', *Journal of Strain Analysis,* Vol. **8**, No. 4, 245 - 252, (1973).

[61] Bathe K. and Dvorkin E. N. 'A formulation of general shell elements - the use of mixed interpolation of tensorial components', *Int. J. Num. Meth. Engn.*, Vol. **22**, 697-722, (1986).

[ 62] Knowles N. C., Razzaque A. and Spooner J. B. 'Experience in finite element analysis of shell structures', *Finite Elements for Thin Shells and Curved Members,* (Eds. Ashwell D. G. and Gallagher R. H.), John Wiley & Sons Ltd., chapter 13, London, (1976).

[63] Morino L., Leech J. W. and Witmer E. A. 'PETROS 2: A new finite-difference method and program for the calculation of large elastic-plastic dynamically-induced deformations of general thin shells', Massachusetts Inst. of Technology, Cambridge Aeroelastic and Structures Research Lab, Technical Rept. ASRL TR 152-1, December (1969).

[64] Morino L., Leech J. W. and Witmer E. A. 'An improved numerical calculation technique for large elastic-plastic transient deformations of thin shells. Part 1 - Background and theoretical formulation', *Journal of Applied Mechanics,* 423-428, June (1971).

[65] Morino L., Leech J. W. and Witmer E. A. 'An improved numerical calculation technique for large elastic-plastic transient deformations of thin shells. Part 2 - Evaluation and Applications', *Journal of Applied Mechanics,* 429-436, June (1971).

[66] Soare M. '*Applications of finite difference equations to shell analysis*', Pergamon Press Ltd., (1976).

[67] Lachance L. and Tremblay M. 'Finite difference solution of Vlasov's general thin shell equations', *2nd Can. Congr. of Appl. Mech.*, Waterloo, Quebec, 20-23, May (1969).

[68] Beg B. I. 'Finite difference approach to hyperbolic paraboloids', *J. Struct. Div., Proc. ASCE,* ST2, 429-445, (1972).

[69] Bhattacharya M. C. 'Static and dynamic deflections of plates of arbitrary geometry by a new finite difference approach', *Journal of Sound and Vibration,* Vol. **107**, No. 3, 507-521, (1986).

[70] Johnson D. E. 'A difference-based variational method for shells', *Int. J. Solids and Structures,* Vol. **6**, 699-724, (1970).

[71] Kwok S. K. 'Geometrically nonlinear analysis of general thin shells using a curvilinear finite difference (CFD) energy approach', *Computers and Structures,* Vol. **20**, No. 4, 683-697, (1985).

[72] Kwok S. K. 'Linear and nonlinear analysis of general thin shells using a curvilinear finite difference energy approach', *Ph.D. Thesis,* University of Western Australia, Nedlands, Australia (1984).

[73] Lau P. C. M. 'Curvilinear finite difference energy method for plate bending analysis', *Seventh Australasian Conference on the Mechanics of Structures and Materials,* University of Western Australia, Nedlands, Australia (1980).

[74] Lau P. C. M. 'Numerical solution of Poisson's equation using curvilinear finite differences', *Appl. Math. Modelling,* Vol. 1, 349-350, (1977).

[75] Christensen P. J. and Lau P. C. M. 'A microcomputer based technique for the analysis of orthotropic thin plates', *Proc. Int. Conf. on Advances in Numerical Methods in Engineering: Theory and Applications,* University College, Swansea, 965-972, (1985).

[76] Barve V. D. and Dey S. S. 'Isoparametric finite difference energy method for plate bending problems', *Computers and Structures,* Vol. **17**, No. 3, 459-465, (1983).

[77] Stein M. 'The effect on the buckling of perfect cylinders of pre-buckling deformations and stresses induced by edge support', *NASA TN D1510,* Dec. (1962).

[78] Budiansky B. and Anderson D. G. M. 'Numerical shell analysis - nodes without elements', *12th Int. Congress of Appl. Mech.,* Stanford Univ., Aug. (1968).

[79] Bushnell D., Almroth B. O. and Brogan F. 'Finite-Difference energy method for nonlinear shell analysis', *Computers and Structures,* Vol. **1**, 361-387, (1971).

[80] Bushnell D. 'Analysis of buckling and vibration of ring-stiffened, segmented shells of revolution', *Int. J. Solids and Structures,* Vol. **6**, 157-181, (1970).

[81] Martins R. A. F. and Owen D. R. J. 'Thin plate semiloof element for structural analysis - including stability and natural vibrations', *Int. J. Num. Meth. Engn.*, Vol. **12**, 1667-1676, (1978).

[82] Weingarten V. I., Ramanathan R. K. and Chen C. N. 'Lanczos eigenvalue algorithm for large structures on a minicomputer', *Computers and Structures,* Vol. **16**., No. 1-4, 253-257, (1983).

[83] Niordson F. I. '*Shell Theory*', North-Holland Series in Applied Mathematics and Mechanics, Vol. **29**, Elsevier Science Publishing Company Inc., (1985).

[84] Mollmann H., '*Introduction to the Theory of Thin Shells*', John Wiley & Sons Ltd., (1981).

[85] Green A. E. and Zerna W. '*Theoretical Elasticity*', Oxford University Press, London, (1954).

[86] Bathe K. J. and Wilson E. L. *'Numerical Methods in Finite Element Analysis'*, Prentice-Hall, Inc., Englewood Cliffs, New Jersey, (1976).

[87] Lowan A. N., Davids N. and Levenson A. 'Table of the zeros of the Legendre polynomials of order 1-16 and the weight coefficients for Gauss' mechanical quadrature formula', Bulletin of the American Mathematical Society, Vol. **48**, 739-743, (1942).

[88] Ralston A. and Rabinowitz P. *'A First Course in Numerical Analysis'*, McGraw-Hill Kogakusha, Ltd., (1978).

[89] Bathe K. J. and Wilson E. L. 'Large eigenvalue problems in dynamic analysis', A.S.C.E., *Journal of Engineering Mechanics Division,* Vol. **98**, 1471-1485, (1972).

[90] Bathe K. J. and Wilson E. L. 'Eigensolution of large structural systems with small bandwidth', A.S.C.E., *Journal of Engineering Mechanics Division,* Vol. **99**, 467-479, (1973).

[91] Bathe K. J. and Wilson E. L. 'Solution methods for eigenvalue problems in structural mechanics', *International Journal for Numerical Methods in Engineering,* Vol. **6**, 213-226, (1973).

[92] Bathe K. J. and Ramaswamy S. 'An accelerated subspace iteration method', *Computer Methods in Applied Mechanics and Engineering,* Vol. **23**, 313-331, (1980).

[93] Nour-Omid B., Parlett B. N. and Taylor R. L. 'Lanczos versus subspace iteration for solution of eigenvalue problems', *International Journal for Numerical Methods in Engineering,* Vol. **19**, 859-871, (1983).

[94] Meirovitch L., *'Computational Methods in Structural Dynamics'*, Sijthoff & Noordhoff International Publishers B.V., Netherlands, (1980).

[95] *'IBM RT PC C Language Guide and Reference'*, International Business Machines Corporation, First Edition, January (1987).

[96] Kernighan B. W. and Ritchie D. M. *'The C Programming Language'*, Englewood Cliffs, Prentice-Hall, (1988).

[97] Huxham F. A., Burnard D. and Takatsuka J. *'Using the Macintosh Toolbox with C'*, Sybex, Berkeley, California, (1986).

[98] *'IBM RT PC AIX Operating System Technical Reference'*, International Business Machines Corporation, First Edition, January (1987).

[99] Ergatoudis I., Irons B. M. and Zienkiewicz O. C. 'Curved isoparametric, "quadrilateral elements" for finite element analysis', *Int. J. Solids and Structures,* Vol. **4**, 31-42, (1968).

[100] Leissa A. W. 'The free vibration of rectangular plates', *Journal of Sound and Vibration,* Vol. **31**, No. 3, 257-293, (1973).

[101] Gorman D. J. *'Free Vibration Analysis of Rectangular Plates'*, Elsevier, North Holland, (1982).

[102] Soedel W. and Dhar M. 'Difficulties in finding modes experimentally when several contribute to a resonance', *Journal of Sound and Vibration,* Vol. **58**, No. 1, 27-38, (1978).

[103] Batoz J. L., Bathe K. J. and Ho L. W. 'A study of three-node triangular plate bending elements', *Int. J. Num. Meth. Engn.*, Vol. **15**, 1771-1812, (1980).

[104] Houlston R.J., Slater J. E., Pegg N. and DesRochers C. G. 'On analysis of structural response of ship panels subjected to air blast loading', *Computers and Structures*, Vol. **21**, No. 1/2, 273-289, (1985).

[105 ] Martins R. A. F. and Owen D. R. J. 'Structural instability and natural vibration analysis of thin arbitrary shells by use of the semiloof element', *Int. J. Num. Meth. Engn.*, Vol. **11**, 481-498, (1977).

[106] Cheung Y. K., Li W. Y. and Tham L.G. 'Free vibration analysis of singly curved shell by spline finite strip method', *Journal of Sound and Vibration,* Vol. **128**, No. 3, 411-422, (1989).

[107] Naomis S. and Lau P.C.M. 'Dynamic analysis of general thin shells', *Proc. Int. Conf. Numerical Methods in Engineering: Theory and Applications,* Vol **1**, D45/1, Swansea, 6-10th July, (1987).

[108] Kielb R. E., Leissa A. W. and MacBain J. C. 'Vibrations of twisted cantilever plates - a comparison of theoretical results', *Int. J. Num. Meth. Engn.*, Vol. **21**, 1365-1380, (1985).

[109] Barton M. V. 'Vibration of rectangular and skew cantilever plates', *J. Appl. Mech.,* Vol. **18**, 129-134, (1951).

[110] Altman W. and Bismarck-Nasr M. N. 'Vibration of thin cylindrical shells based on a mixed finite element formulation', *Computers and Structures,* Vol. **23**, No 3., 291-303, (1986).

[111] Sander G. and Idelsohn S. 'A family of conforming finite elements for deep shell analysis', *Int. J. Num. Meth. Engn.*, Vol. **18**, 363-380, (1982).

[112] Lau P. C. M. and Brebbia C.A. 'The Cell Collocation Method in Continuum Mechanics.', *Int. J. Mech. Sci,* Vol. 20, 83-95, (1978).

[113] Lau P. C. M. 'Curvilinear Finite Difference Method for Three Dimensional Potential Problems.', *J. Computational Physics* Vol. 32, No 3,325-344,Sept. (1979).

[114] Lau P. C. M. 'Curvilinear Finite Difference Method for Biharmonic Equations.', *Int. J. Num. Meth. Engrg.* Vol. 14, 791-798,( 1979).

# Appendix A

# DISPLACEMENT TRANSFORMATIONS

An important aspect in the application of the curvilinear finite difference method to the analysis of shell structures is the selection of the displacement field. In this proposed method, the displacement field is described with respect to a common global coordinate system. For an internal or boundary mesh, the displacement components $u_x$, $u_y$ and $u_z$ are chosen as the nodal unknowns. However, Flugge's [1] general thin shell equations are defined with respect to a local curvilinear coordinate system. It is therefore necessary to be able to approximate the local displacement components $u_\alpha$, $u_\beta$ and w and their corresponding derivatives within a computational mesh.

In this appendix, the transformation equations defined in section 3.4 are expanded and assembled in matrix form.

The covariant components of the local displacements $u_\alpha$, $u_\beta$ and w can be computed from the following set of equations.

$$u_\alpha = \beta^i_\alpha u_i \tag{A.1}$$

$$w = \beta^i_3 u_i \tag{A.2}$$

In matrix form, equations A.1 and A.2 can be combined to yield

$$\{u_\gamma\} = [T_u] \{u_i\} \tag{A.3a}$$

where

$$\{u_\gamma\}^t = \{ u_\alpha \ \ u_\beta \ \ w \} \tag{A.3b}$$

$$\{u_i\}^t = \{ u_x \ \ u_y \ \ u_z \} \tag{A.3c}$$

$$[T_u] = \begin{bmatrix} x,_{\alpha} & y,_{\alpha} & z,_{\alpha} \\ x,_{\beta} & y,_{\beta} & z,_{\beta} \\ d_1/H & d_2/H & d_3/H \end{bmatrix} \tag{A.3d}$$

Differentiating equations A.1 and A.2, the first order inplane displacement derivatives are

$$u_{\alpha'\beta} = \beta^{i}_{\alpha'\beta}\, u_i + \beta^{i}_{\alpha}\, u_{i'\beta} \tag{A.4}$$

which can be expressed in matrix form as

$$\{u_{\gamma'\delta}\} = [T_{\gamma'\delta}]\,\{u_i\} + [T_{\gamma}]\,\{u_{i'\delta}\} \tag{A.5a}$$

where $\{u_{\gamma'\delta}\}^t = \{\, u_{\alpha'\alpha} \;\; u_{\alpha'\beta} \;\; u_{\beta'\alpha} \;\; u_{\beta'\beta} \,\}$ (A.5b)

$$\{u_{i'\delta}\}^t = \{\, u_{x'\alpha} \;\; u_{x'\beta} \;\; u_{y'\alpha} \;\; u_{y'\beta} \;\; u_{z'\alpha} \;\; u_{z'\beta} \} \tag{A.5c}$$

$$[T_{\gamma'\delta}] = \begin{bmatrix} x,_{\alpha\alpha} & y,_{\alpha\alpha} & z,_{\alpha\alpha} \\ x,_{\alpha\beta} & y,_{\alpha\beta} & z,_{\alpha\beta} \\ x,_{\beta\alpha} & y,_{\beta\alpha} & z,_{\beta\alpha} \\ x,_{\beta\beta} & y,_{\beta\beta} & z,_{\beta\beta} \end{bmatrix} \tag{A.5d}$$

$$[T_{\gamma}] = \begin{bmatrix} x,_{\alpha} & 0 & y,_{\alpha} & 0 & z,_{\alpha} & 0 \\ 0 & x,_{\alpha} & 0 & y,_{\alpha} & 0 & z,_{\alpha} \\ x,_{\beta} & 0 & y,_{\beta} & 0 & z,_{\beta} & 0 \\ 0 & x,_{\beta} & 0 & y,_{\beta} & 0 & z,_{\beta} \end{bmatrix} \tag{A.5e}$$

Similarly the normal displacement, w, can be differentiated and represented by the equation

$$w_{,\alpha} = \beta^{i}_{3,\alpha}\, u_i + \beta^{i}_{3}\, u_{i,\alpha} \tag{A.6}$$

Expressed in matrix form

$$\{w_{,\delta}\} = [T_{3,\delta}]\,\{u_i\} + [T_3]\,\{u_{i,\delta}\} \tag{A.7a}$$

where $\{w_{,\delta}\}^t = \{\, w_{,\alpha} \quad w_{,\beta} \,\}$ (A.7b)

$$[T_{3,\delta}] = \begin{bmatrix} (d_1/H)_{,\alpha} & (d_2/H)_{,\alpha} & (d_3/H)_{,\alpha} \\ (d_1/H)_{,\beta} & (d_2/H)_{,\beta} & (d_3/H)_{,\beta} \end{bmatrix} \tag{A.7c}$$

$$[T_3] = \begin{bmatrix} d_1/H & 0 & d_2/H & 0 & d_3/H & 0 \\ 0 & d_1/H & 0 & d_2/H & 0 & d_3/H \end{bmatrix} \tag{A.7d}$$

To obtain the second order derivatives of the normal displacement, equations A.6 is differentiated with respect to $x^{\beta}$:

$$w_{,\alpha\beta} = \beta^{i}_{3,\alpha\beta}\, u_i + \beta^{i}_{3,\alpha}\, u_{i,\beta} + \beta^{i}_{3,\beta}\, u_{i,\alpha} + \beta^{i}_{3}\, u_{i,\alpha\beta} \tag{A.8}$$

In matrix form, equation A.8 becomes

$$\{w_{,\gamma\delta}\} = [T_{3,\gamma\delta}]\,\{u_i\} + [T_{3,\gamma}]\,\{u_{i,\delta}\} + [T_3]\{u_{i,\gamma\delta}\} \tag{A.9a}$$

where $\{w_{,\gamma\delta}\}^t = \{\, w_{,\alpha\alpha} \quad w_{,\alpha\beta} \quad w_{,\beta\alpha} \quad w_{,\beta\beta} \,\}$ (A.9b)

$$\{u_{i',\gamma\delta}\}^t = \{\, u_{x',\alpha\alpha}\ u_{x',\alpha\beta}\ u_{x',\beta\alpha}\ u_{x',\beta\beta}\ u_{y',\alpha\alpha}\ u_{y',\alpha\beta}\ u_{y',\beta\alpha}\ u_{y',\beta\beta}\ u_{z',\alpha\alpha}\ u_{z',\alpha\beta}\ u_{z',\beta\alpha}\ u_{z',\beta\beta}\,\} \tag{A.9c}$$

$$[T_{3',\gamma\delta}] = \begin{bmatrix} (d_1/H)_{,\alpha\alpha} & (d_2/H)_{,\alpha\alpha} & (d_3/H)_{,\alpha\alpha} \\ (d_1/H)_{,\alpha\beta} & (d_2/H)_{,\alpha\beta} & (d_3/H)_{,\alpha\beta} \\ (d_1/H)_{,\beta\alpha} & (d_2/H)_{,\beta\alpha} & (d_3/H)_{,\beta\alpha} \\ (d_1/H)_{,\beta\beta} & (d_2/H)_{,\beta\beta} & (d_3/H)_{,\beta\beta} \end{bmatrix} \tag{A.9d}$$

$$[T_{3',\gamma}] = \begin{bmatrix} 2(d_1/H)_{,\alpha} & 0 & 2(d_2/H)_{,\alpha} & 0 & 2(d_3/H)_{,\alpha} & 0 \\ (d_1/H)_{,\beta} & (d_1/H)_{,\alpha} & (d_2/H)_{,\beta} & (d_2/H)_{,\alpha} & (d_3/H)_{,\beta} & (d_3/H)_{,\alpha} \\ (d_1/H)_{,\beta} & (d_1/H)_{,\alpha} & (d_2/H)_{,\beta} & (d_2/H)_{,\alpha} & (d_3/H)_{,\beta} & (d_3/H)_{,\alpha} \\ 0 & 2(d_1/H)_{,\beta} & 0 & 2(d_2/H)_{,\beta} & 0 & 2(d_3/H)_{,\beta} \end{bmatrix} \tag{A.9e}$$

$$[T_3] = \begin{bmatrix} d_1/H & 0 & 0 & 0 & d_2/H & 0 & 0 & 0 & d_3/H & 0 & 0 & 0 \\ 0 & d_1/H & 0 & 0 & 0 & d_2/H & 0 & 0 & 0 & d_3/H & 0 & 0 \\ 0 & 0 & d_1/H & 0 & 0 & 0 & d_2/H & 0 & 0 & 0 & d_3/H & 0 \\ 0 & 0 & 0 & d_1/H & 0 & 0 & 0 & d_2/H & 0 & 0 & 0 & d_3/H \end{bmatrix} \tag{A.9f}$$

In equations A.9a - A.9f, expansions for the terms $(d_i/H)_{,\alpha}$ and $(d_i/H)_{,\alpha\beta}$ are obtained by applying the chain rule for partial differentiation. i.e.,

$$(d_i/H)_{,\alpha} = (\, d_{i',\alpha}\, H \;-\; d_i\, H_{,\alpha}\,)/\, H^2 \tag{A.10}$$

where $$H,_{\alpha} = ( d_1 d_{1},_{\alpha} + d_2 d_{2},_{\alpha} + d_3 d_{3},_{\alpha} )/H \tag{A.11}$$

Similarly,

$$\begin{aligned}(d_i/H),_{\alpha\beta} &= ( d_{i},_{\alpha\beta} H^2 - d_{i},_{\alpha} H,_{\beta} H - d_{i},_{\beta} H,_{\alpha} H \\ &\quad - d_i H,_{\alpha\beta} H + 2 d_i H,_{\alpha} H,_{\beta} )/ H^3\end{aligned} \tag{A.12}$$

where
$$\begin{aligned}H,_{\alpha\beta} &= ( d_{1},_{\alpha} d_{1},_{\beta} + d_{2},_{\alpha} d_{2},_{\beta} + d_{3},_{\alpha} d_{3},_{\beta} )/H \\ &\quad + ( d_1 d_{1},_{\alpha\beta} + d_2 d_{2},_{\alpha\beta} + d_3 d_{3},_{\alpha\beta} )/H \\ &\quad - H,_{\alpha} H,_{\beta}/H\end{aligned} \tag{A.13}$$

## Appendix B

## FINITE DIFFERENCE EXPRESSIONS

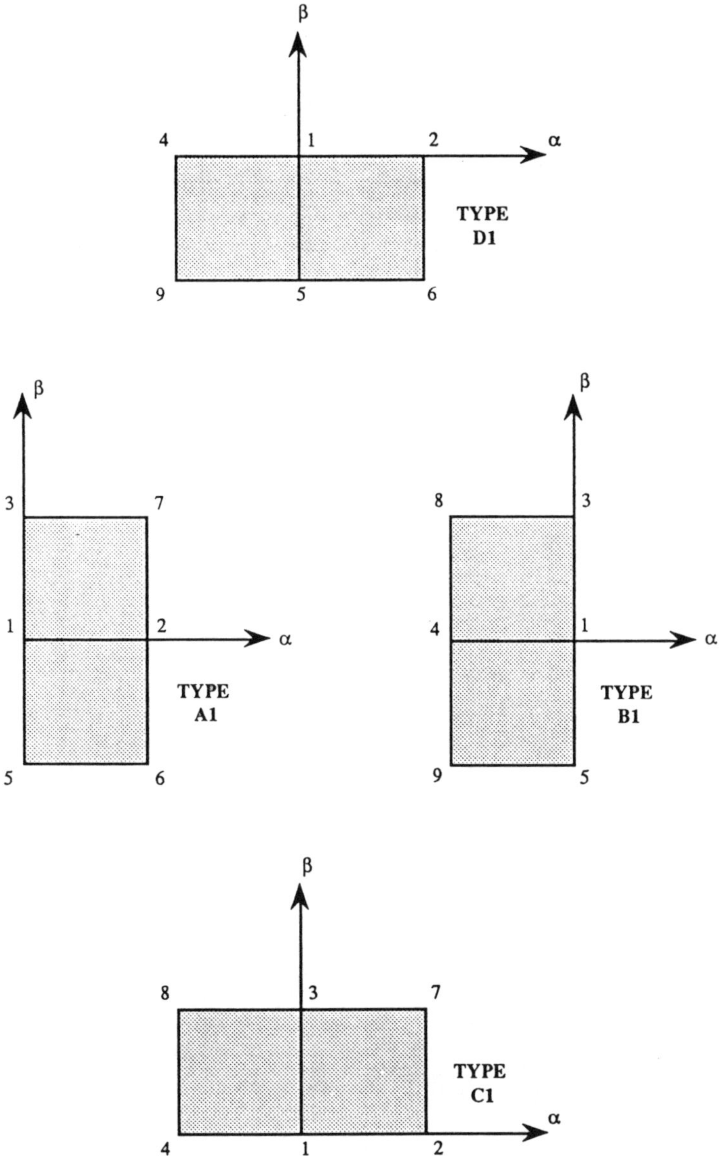


Figure B1: Edge Meshes.

| $\phi_i$ | $\phi$ | $\phi_{,\alpha}$ | $\phi_{,\beta}$ | $\phi_{,\alpha\alpha}$ | $\phi_{,\alpha\beta}$ | $\phi_{,\beta\beta}$ |
|---|---|---|---|---|---|---|
| $\phi_1$ | $1 - \alpha - \beta^2 + \alpha\beta^2$ | $-1 + \beta^2$ | $2(-\beta + \alpha\beta)$ | 0 | $2\beta$ | $2(-1 + \alpha)$ |
| $\phi_2$ | $\alpha - \alpha\beta^2$ | $1 - \beta^2$ | $-2\alpha\beta$ | 0 | $-2\beta$ | $-2\alpha$ |
| $\phi_3$ | $\frac{1}{2}(\beta - \alpha\beta + \beta^2 - \alpha\beta^2)$ | $\frac{1}{2}(-\beta - \beta^2)$ | $\frac{1}{2} - \frac{1}{2}\alpha + \beta - \alpha\beta$ | 0 | $-\frac{1}{2} - \beta$ | $1 - \alpha$ |
| $\phi_4$ | 0 | 0 | 0 | 0 | 0 | 0 |
| $\phi_5$ | $\frac{1}{2}(-\beta + \alpha\beta + \beta^2 - \alpha\beta^2)$ | $\frac{1}{2}(\beta - \beta^2)$ | $-\frac{1}{2} + \frac{1}{2}\alpha + \beta - \alpha\beta$ | 0 | $\frac{1}{2} - \beta$ | $1 - \alpha$ |
| $\phi_6$ | $\frac{1}{2}(-\alpha\beta + \alpha\beta^2)$ | $\frac{1}{2}(-\beta + \beta^2)$ | $-\frac{1}{2}\alpha + \alpha\beta$ | 0 | $-\frac{1}{2} + \beta$ | $\alpha$ |
| $\phi_7$ | $\frac{1}{2}(\alpha\beta + \alpha\beta^2)$ | $\frac{1}{2}(\beta + \beta^2)$ | $\frac{1}{2}\alpha + \alpha\beta$ | 0 | $\frac{1}{2} + \beta$ | $\alpha$ |
| $\phi_8$ | 0 | 0 | 0 | 0 | 0 | 0 |
| $\phi_9$ | 0 | 0 | 0 | 0 | 0 | 0 |

Table B1a: Interpolation functions for Mesh Type A1.

| $\phi_i$ | $\phi_{,\alpha\alpha\alpha}$ | $\phi_{,\alpha\beta\alpha}$ | $\phi_{,\beta\beta\alpha}$ | $\phi_{,\beta\beta\beta}$ |
|---|---|---|---|---|
| $\phi_1$ | 0 | 0 | 2 | 0 |
| $\phi_2$ | 0 | 0 | -2 | 0 |
| $\phi_3$ | 0 | 0 | -1 | 0 |
| $\phi_4$ | 0 | 0 | 0 | 0 |
| $\phi_5$ | 0 | 0 | -1 | 0 |
| $\phi_6$ | 0 | 0 | 1 | 0 |
| $\phi_7$ | 0 | 0 | 1 | 0 |
| $\phi_8$ | 0 | 0 | 0 | 0 |
| $\phi_9$ | 0 | 0 | 0 | 0 |

Table B1b: Interpolation functions for Mesh Type A1.

| $\phi_i$ | $\phi$ | $\phi_{,\alpha}$ | $\phi_{,\beta}$ | $\phi_{,\alpha\alpha}$ | $\phi_{,\alpha\beta}$ | $\phi_{,\beta\beta}$ |
|---|---|---|---|---|---|---|
| $\phi_1$ | $1 + \alpha - \beta^2 - \alpha\beta^2$ | $1 - \beta^2$ | $2(-\beta - \alpha\beta)$ | 0 | $-2\beta$ | $2(-1 - \alpha)$ |
| $\phi_2$ | 0 | 0 | 0 | 0 | 0 | 0 |
| $\phi_3$ | $\frac{1}{2}(\beta + \alpha\beta + \beta^2 + \alpha\beta^2)$ | $\frac{1}{2}(\beta + \beta^2)$ | $\frac{1}{2} + \frac{1}{2}\alpha + \beta + \alpha\beta$ | 0 | $\frac{1}{2} + \beta$ | $1 + \alpha$ |
| $\phi_4$ | $-\alpha + \alpha\beta^2$ | $-1 + \beta^2$ | $2\alpha\beta$ | 0 | $2\beta$ | $2\alpha$ |
| $\phi_5$ | $\frac{1}{2}(-\beta - \alpha\beta + \beta^2 + \alpha\beta^2)$ | $\frac{1}{2}(-\beta + \beta^2)$ | $-\frac{1}{2} - \frac{1}{2}\alpha + \beta + \alpha\beta$ | 0 | $-\frac{1}{2} + \beta$ | $1 + \alpha$ |
| $\phi_6$ | 0 | 0 | 0 | 0 | 0 | 0 |
| $\phi_7$ | 0 | 0 | 0 | 0 | 0 | 0 |
| $\phi_8$ | $\frac{1}{2}(-\alpha\beta - \alpha\beta^2)$ | $\frac{1}{2}(-\beta - \beta^2)$ | $-\frac{1}{2}\alpha - \alpha\beta$ | 0 | $-\frac{1}{2} - \beta$ | $-\alpha$ |
| $\phi_9$ | $\frac{1}{2}(\alpha\beta - \alpha\beta^2)$ | $\frac{1}{2}(\beta - \beta^2)$ | $\frac{1}{2}\alpha - \alpha\beta$ | 0 | $\frac{1}{2} - \beta$ | $-\alpha$ |

Table B2a: Interpolation functions for Mesh Type B1.

| $\phi_i$ | $\phi_{,\alpha\alpha\alpha}$ | $\phi_{,\alpha\beta\alpha}$ | $\phi_{,\beta\beta\alpha}$ | $\phi_{,\beta\beta\beta}$ |
|---|---|---|---|---|
| $\phi_1$ | 0 | 0 | -2 | 0 |
| $\phi_2$ | 0 | 0 | 0 | 0 |
| $\phi_3$ | 0 | 0 | 1 | 0 |
| $\phi_4$ | 0 | 0 | 2 | 0 |
| $\phi_5$ | 0 | 0 | 1 | 0 |
| $\phi_6$ | 0 | 0 | 0 | 0 |
| $\phi_7$ | 0 | 0 | 0 | 0 |
| $\phi_8$ | 0 | 0 | -1 | 0 |
| $\phi_9$ | 0 | 0 | -1 | 0 |

Table B2b: Interpolation functions for Mesh Type B1.

| $\phi_i$ | $\phi$ | $\phi_{,\alpha}$ | $\phi_{,\beta}$ | $\phi_{,\alpha\alpha}$ | $\phi_{,\alpha\beta}$ | $\phi_{,\beta\beta}$ |
|---|---|---|---|---|---|---|
| $\phi_1$ | $1 - \beta - \alpha^2 + \alpha^2\beta$ | $2(-\alpha + \alpha\beta)$ | $-1 + \alpha^2$ | $2(-1 + \beta)$ | $-2\alpha$ | 0 |
| $\phi_2$ | $\frac{1}{2}(\alpha + \alpha^2 - \alpha\beta - \alpha^2\beta)$ | $\frac{1}{2} + \alpha - \frac{1}{2}\beta - \alpha\beta$ | $\frac{1}{2}(-\alpha - \alpha^2)$ | $1 - \beta$ | $-\frac{1}{2} - \alpha$ | 0 |
| $\phi_3$ | $\beta - \alpha^2\beta$ | $-2\alpha\beta$ | $1 - \alpha^2$ | $-2\beta$ | $-2\alpha$ | 0 |
| $\phi_4$ | $\frac{1}{2}(-\alpha + \alpha^2 + \alpha\beta - \alpha^2\beta)$ | $-\frac{1}{2} + \alpha + \frac{1}{2}\beta - \alpha\beta$ | $\frac{1}{2}(\alpha - \alpha^2)$ | $1 - \beta$ | $\frac{1}{2} - \alpha$ | 0 |
| $\phi_5$ | 0 | 0 | 0 | 0 | 0 | 0 |
| $\phi_6$ | 0 | 0 | 0 | 0 | 0 | 0 |
| $\phi_7$ | $\frac{1}{2}(\alpha\beta + \alpha^2\beta)$ | $\frac{1}{2}\beta + \alpha\beta$ | $\frac{1}{2}(\alpha + \alpha^2)$ | $\beta$ | $\frac{1}{2} + \alpha$ | 0 |
| $\phi_8$ | $\frac{1}{2}(-\alpha\beta + \alpha^2\beta)$ | $-\frac{1}{2}\beta + \alpha\beta$ | $\frac{1}{2}(-\alpha + \alpha^2)$ | $\beta$ | $-\frac{1}{2} + \alpha$ | 0 |
| $\phi_9$ | 0 | 0 | 0 | 0 | 0 | 0 |

Table B3a: Interpolation functions for Mesh Type C1.

| $\phi_i$ | $\phi_{,\alpha\alpha\alpha}$ | $\phi_{,\alpha\beta\alpha}$ | $\phi_{,\beta\beta\alpha}$ | $\phi_{,\beta\beta\beta}$ |
|---|---|---|---|---|
| $\phi_1$ | 0 | 2 | 0 | 0 |
| $\phi_2$ | 0 | -1 | 0 | 0 |
| $\phi_3$ | 0 | -2 | 0 | 0 |
| $\phi_4$ | 0 | -1 | 0 | 0 |
| $\phi_5$ | 0 | 0 | 0 | 0 |
| $\phi_6$ | 0 | 0 | 0 | 0 |
| $\phi_7$ | 0 | 1 | 0 | 0 |
| $\phi_8$ | 0 | 1 | 0 | 0 |
| $\phi_9$ | 0 | 0 | 0 | 0 |

Table B3b: Interpolation functions for Mesh Type C1.

| $\phi_i$ | $\phi$ | $\phi_{,\alpha}$ | $\phi_{,\beta}$ | $\phi_{,\alpha\alpha}$ | $\phi_{,\alpha\beta}$ | $\phi_{,\beta\beta}$ |
|---|---|---|---|---|---|---|
| $\phi_1$ | $1+\beta-\alpha^2-\alpha^2\beta$ | $2(-\alpha-\alpha\beta)$ | $1-\alpha^2$ | $2(-1-\beta)$ | $-2\alpha$ | 0 |
| $\phi_2$ | $\frac{1}{2}(\alpha+\alpha^2+\alpha\beta+\alpha^2\beta)$ | $\frac{1}{2}+\alpha+\frac{1}{2}\beta+\alpha\beta$ | $\frac{1}{2}(\alpha+\alpha^2)$ | $1+\beta$ | $\frac{1}{2}+\alpha$ | 0 |
| $\phi_3$ | 0 | 0 | 0 | 0 | 0 | 0 |
| $\phi_4$ | $\frac{1}{2}(-\alpha+\alpha^2-\alpha\beta-\alpha^2\beta)$ | $-\frac{1}{2}+\alpha-\frac{1}{2}\beta+\alpha\beta$ | $\frac{1}{2}(-\alpha+\alpha^2)$ | $1+\beta$ | $-\frac{1}{2}+\alpha$ | 0 |
| $\phi_5$ | $-\beta+\alpha^2\beta$ | $2\alpha\beta$ | $-1+\alpha^2$ | $2\beta$ | $2\alpha$ | 0 |
| $\phi_6$ | $\frac{1}{2}(-\alpha\beta-\alpha^2\beta)$ | $-\frac{1}{2}\beta-\alpha\beta$ | $\frac{1}{2}(-\alpha-\alpha^2)$ | $-\beta$ | $-\frac{1}{2}-\alpha$ | 0 |
| $\phi_7$ | 0 | 0 | 0 | 0 | 0 | 0 |
| $\phi_8$ | 0 | 0 | 0 | 0 | 0 | 0 |
| $\phi_9$ | $\frac{1}{2}(\alpha\beta-\alpha^2\beta)$ | $\frac{1}{2}\beta-\alpha\beta$ | $\frac{1}{2}(\alpha-\alpha^2)$ | $-\beta$ | $\frac{1}{2}-\alpha$ | 0 |

Table B4a: Interpolation functions for Mesh Type D1.

| $\phi_i$ | $\phi_{,\alpha\alpha\alpha}$ | $\phi_{,\alpha\beta\alpha}$ | $\phi_{,\beta\beta\alpha}$ | $\phi_{,\beta\beta\beta}$ |
|---|---|---|---|---|
| $\phi_1$ | 0 | -2 | 0 | 0 |
| $\phi_2$ | 0 | 1 | 0 | 0 |
| $\phi_3$ | 0 | 0 | 0 | 0 |
| $\phi_4$ | 0 | 1 | 0 | 0 |
| $\phi_5$ | 0 | 2 | 0 | 0 |
| $\phi_6$ | 0 | -1 | 0 | 0 |
| $\phi_7$ | 0 | 0 | 0 | 0 |
| $\phi_8$ | 0 | 0 | 0 | 0 |
| $\phi_9$ | 0 | -1 | 0 | 0 |

Table B4b: Interpolation functions for Mesh Type D1.

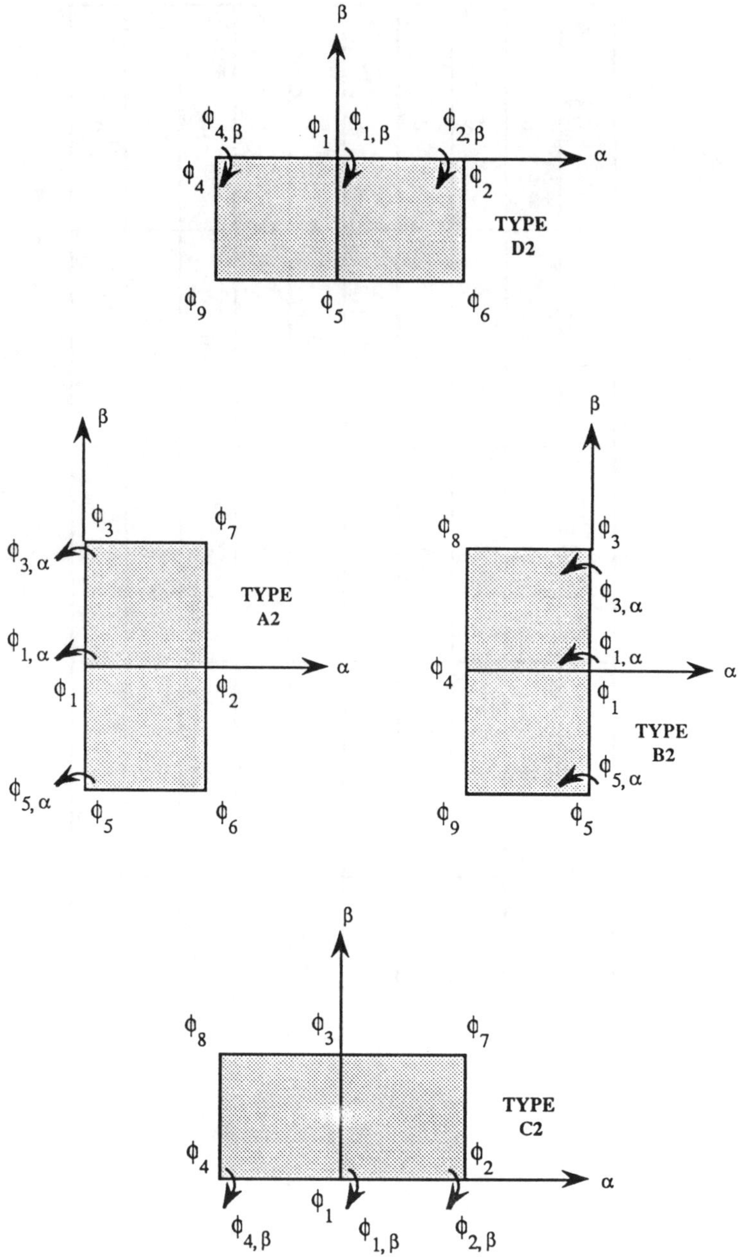


Figure B2: Edge Meshes.

| $\phi_i$ | $\phi$ | $\phi_{,\alpha}$ | $\phi_{,\beta}$ | $\phi_{,\alpha\alpha}$ | $\phi_{,\alpha\beta}$ | $\phi_{,\beta\beta}$ |
|---|---|---|---|---|---|---|
| $\phi_1$ | $1 - \alpha^2 - \beta^2 + \alpha^2\beta^2$ | $2(-\alpha + \alpha\beta^2)$ | $2(-\beta + \alpha^2\beta)$ | $2(-1+\beta^2)$ | $4\alpha\beta$ | $-2 + 2\alpha^2$ |
| $\phi_2$ | $\alpha^2 - \alpha^2\beta^2$ | $2(\alpha - \alpha\beta^2)$ | $2(-\alpha\beta - \alpha^2\beta)$ | $2(1 - \beta^2)$ | $-2\beta - 4\alpha\beta$ | $-2\alpha - 2\alpha^2$ |
| $\phi_3$ | $\frac{1}{2}(\beta + \beta^2 - \alpha^2\beta - \alpha^2\beta^2)$ | $-\alpha\beta - \alpha\beta^2$ | $\frac{1}{2} + \beta - \frac{1}{2}\alpha^2 - \alpha^2\beta$ | $-\beta - \beta^2$ | $-\alpha - 2\alpha\beta$ | $1 - \alpha^2$ |
| $\phi_{1,\alpha}$ | $\alpha - \alpha^2 - \alpha\beta^2 + \alpha^2\beta^2$ | $1 - 2\alpha - \beta^2 + 2\alpha\beta^2$ | $-2\alpha^2\beta$ | $2(1 - \beta^2)$ | $-4\alpha\beta$ | $-2\alpha^2$ |
| $\phi_5$ | $\frac{1}{2}(-\beta + \beta^2 + \alpha^2\beta - \alpha^2\beta^2)$ | $\alpha\beta - \alpha\beta^2$ | $-\frac{1}{2} + \beta + \frac{1}{2}\alpha^2 - \alpha^2\beta$ | $\beta - \beta^2$ | $\alpha - 2\alpha\beta$ | $1 - \alpha^2$ |
| $\phi_6$ | $\frac{1}{2}(-\alpha^2\beta + \alpha^2\beta^2)$ | $-\alpha\beta + \alpha\beta^2$ | $-\frac{1}{2}\alpha - \frac{1}{2}\alpha^2 + \alpha\beta + \alpha^2\beta$ | $-\beta + \beta^2$ | $-\frac{1}{2} - \alpha + \beta + 2\alpha\beta$ | $\alpha + \alpha^2$ |
| $\phi_7$ | $\frac{1}{2}(\alpha^2\beta + \alpha^2\beta^2)$ | $\alpha\beta + \alpha\beta^2$ | $\frac{1}{2}\alpha + \frac{1}{2}\alpha^2 + \alpha\beta + \alpha^2\beta$ | $\beta + \beta^2$ | $\frac{1}{2} + \alpha + \beta + 2\alpha\beta$ | $\alpha + \alpha^2$ |
| $\phi_{3,\alpha}$ | $\frac{1}{2}(\alpha\beta - \alpha^2\beta + \alpha\beta^2 - \alpha^2\beta^2)$ | $\frac{1}{2}\beta - \alpha\beta + \frac{1}{2}\beta^2 - \alpha\beta^2$ | $\frac{1}{2}\alpha^2 + \alpha^2\beta$ | $\beta + \beta^2$ | $\alpha + 2\alpha\beta$ | $\alpha^2$ |
| $\phi_{5,\alpha}$ | $\frac{1}{2}(-\alpha\beta + \alpha^2\beta + \alpha\beta^2 - \alpha^2\beta^2)$ | $-\frac{1}{2}\beta + \alpha\beta + \frac{1}{2}\beta^2 - \alpha\beta^2$ | $-\frac{1}{2}\alpha^2 + \alpha^2\beta$ | $-\beta + \beta^2$ | $-\alpha + 2\alpha\beta$ | $\alpha^2$ |

Table B5a: Interpolation functions for Mesh Type A2.

| $\phi_i$ | $\phi_{,\alpha\alpha\alpha}$ | $\phi_{,\alpha\beta\alpha}$ | $\phi_{,\beta\beta\alpha}$ | $\phi_{,\beta\beta\beta}$ |
|---|---|---|---|---|
| $\phi_1$ | 0 | $4\beta$ | $4\alpha$ | 0 |
| $\phi_2$ | 0 | $-4\beta$ | $-2 - 4\alpha$ | 0 |
| $\phi_3$ | 0 | $-1 - 2\beta$ | $-2\alpha$ | 0 |
| $\phi_{1',\alpha}$ | 0 | $-4\beta$ | $-4\alpha$ | 0 |
| $\phi_5$ | 0 | $1 - 2\beta$ | $-2\alpha$ | 0 |
| $\phi_6$ | 0 | $-1 + 2\beta$ | $1 + 2\alpha$ | 0 |
| $\phi_7$ | 0 | $1 + 2\beta$ | $1 + 2\alpha$ | 0 |
| $\phi_{3',\alpha}$ | 0 | $1 + 2\beta$ | $2\alpha$ | 0 |
| $\phi_{5',\alpha}$ | 0 | $-1 + 2\beta$ | $2\alpha$ | 0 |

Table B5b: Interpolation functions for Mesh Type A2.

| $\phi_i$ | $\phi$ | $\phi_{,\alpha}$ | $\phi_{,\beta}$ | $\phi_{,\alpha\alpha}$ | $\phi_{,\alpha\beta}$ | $\phi_{,\beta\beta}$ |
|---|---|---|---|---|---|---|
| $\phi_1$ | $1 - \alpha^2 - \beta^2 + \alpha^2\beta^2$ | $2(-\alpha + \alpha\beta^2)$ | $2(-\beta + \alpha^2\beta)$ | $2(-1+\beta^2)$ | $4\alpha\beta$ | $-2 + 2\alpha^2$ |
| $\phi_{1,\alpha}$ | $\alpha + \alpha^2 - \alpha\beta^2 - \alpha^2\beta^2$ | $1+ 2\alpha - \beta^2 - 2\alpha\beta^2$ | $2(-\alpha\beta - \alpha^2\beta)$ | $2(1 - \beta^2)$ | $-2\beta - 4\alpha\beta$ | $-2\alpha - 2\alpha^2$ |
| $\phi_3$ | $\frac{1}{2}(\beta + \beta^2 - \alpha^2\beta - \alpha^2\beta^2)$ | $-\alpha\beta - \alpha\beta^2$ | $\frac{1}{2} + \beta - \frac{1}{2}\alpha^2 - \alpha^2\beta$ | $-\beta - \beta^2$ | $-\alpha - 2\alpha\beta$ | $1 - \alpha^2$ |
| $\phi_4$ | $\alpha^2 - \alpha^2\beta^2$ | $2(\alpha - \alpha\beta^2)$ | $-2\alpha^2\beta$ | $2(1 - \beta^2)$ | $-4\alpha\beta$ | $-2\alpha^2$ |
| $\phi_5$ | $\frac{1}{2}(-\beta + \beta^2 + \alpha^2\beta - \alpha^2\beta^2)$ | $\alpha\beta - \alpha\beta^2$ | $-\frac{1}{2} + \beta + \frac{1}{2}\alpha^2 - \alpha^2\beta$ | $\beta - \beta^2$ | $\alpha - 2\alpha\beta$ | $1 - \alpha^2$ |
| $\phi_{5,\alpha}$ | $\frac{1}{2}(-\alpha\beta - \alpha^2\beta + \alpha\beta^2 + \alpha^2\beta^2)$ | $-\frac{1}{2}\beta - \alpha\beta + \frac{1}{2}\beta^2 + \alpha\beta^2$ | $-\frac{1}{2}\alpha - \frac{1}{2}\alpha^2 + \alpha\beta + \alpha^2\beta$ | $-\beta + \beta^2$ | $-\frac{1}{2} - \alpha + \beta + 2\alpha\beta$ | $\alpha + \alpha^2$ |
| $\phi_{3,\alpha}$ | $\frac{1}{2}(\alpha\beta + \alpha^2\beta + \alpha\beta^2 + \alpha^2\beta^2)$ | $\frac{1}{2}\beta + \alpha\beta + \frac{1}{2}\beta^2 + \alpha\beta^2$ | $\frac{1}{2}\alpha + \frac{1}{2}\alpha^2 + \alpha\beta + \alpha^2\beta$ | $\beta + \beta^2$ | $\frac{1}{2} + \alpha + \beta + 2\alpha\beta$ | $\alpha + \alpha^2$ |
| $\phi_8$ | $\frac{1}{2}(\alpha^2\beta + \alpha^2\beta^2)$ | $\alpha\beta + \alpha\beta^2$ | $\frac{1}{2}\alpha^2 + \alpha^2\beta$ | $\beta + \beta^2$ | $\alpha + 2\alpha\beta$ | $\alpha^2$ |
| $\phi_9$ | $\frac{1}{2}(-\alpha^2\beta + \alpha^2\beta^2)$ | $-\alpha\beta + \alpha\beta^2$ | $-\frac{1}{2}\alpha^2 + \alpha^2\beta$ | $-\beta + \beta^2$ | $-\alpha + 2\alpha\beta$ | $\alpha^2$ |

Table B6a: Interpolation functions for Mesh Type B2.

| $\phi_i$ | $\phi_{,\alpha\alpha\alpha}$ | $\phi_{,\alpha\beta\alpha}$ | $\phi_{,\beta\beta\alpha}$ | $\phi_{,\beta\beta\beta}$ |
|---|---|---|---|---|
| $\phi_1$ | 0 | $4\beta$ | $4\alpha$ | 0 |
| $\phi_{1,\alpha}$ | 0 | $-4\beta$ | $-2 - 4\alpha$ | 0 |
| $\phi_3$ | 0 | $-1 - 2\beta$ | $-2\alpha$ | 0 |
| $\phi_4$ | 0 | $-4\beta$ | $-4\alpha$ | 0 |
| $\phi_5$ | 0 | $1 - 2\beta$ | $-2\alpha$ | 0 |
| $\phi_{5,\alpha}$ | 0 | $-1 + 2\beta$ | $1 + 2\alpha$ | 0 |
| $\phi_{3,\alpha}$ | 0 | $1 + 2\beta$ | $1 + 2\alpha$ | 0 |
| $\phi_8$ | 0 | $1 + 2\beta$ | $2\alpha$ | 0 |
| $\phi_9$ | 0 | $-1 + 2\beta$ | $2\alpha$ | 0 |

Table B6b: Interpolation functions for Mesh Type B2.

| $\phi_i$ | $\phi$ | $\phi_{,\alpha}$ | $\phi_{,\beta}$ | $\phi_{,\alpha\alpha}$ | $\phi_{,\alpha\beta}$ | $\phi_{,\beta\beta}$ |
|---|---|---|---|---|---|---|
| $\phi_1$ | $1-\alpha^2-\beta^2+\alpha^2\beta^2$ | $-2\alpha+2\alpha\beta^2$ | $-2\beta+2\alpha^2\beta$ | $-2+2\beta^2$ | $4\alpha\beta$ | $-2+2\alpha^2$ |
| $\phi_2$ | $\frac{1}{2}(\alpha+\alpha^2-\alpha\beta^2-\alpha^2\beta^2)$ | $\frac{1}{2}+\alpha-\frac{1}{2}\beta^2-\alpha\beta^2$ | $-\alpha\beta-\alpha^2\beta$ | $1-\beta^2$ | $-\beta-2\alpha\beta$ | $-\alpha-\alpha^2$ |
| $\phi_3$ | $\beta^2-\alpha^2\beta^2$ | $-2\alpha\beta^2$ | $2\beta-2\alpha^2\beta$ | $-2\beta^2$ | $-4\alpha\beta$ | $2-2\alpha^2$ |
| $\phi_4$ | $\frac{1}{2}(-\alpha+\alpha^2+\alpha\beta^2-\alpha^2\beta^2)$ | $-\frac{1}{2}+\alpha+\frac{1}{2}\beta^2-\alpha\beta^2$ | $\alpha\beta-\alpha^2\beta$ | $1-\beta^2$ | $\beta-2\alpha\beta$ | $\alpha-\alpha^2$ |
| $\phi_{1,\beta}$ | $\beta-\beta^2-\alpha^2\beta+\alpha^2\beta^2$ | $-2\alpha\beta+2\alpha\beta^2$ | $1-2\beta-\alpha^2+2\alpha^2\beta$ | $-2\beta+2\beta^2$ | $-2\alpha+4\alpha\beta$ | $-2+2\alpha^2$ |
| $\phi_{2,\beta}$ | $\frac{1}{2}(\alpha\beta+\alpha^2\beta-\alpha\beta^2-\alpha^2\beta^2)$ | $\frac{1}{2}\beta+\alpha\beta-\frac{1}{2}\beta^2-\alpha\beta^2$ | $\frac{1}{2}\alpha+\frac{1}{2}\alpha^2-\alpha\beta-\alpha^2\beta$ | $\beta-\beta^2$ | $\frac{1}{2}+\alpha-\beta-2\alpha\beta$ | $-\alpha-\alpha^2$ |
| $\phi_7$ | $\frac{1}{2}(\alpha\beta^2+\alpha^2\beta^2)$ | $\frac{1}{2}\beta^2+\alpha\beta^2$ | $\alpha\beta+\alpha^2\beta$ | $\beta^2$ | $\beta+2\alpha\beta$ | $\alpha+\alpha^2$ |
| $\phi_8$ | $\frac{1}{2}(-\alpha\beta^2+\alpha^2\beta^2)$ | $-\frac{1}{2}\beta^2+\alpha\beta^2$ | $-\alpha\beta+\alpha^2\beta$ | $\beta^2$ | $-\beta+2\alpha\beta$ | $-\alpha+\alpha^2$ |
| $\phi_{4,\beta}$ | $\frac{1}{2}(-\alpha\beta+\alpha^2\beta+\alpha\beta^2-\alpha^2\beta^2)$ | $-\frac{1}{2}\beta+\alpha\beta+\frac{1}{2}\beta^2-\alpha\beta^2$ | $-\frac{1}{2}\alpha+\frac{1}{2}\alpha^2+\alpha\beta-\alpha^2\beta$ | $\beta-\beta^2$ | $-\frac{1}{2}+\alpha+\beta-2\alpha\beta$ | $\alpha-\alpha^2$ |

Table B7a: Interpolation functions for Mesh Type C2.

| $\phi_i$ | $\phi,_{\alpha\alpha\alpha}$ | $\phi,_{\alpha\beta\alpha}$ | $\phi,_{\beta\beta\alpha}$ | $\phi,_{\beta\beta\beta}$ |
|---|---|---|---|---|
| $\phi_1$ | 0 | $4\beta$ | $4\alpha$ | 0 |
| $\phi_2$ | 0 | $-2\beta$ | $-1 - 2\alpha$ | 0 |
| $\phi_3$ | 0 | $-4\beta$ | $-4\alpha$ | 0 |
| $\phi_4$ | 0 | $-2\beta$ | $1 - 2\alpha$ | 0 |
| $\phi_{1',\beta}$ | 0 | $-2 + 4\beta$ | $4\alpha$ | 0 |
| $\phi_{2',\beta}$ | 0 | $1 - 2\beta$ | $-1 - 2\alpha$ | 0 |
| $\phi_7$ | 0 | $2\beta$ | $1 + 2\alpha$ | 0 |
| $\phi_8$ | 0 | $2\beta$ | $-1 + 2\alpha$ | 0 |
| $\phi_{4',\beta}$ | 0 | $1 - 2\beta$ | $1 - 2\alpha$ | 0 |

Table B7b: Interpolation functions for Mesh Type C2.

| $\phi_i$ | $\phi$ | $\phi_{,\alpha}$ | $\phi_{,\beta}$ | $\phi_{,\alpha\alpha}$ | $\phi_{,\alpha\beta}$ | $\phi_{,\beta\beta}$ |
|---|---|---|---|---|---|---|
| $\phi_1$ | $1-\alpha^2-\beta^2+\alpha^2\beta^2$ | $-2\alpha+2\alpha\beta^2$ | $-2\beta+2\alpha^2\beta$ | $-2+2\beta^2$ | $4\alpha\beta$ | $-2+2\alpha^2$ |
| $\phi_2$ | $\frac{1}{2}(\alpha+\alpha^2-\alpha\beta^2-\alpha^2\beta^2)$ | $\frac{1}{2}+\alpha-\frac{1}{2}\beta^2-\alpha\beta^2$ | $-\alpha\beta-\alpha^2\beta$ | $1-\beta^2$ | $-\beta-2\alpha\beta$ | $-\alpha-\alpha^2$ |
| $\phi_{1,\beta}$ | $\beta+\beta^2-\alpha^2\beta-\alpha^2\beta^2$ | $-2\alpha\beta-2\alpha\beta^2$ | $1+2\beta-\alpha^2-2\alpha^2\beta$ | $-2\beta-2\beta^2$ | $-2\alpha-4\alpha\beta$ | $2-2\alpha^2$ |
| $\phi_4$ | $\frac{1}{2}(-\alpha+\alpha^2+\alpha\beta^2-\alpha^2\beta^2)$ | $-\frac{1}{2}+\alpha+\frac{1}{2}\beta^2-\alpha\beta^2$ | $\alpha\beta-\alpha^2\beta$ | $1-\beta^2$ | $\beta-2\alpha\beta$ | $\alpha-\alpha^2$ |
| $\phi_5$ | $\beta^2-\alpha^2\beta^2$ | $-2\alpha\beta^2$ | $2\beta-2\alpha^2\beta$ | $-2\beta^2$ | $-4\alpha\beta$ | $2-2\alpha^2$ |
| $\phi_6$ | $\frac{1}{2}(\alpha\beta^2+\alpha^2\beta^2)$ | $\frac{1}{2}\beta^2+\alpha\beta^2$ | $\alpha\beta+\alpha^2\beta$ | $\beta^2$ | $\beta+2\alpha\beta$ | $\alpha+\alpha^2$ |
| $\phi_{2,\beta}$ | $\frac{1}{2}(\alpha\beta+\alpha^2\beta+\alpha\beta^2+\alpha^2\beta^2)$ | $\frac{1}{2}\beta+\alpha\beta+\frac{1}{2}\beta^2+\alpha\beta^2$ | $\frac{1}{2}\alpha+\frac{1}{2}\alpha^2+\alpha\beta+\alpha^2\beta$ | $\beta+\beta^2$ | $\frac{1}{2}+\alpha+\beta+2\alpha\beta$ | $\alpha+\alpha^2$ |
| $\phi_{4,\beta}$ | $\frac{1}{2}(-\alpha\beta+\alpha^2\beta-\alpha\beta^2+\alpha^2\beta^2)$ | $-\frac{1}{2}\beta+\alpha\beta-\frac{1}{2}\beta^2+\alpha\beta^2$ | $-\frac{1}{2}\alpha+\frac{1}{2}\alpha^2-\alpha\beta+\alpha^2\beta$ | $\beta+\beta^2$ | $-\frac{1}{2}+\alpha-\beta+2\alpha\beta$ | $-\alpha+\alpha^2$ |
| $\phi_9$ | $\frac{1}{2}(-\alpha\beta^2+\alpha^2\beta^2)$ | $-\frac{1}{2}\beta^2+\alpha\beta^2$ | $-\alpha\beta+\alpha^2\beta$ | $\beta^2$ | $-\beta+2\alpha\beta$ | $-\alpha+\alpha^2$ |

Table B8a: Interpolation functions for Mesh Type D2.

| $\phi_i$ | $\phi,_{\alpha\alpha\alpha}$ | $\phi,_{\alpha\beta\alpha}$ | $\phi,_{\beta\beta\alpha}$ | $\phi,_{\beta\beta\beta}$ |
|---|---|---|---|---|
| $\phi_1$ | 0 | $4\beta$ | $4\alpha$ | 0 |
| $\phi_2$ | 0 | $-2\beta$ | $-1-2\alpha$ | 0 |
| $\phi_{1},_{\beta}$ | 0 | $-2-4\beta$ | $-4\alpha$ | 0 |
| $\phi_4$ | 0 | $-2\beta$ | $1-2\alpha$ | 0 |
| $\phi_5$ | 0 | $-4\beta$ | $-4\alpha$ | 0 |
| $\phi_6$ | 0 | $2\beta$ | $1+2\alpha$ | 0 |
| $\phi_{2},_{\beta}$ | 0 | $1+2\beta$ | $1+2\alpha$ | 0 |
| $\phi_{4},_{\beta}$ | 0 | $1+2\beta$ | $-1+2\alpha$ | 0 |
| $\phi_9$ | 0 | $2\beta$ | $-1+2\alpha$ | 0 |

Table B8b: Interpolation functions for Mesh Type D2.

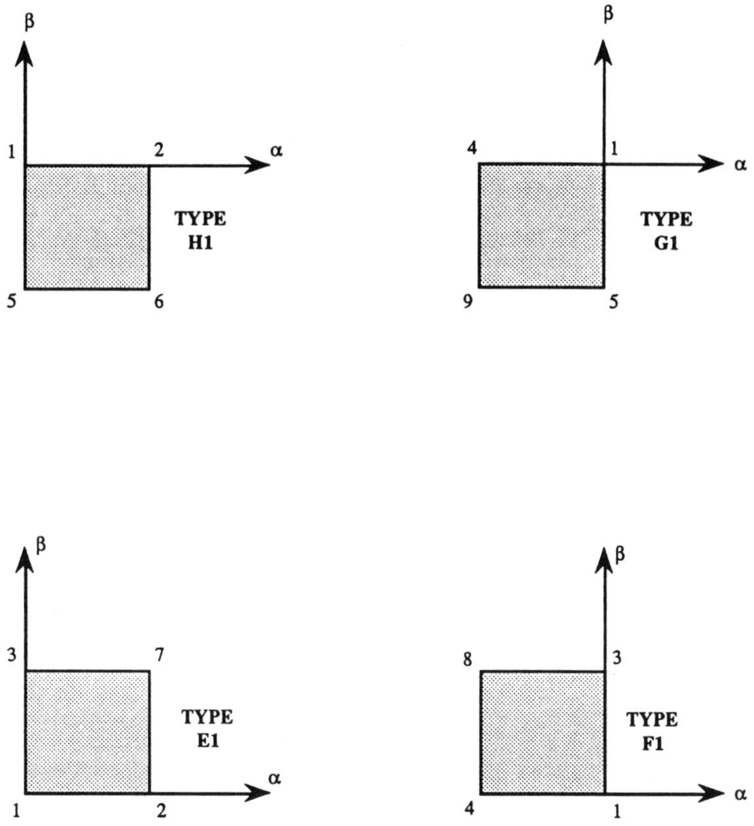


Figure B3: Corner Meshes.

| $\phi_i$ | $\phi$ | $\phi_{,\alpha}$ | $\phi_{,\beta}$ | $\phi_{,\alpha\alpha}$ | $\phi_{,\alpha\beta}$ | $\phi_{,\beta\beta}$ |
|---|---|---|---|---|---|---|
| $\phi_1$ | $1-\alpha-\beta+\alpha\beta$ | $-1+\beta$ | $-1+\alpha$ | 0 | 1 | 0 |
| $\phi_2$ | $\alpha-\alpha\beta$ | $1-\beta$ | $-\alpha$ | 0 | $-1$ | 0 |
| $\phi_3$ | $\beta-\alpha\beta$ | $-\beta$ | $1-\alpha$ | 0 | $-1$ | 0 |
| $\phi_4$ | 0 | 0 | 0 | 0 | 0 | 0 |
| $\phi_5$ | 0 | 0 | 0 | 0 | 0 | 0 |
| $\phi_6$ | 0 | 0 | 0 | 0 | 0 | 0 |
| $\phi_7$ | $\alpha\beta$ | $\beta$ | $\alpha$ | 0 | 1 | 0 |
| $\phi_8$ | 0 | 0 | 0 | 0 | 0 | 0 |
| $\phi_9$ | 0 | 0 | 0 | 0 | 0 | 0 |

Table B9: Interpolation functions for Mesh Type E1.

| $\phi_i$ | $\phi$ | $\phi,_{\alpha}$ | $\phi,_{\beta}$ | $\phi,_{\alpha\alpha}$ | $\phi,_{\alpha\beta}$ | $\phi,_{\beta\beta}$ |
|---|---|---|---|---|---|---|
| $\phi_1$ | $1 + \alpha - \beta - \alpha\beta$ | $1 - \beta$ | $-1 - \alpha$ | 0 | $-1$ | 0 |
| $\phi_2$ | 0 | 0 | 0 | 0 | 0 | 0 |
| $\phi_3$ | $\beta + \alpha\beta$ | $\beta$ | $1 + \alpha$ | 0 | 1 | 0 |
| $\phi_4$ | $-\alpha + \alpha\beta$ | $-1 + \beta$ | $\alpha$ | 0 | 1 | 0 |
| $\phi_5$ | 0 | 0 | 0 | 0 | 0 | 0 |
| $\phi_6$ | 0 | 0 | 0 | 0 | 0 | 0 |
| $\phi_7$ | 0 | 0 | 0 | 0 | 0 | 0 |
| $\phi_8$ | $-\alpha\beta$ | $-\beta$ | $-\alpha$ | 0 | $-1$ | 0 |
| $\phi_9$ | 0 | 0 | 0 | 0 | 0 | 0 |

Table B10: Interpolation functions for Mesh Type F1.

| $\phi_i$ | $\phi$ | $\phi_{,\alpha}$ | $\phi_{,\beta}$ | $\phi_{,\alpha\alpha}$ | $\phi_{,\alpha\beta}$ | $\phi_{,\beta\beta}$ |
|---|---|---|---|---|---|---|
| $\phi_1$ | $1+\alpha+\beta+\alpha\beta$ | $1+\beta$ | $1+\alpha$ | 0 | 1 | 0 |
| $\phi_2$ | 0 | 0 | 0 | 0 | 0 | 0 |
| $\phi_3$ | 0 | 0 | 0 | 0 | 0 | 0 |
| $\phi_4$ | $-\alpha-\alpha\beta$ | $-1-\beta$ | $-\alpha$ | 0 | $-1$ | 0 |
| $\phi_5$ | $-\beta-\alpha\beta$ | $-\beta$ | $-1-\alpha$ | 0 | $-1$ | 0 |
| $\phi_6$ | 0 | 0 | 0 | 0 | 0 | 0 |
| $\phi_7$ | 0 | 0 | 0 | 0 | 0 | 0 |
| $\phi_8$ | 0 | 0 | 0 | 0 | 0 | 0 |
| $\phi_9$ | $\alpha\beta$ | $\beta$ | $\alpha$ | 0 | 1 | 0 |

Table B11: Interpolation functions for Mesh Type G1.

| $\phi_i$ | $\phi$ | $\phi_{,\alpha}$ | $\phi_{,\beta}$ | $\phi_{,\alpha\alpha}$ | $\phi_{,\alpha\beta}$ | $\phi_{,\beta\beta}$ |
|---|---|---|---|---|---|---|
| $\phi_1$ | $1-\alpha+\beta-\alpha\beta$ | $-1-\beta$ | $1-\alpha$ | 0 | $-1$ | 0 |
| $\phi_2$ | $\alpha+\alpha\beta$ | $1+\beta$ | $\alpha$ | 0 | 1 | 0 |
| $\phi_3$ | 0 | 0 | 0 | 0 | 0 | 0 |
| $\phi_4$ | 0 | 0 | 0 | 0 | 0 | 0 |
| $\phi_5$ | $-\beta+\alpha\beta$ | $\beta$ | $-1+\alpha$ | 0 | 1 | 0 |
| $\phi_6$ | $-\alpha\beta$ | $-\beta$ | $-\alpha$ | 0 | $-1$ | 0 |
| $\phi_7$ | 0 | 0 | 0 | 0 | 0 | 0 |
| $\phi_8$ | 0 | 0 | 0 | 0 | 0 | 0 |
| $\phi_9$ | 0 | 0 | 0 | 0 | 0 | 0 |

Table B12: Interpolation functions for Mesh Type H1.

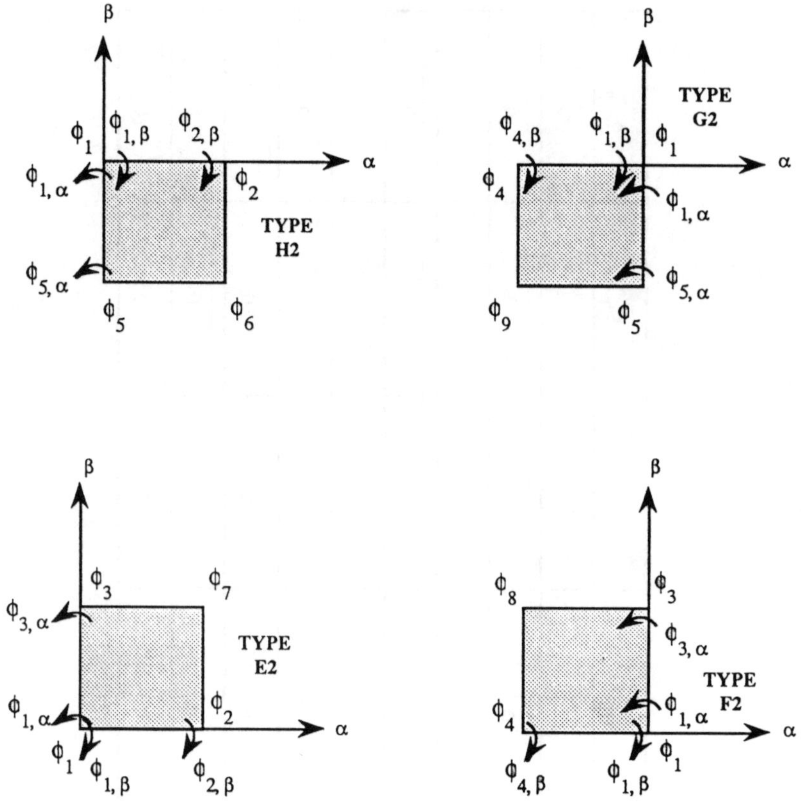


Figure B4: Corner Meshes.

| $\phi_i$ | $\phi$ | $\phi_{,\alpha}$ | $\phi_{,\beta}$ | $\phi_{,\alpha\alpha}$ | $\phi_{,\alpha\beta}$ | $\phi_{,\beta\beta}$ |
|---|---|---|---|---|---|---|
| $\phi_1$ | $1-\alpha^2-\alpha\beta-\beta^2+\alpha^2\beta+\alpha\beta^2$ | $-2\alpha-\beta+2\alpha\beta+\beta^2$ | $-\alpha-2\beta+\alpha^2+2\alpha\beta$ | $-2+2\beta$ | $-1+2\alpha+2\beta$ | $-2+2\alpha$ |
| $\phi_2$ | $\alpha^2+\alpha\beta-\alpha^2\beta-\alpha\beta^2$ | $2\alpha+\beta-2\alpha\beta-\beta^2$ | $\alpha-\alpha^2-2\alpha\beta$ | $2-2\beta$ | $1-2\alpha-2\beta$ | $-2\alpha$ |
| $\phi_3$ | $\alpha\beta+\beta^2-\alpha^2\beta-\alpha\beta^2$ | $\beta-2\alpha\beta-\beta^2$ | $\alpha+2\beta-\alpha^2-2\alpha\beta$ | $-2\beta$ | $1-2\alpha-2\beta$ | $2-2\alpha$ |
| $\phi_{1,\alpha}$ | $\alpha-\alpha^2-\alpha\beta+\alpha^2\beta$ | $1-2\alpha-\beta+2\alpha\beta$ | $-\alpha+\alpha^2$ | $-2+2\beta$ | $-1+2\alpha$ | 0 |
| $\phi_{1,\beta}$ | $\beta-\alpha\beta-\beta^2+\alpha\beta^2$ | $-\beta+\beta^2$ | $1-\alpha-2\beta+2\alpha\beta$ | 0 | $-1+2\beta$ | $-2+2\alpha$ |
| $\phi_{2,\beta}$ | $\alpha\beta-\alpha\beta^2$ | $\beta-\beta^2$ | $\alpha-2\alpha\beta$ | 0 | $1-2\beta$ | $-2\alpha$ |
| $\phi_7$ | $-\alpha\beta+\alpha^2\beta+\alpha\beta^2$ | $-\beta+2\alpha\beta+\beta^2$ | $-\alpha+\alpha^2+2\alpha\beta$ | $2\beta$ | $-1+2\alpha+2\beta$ | $2\alpha$ |
| $\phi_{3,\alpha}$ | $\alpha\beta-\alpha^2\beta$ | $\beta-2\alpha\beta$ | $\alpha-\alpha^2$ | $-2\beta$ | $1-2\alpha$ | 0 |
| $\phi_9$ | 0 | 0 | 0 | 0 | 0 | 0 |

Table B13a: Interpolation functions for Mesh Type E2.

| $\phi_i$ | $\phi_{,\alpha\alpha\alpha}$ | $\phi_{,\alpha\beta\alpha}$ | $\phi_{,\beta\beta\alpha}$ | $\phi_{,\beta\beta\beta}$ |
|---|---|---|---|---|
| $\phi_1$ | 0 | 2 | 2 | 0 |
| $\phi_2$ | 0 | –2 | –2 | 0 |
| $\phi_3$ | 0 | –2 | –2 | 0 |
| $\phi_{1,\alpha}$ | 0 | 2 | 0 | 0 |
| $\phi_{1,\beta}$ | 0 | 0 | 2 | 0 |
| $\phi_{2,\beta}$ | 0 | 0 | -2 | 0 |
| $\phi_7$ | 0 | 2 | 2 | 0 |
| $\phi_{3,\alpha}$ | 0 | -2 | 0 | 0 |
| $\phi_9$ | 0 | 0 | 0 | 0 |

Table B13b: Interpolation functions for Mesh Type E2.

| $\phi_i$ | $\phi$ | $\phi,_{\alpha}$ | $\phi,_{\beta}$ | $\phi,_{\alpha\alpha}$ | $\phi,_{\alpha\beta}$ | $\phi,_{\beta\beta}$ |
|---|---|---|---|---|---|---|
| $\phi_1$ | $1-\alpha^2+\alpha\beta-\beta^2+\alpha^2\beta-\alpha\beta^2$ | $-2\alpha+\beta+2\alpha\beta-\beta^2$ | $\alpha-2\beta+\alpha^2-2\alpha\beta$ | $-2+2\beta$ | $1+2\alpha-2\beta$ | $-2-2\alpha$ |
| $\phi_{1},_{\alpha}$ | $\alpha+\alpha^2-\alpha\beta-\alpha^2\beta$ | $1+2\alpha-\beta-2\alpha\beta$ | $-\alpha-\alpha^2$ | $2-2\beta$ | $-1-2\alpha$ | $0$ |
| $\phi_3$ | $-\alpha\beta+\beta^2-\alpha^2\beta+\alpha\beta^2$ | $-\beta-2\alpha\beta+\beta^2$ | $-\alpha+2\beta-\alpha^2+2\alpha\beta$ | $-2\beta$ | $-1-2\alpha+2\beta$ | $2+2\alpha$ |
| $\phi_4$ | $\alpha^2-\alpha\beta-\alpha^2\beta+\alpha\beta^2$ | $2\alpha-\beta-2\alpha\beta+\beta^2$ | $-\alpha-\alpha^2+2\alpha\beta$ | $2-2\beta$ | $-1-2\alpha+2\beta$ | $2\alpha$ |
| $\phi_{1},_{\beta}$ | $\beta+\alpha\beta-\beta^2-\alpha\beta^2$ | $\beta-\beta^2$ | $1+\alpha-2\beta-2\alpha\beta$ | $0$ | $1-2\beta$ | $-2-2\alpha$ |
| $\phi_6$ | $0$ | $0$ | $0$ | $0$ | $0$ | $0$ |
| $\phi_{3},_{\alpha}$ | $\alpha\beta+\alpha^2\beta$ | $\beta+2\alpha\beta$ | $\alpha+\alpha^2$ | $2\beta$ | $1+2\alpha$ | $0$ |
| $\phi_8$ | $\alpha\beta+\alpha^2\beta-\alpha\beta^2$ | $\beta+2\alpha\beta-\beta^2$ | $\alpha+\alpha^2-2\alpha\beta$ | $2\beta$ | $1+2\alpha-2\beta$ | $-2\alpha$ |
| $\phi_{4},_{\beta}$ | $-\alpha\beta+\alpha\beta^2$ | $-\beta+\beta^2$ | $-\alpha+2\alpha\beta$ | $0$ | $-1+2\beta$ | $2\alpha$ |

Table B14a: Interpolation functions for Mesh Type F2.

| $\phi_i$ | $\phi_{,\alpha\alpha\alpha}$ | $\phi_{,\alpha\beta\alpha}$ | $\phi_{,\beta\beta\alpha}$ | $\phi_{,\beta\beta\beta}$ |
|---|---|---|---|---|
| $\phi_1$ | 0 | 2 | −2 | 0 |
| $\phi_{1,\alpha}$ | 0 | −2 | 0 | 0 |
| $\phi_3$ | 0 | −2 | 2 | 0 |
| $\phi_4$ | 0 | −2 | 2 | 0 |
| $\phi_{1,\beta}$ | 0 | 0 | −2 | 0 |
| $\phi_6$ | 0 | 0 | 0 | 0 |
| $\phi_{3,\alpha}$ | 0 | 2 | 0 | 0 |
| $\phi_8$ | 0 | 2 | −2 | 0 |
| $\phi_{4,\beta}$ | 0 | 0 | 2 | 0 |

Table B14b: Interpolation functions for Mesh Type F2.

| $\phi_i$ | $\phi$ | $\phi_{,\alpha}$ | $\phi_{,\beta}$ | $\phi_{,\alpha\alpha}$ | $\phi_{,\alpha\beta}$ | $\phi_{,\beta\beta}$ |
|---|---|---|---|---|---|---|
| $\phi_1$ | $1-\alpha^2-\alpha\beta-\beta^2-\alpha^2\beta-\alpha\beta^2$ | $-2\alpha-\beta-2\alpha\beta-\beta^2$ | $-\alpha-2\beta-\alpha^2-2\alpha\beta$ | $-2-2\beta$ | $-1-2\alpha-2\beta$ | $-2-2\alpha$ |
| $\phi_{1,\alpha}$ | $\alpha+\alpha^2+\alpha\beta+\alpha^2\beta$ | $1+2\alpha+\beta+2\alpha\beta$ | $\alpha+\alpha^2$ | $2+2\beta$ | $1+2\alpha$ | 0 |
| $\phi_{1,\beta}$ | $\beta+\alpha\beta+\beta 2+\alpha\beta^2$ | $\beta+\beta^2$ | $1+\alpha+2\beta+2\alpha\beta$ | 0 | $1+2\beta$ | $2+2\alpha$ |
| $\phi_4$ | $\alpha^2+\alpha\beta+\alpha^2\beta+\alpha\beta^2$ | $2\alpha+\beta+2\alpha\beta+\beta^2$ | $\alpha+\alpha^2+2\alpha\beta$ | $2+2\beta$ | $1+2\alpha+2\beta$ | $2\alpha$ |
| $\phi_5$ | $\alpha\beta+\beta^2+\alpha^2\beta+\alpha\beta^2$ | $\beta+2\alpha\beta+\beta^2$ | $\alpha+2\beta+\alpha^2+2\alpha\beta$ | $2\beta$ | $1+2\alpha+2\beta$ | $2+2\alpha$ |
| $\phi_{5,\alpha}$ | $-\alpha\beta-\alpha^2\beta$ | $-\beta-2\alpha\beta$ | $-\alpha-\alpha^2$ | $-2\beta$ | $-1-2\alpha$ | 0 |
| $\phi_7$ | 0 | 0 | 0 | 0 | 0 | 0 |
| $\phi_{4,\beta}$ | $-\alpha\beta-\alpha\beta^2$ | $-\beta-\beta^2$ | $-\alpha-2\alpha\beta$ | 0 | $-1-2\beta$ | $-2\alpha$ |
| $\phi_9$ | $-\alpha\beta-\alpha^2\beta-\alpha\beta^2$ | $-\beta-2\alpha\beta-\beta^2$ | $-\alpha-\alpha^2-2\alpha\beta$ | $-2\beta$ | $-1-2\alpha-2\beta$ | $-2\alpha$ |

Table B15a: Interpolation functions for Mesh Type G2.

| $\phi_i$ | $\phi_{,\alpha\alpha\alpha}$ | $\phi_{,\alpha\beta\alpha}$ | $\phi_{,\beta\beta\alpha}$ | $\phi_{,\beta\beta\beta}$ |
|---|---|---|---|---|
| $\phi_1$ | 0 | −2 | −2 | 0 |
| $\phi_{1,\alpha}$ | 0 | 2 | 0 | 0 |
| $\phi_{1,\beta}$ | 0 | 0 | 2 | 0 |
| $\phi_4$ | 0 | 2 | 2 | 0 |
| $\phi_5$ | 0 | 2 | 2 | 0 |
| $\phi_{5,\alpha}$ | 0 | −2 | 0 | 0 |
| $\phi_7$ | 0 | 0 | 0 | 0 |
| $\phi_{4,\beta}$ | 0 | 0 | −2 | 0 |
| $\phi_9$ | 0 | −2 | −2 | 0 |

Table B15b: Interpolation functions for Mesh Type G2.

| $\phi_i$ | $\phi$ | $\phi_{,\alpha}$ | $\phi_{,\beta}$ | $\phi_{,\alpha\alpha}$ | $\phi_{,\alpha\beta}$ | $\phi_{,\beta\beta}$ |
|---|---|---|---|---|---|---|
| $\phi_1$ | $1-\alpha^2+\alpha\beta-\beta^2-\alpha^2\beta+\alpha\beta^2$ | $-2\alpha+\beta-2\alpha\beta+\beta^2$ | $\alpha-2\beta-\alpha^2+2\alpha\beta$ | $-2-2\beta$ | $1-2\alpha+2\beta$ | $-2+2\alpha$ |
| $\phi_2$ | $\alpha^2-\alpha\beta+\alpha^2\beta-\alpha\beta^2$ | $2\alpha-\beta+2\alpha\beta-\beta^2$ | $-\alpha+\alpha^2-2\alpha\beta$ | $2+2\beta$ | $-1+2\alpha-2\beta$ | $-2\alpha$ |
| $\phi_{1,\beta}$ | $\beta-\alpha\beta+\beta^2-\alpha\beta^2$ | $-\beta-\beta^2$ | $1-\alpha+2\beta-2\alpha\beta$ | $0$ | $-1-2\beta$ | $2-2\alpha$ |
| $\phi_{1,\alpha}$ | $\alpha-\alpha^2+\alpha\beta-\alpha^2\beta$ | $1-2\alpha+\beta-2\alpha\beta$ | $\alpha-\alpha^2$ | $-2-2\beta$ | $1-2\alpha$ | $0$ |
| $\phi_5$ | $-\alpha\beta+\beta^2+\alpha^2\beta-\alpha\beta^2$ | $-\beta+2\alpha\beta-\beta^2$ | $-\alpha+2\beta+\alpha^2-2\alpha\beta$ | $2\beta$ | $-1+2\alpha-2\beta$ | $2-2\alpha$ |
| $\phi_6$ | $\alpha\beta-\alpha^2\beta+\alpha\beta^2$ | $\beta-2\alpha\beta+\beta^2$ | $\alpha-\alpha^2+2\alpha\beta$ | $-2\beta$ | $1-2\alpha+2\beta$ | $2\alpha$ |
| $\phi_{2,\beta}$ | $\alpha\beta+\alpha\beta^2$ | $\beta+\beta^2$ | $\alpha+2\alpha\beta$ | $0$ | $1+2\beta$ | $2\alpha$ |
| $\phi_8$ | $0$ | $0$ | $0$ | $0$ | $0$ | $0$ |
| $\phi_{5,\alpha}$ | $-\alpha\beta+\alpha^2\beta$ | $-\beta+2\alpha\beta$ | $-\alpha+\alpha^2$ | $2\beta$ | $-1+2\alpha$ | $0$ |

Table B16a: Interpolation functions for Mesh Type H2.

| $\phi_i$ | $\phi_{,\alpha\alpha\alpha}$ | $\phi_{,\alpha\beta\alpha}$ | $\phi_{,\beta\beta\alpha}$ | $\phi_{,\beta\beta\beta}$ |
|---|---|---|---|---|
| $\phi_1$ | 0 | −2 | 2 | 0 |
| $\phi_2$ | 0 | 2 | −2 | 0 |
| $\phi_{1,\beta}$ | 0 | 0 | −2 | 0 |
| $\phi_{1,\alpha}$ | 0 | −2 | 0 | 0 |
| $\phi_5$ | 0 | 2 | −2 | 0 |
| $\phi_6$ | 0 | −2 | 2 | 0 |
| $\phi_{2,\beta}$ | 0 | 0 | 2 | 0 |
| $\phi_8$ | 0 | 0 | 0 | 0 |
| $\phi_{5,\alpha}$ | 0 | 2 | 0 | 0 |

Table B16b: Interpolation functions for Mesh Type H2.

## Appendix C

# NUMERICAL INTEGRATION OF THE STIFFNESS MATRIX

The application of the curvilinear finite difference method to the analysis of shell structures enables the Principle of Virtual Displacements to be written in discretized form. i.e.,

$$\sum_{i=1}^{n_m}\Big[[M]_i \{\ddot{U}_n\}_i + [K]_i \{U_n\}_i\Big] = \{0\} \tag{C.1}$$

The matrices $[K]_i$ and $[M]_i$ are referred to as the stiffness and mass matrices respectively and are evaluated for each subdomain by the expressions

$$[K]_i = \int_v [H_{\alpha\beta}]^t [\Sigma^{\alpha\beta}] \, dV \tag{C.2}$$

and

$$[M]_i = \int_v \rho \, [D^{cn}]^t [D_{cv}] \, dV \tag{C.3}$$

In previous work [71, 72], the integrand described by equation C.2 was assumed to be constant over a subdomain. Based on this assumption, the stiffness matrix was evaluated using a one-point integration scheme. The results generated by adopting such an approximation were shown to be acceptable when applied to a series of shell structures subjected to normal static loads.

However, in section 4.1.3 a plane stress example was presented which illustrated that a one-point integration scheme may produce inaccurate results.

This appendix examines in detail the effects of evaluating the shell's stiffness matrix using a low order integration scheme. Based on the example described in section 4.1.3, the general shell equations are simplified to represent a flat plate in plane stress. Thus, the strain-displacement and the stress-strain relationships defined by equations 2.6.50, 2.6.69 and 3.5.13 may written in the form

$$\eta_{\xi\nu} = \frac{1}{2}[\, u_\nu\|_\xi + u_\xi\|_\nu ] \tag{C.4}$$

$$\sigma_{\alpha\beta} = \frac{E}{(1-\nu^2)}[\,(1-\nu)\, a^{\alpha\nu}\, a^{\xi\beta} + \nu\, a^{\xi\nu}\, a^{\alpha\beta}\,]\,\eta_{\xi\nu} \tag{C.5}$$

Assuming the problem is discretized using a square orthogonal mesh, as illustrated in figure C.1, the following observations can be made.

$$\begin{aligned} x,_\beta &= y,_\alpha = 0.0 \\ x,_{\alpha\alpha} &= x,_{\alpha\beta} = x,_{\beta\beta} = 0.0 \\ y,_{\alpha\alpha} &= y,_{\alpha\beta} = y,_{\beta\beta} = 0.0 \\ x,_\alpha &= y,_\beta \end{aligned} \tag{C.6}$$

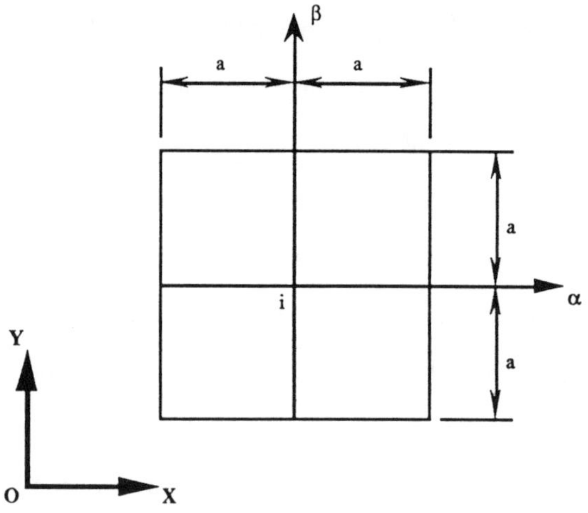


Figure C.1: Mesh Discretization.

For a nine node mesh the displacement fields $U_x$ and $U_y$ may be represented by the second order polynomial

$$U_i = a_1 + a_2\alpha + a_3\beta + a_4\alpha^2 + a_5\beta^2 + a_6\alpha\beta + a_7\alpha^2\beta + a_8\alpha\beta^2 + a_9\alpha^2\beta^2 \tag{C.7}$$

where $a_1, a_2, a_3, \ldots, a_9$ are constant coefficients. Equation C.7 may be differentiated with respect to the local coordinate system to form a series of expressions which approximate the displacements partial derivatives. i.e,

$$\{U_{i,\alpha}\} = [\phi^{u,\alpha}]\{A\} \tag{C.8}$$

where $\{U_{i,\alpha}\}^t = \{u_{x,\alpha} \quad u_{x,\beta} \quad u_{y,\alpha} \quad u_{y,\beta}\}$ (C.9)

$$[\phi^{u,\alpha}]^t = \begin{bmatrix} 0 & 0 & 0 & 0 \\ 0 & 0 & 0 & 0 \\ 1 & 0 & 0 & 0 \\ 0 & 0 & 1 & 0 \\ 0 & 1 & 0 & 0 \\ 0 & 0 & 0 & 1 \\ 2\alpha & 0 & 0 & 0 \\ 0 & 0 & 2\alpha & 0 \\ \beta & \alpha & 0 & 0 \\ 0 & 0 & \beta & \alpha \\ 0 & 2\beta & 0 & 0 \\ 0 & 0 & 0 & 2\beta \\ 2\alpha\beta & \alpha^2 & 0 & 0 \\ 0 & 0 & 2\alpha\beta & \alpha^2 \\ \beta^2 & 2\alpha\beta & 0 & 0 \\ 0 & 0 & \beta^2 & 2\alpha\beta \\ 2\alpha\beta^2 & 2\alpha^2\beta & 0 & 0 \\ 0 & 0 & 2\alpha\beta^2 & 2\alpha^2\beta \end{bmatrix} \tag{C.10}$$

and $[A]^t = \{a_1 \quad a_2 \quad a_3 \quad \ldots \quad a_9\}$ (C.11)

Equation C.7 can be substituted into the transformation equations defined in appendix A, to enable the local displacement partial derivatives to be written in the form

$$\begin{bmatrix} u_{\alpha,\alpha} \\ u_{\alpha,\beta} \\ u_{\beta,\alpha} \\ u_{\beta,\beta} \end{bmatrix} = x,_{\alpha} \begin{bmatrix} 1 & 0 & 0 & 0 \\ 0 & 1 & 0 & 0 \\ 0 & 0 & 1 & 0 \\ 0 & 0 & 0 & 1 \end{bmatrix} \begin{bmatrix} u_{x,\alpha} \\ u_{x,\beta} \\ u_{y,\alpha} \\ u_{y,\beta} \end{bmatrix} \tag{C.12}$$

This above expression is then combined with equations C.2, C.4 and C.5 to yield

$$[K]_i = \frac{E\, x,_{\alpha}\, x,_{\alpha}\, a^{\alpha\alpha}\, a^{\alpha\alpha}}{(1-v^2)} \int_v \{U_{x,\alpha}\}^t \begin{bmatrix} 1 & 0 & 0 & v \\ 0 & \frac{1}{2}(1-v) & \frac{1}{2}(1-v) & 0 \\ 0 & \frac{1}{2}(1-v) & \frac{1}{2}(1-v) & 0 \\ v & 0 & 0 & 1 \end{bmatrix} \{U_{x,\alpha}\}\, dV \tag{C.13}$$

Substituting equation C.8 into C.13 and expanding the volume integral allows the stiffness matrix to be written in the form

$$[K]_i = \frac{E\,(x,_{\alpha})^4\, a^{\alpha\alpha}\, a^{\alpha\alpha}\, t}{(1-v^2)} \int_{-0.5}^{0.5} [F]\, d\alpha\, d\beta \tag{C.14}$$

where the matrix [F] contains a series of algebraic expressions in terms of the local coordinates $\alpha$ and $\beta$ (see equation C.15).

$$[F] = \begin{bmatrix} [F_1] & [F_2] & [F_3] \\ & [F_4] & [F_5] \\ & & [F_6] \end{bmatrix} \tag{C.15a}$$

$$[F_1] = \begin{bmatrix} 0 & 0 & 0 & 0 & 0 & 0 \\ & 0 & 0 & 0 & 0 & 0 \\ & & 1 & 0 & 0 & \nu \\ & \mathbb{S} & & \frac{1}{2}(1-\nu) & \frac{1}{2}(1-\nu) & 0 \\ & & & & \frac{1}{2}(1-\nu) & 0 \\ & & & & & 1 \end{bmatrix} \tag{C.15b}$$

$$[F_2] = \begin{bmatrix} 0 & 0 & 0 & 0 & 0 & 0 \\ 0 & 0 & 0 & 0 & 0 & 0 \\ 2\alpha & 0 & \beta & \nu\alpha & 0 & 2\nu\beta \\ 0 & (1-\nu)\alpha & \frac{1}{2}(1-\nu)\alpha & \frac{1}{2}(1-\nu)\beta & (1-\nu)\beta & 0 \\ 0 & (1-\nu)\alpha & \frac{1}{2}(1-\nu)\alpha & \frac{1}{2}(1-\nu)\beta & (1-\nu)\beta & 0 \\ 2\nu\alpha & 0 & \nu\beta & \alpha & 0 & 2\beta \end{bmatrix} \tag{C.15c}$$

$$[F_3] = \begin{bmatrix} 0 & 0 & 0 & 0 & 0 & 0 \\ 0 & 0 & 0 & 0 & 0 & 0 \\ 2\alpha\beta & \nu\alpha^2 & \beta^2 & 2\nu\alpha\beta & 2\alpha\beta^2 & 2\nu\alpha^2\beta \\ \frac{1}{2}(1-\nu)\alpha^2 & (1-\nu)\alpha\beta & (1-\nu)\alpha\beta & \frac{1}{2}(1-\nu)\beta^2 & (1-\nu)\alpha^2\beta & (1-\nu)\alpha\beta^2 \\ \frac{1}{2}(1-\nu)\alpha^2 & (1-\nu)\alpha\beta & (1-\nu)\alpha\beta & \frac{1}{2}(1-\nu)\beta^2 & (1-\nu)\alpha^2\beta & (1-\nu)\alpha\beta^2 \\ 2\nu\alpha\beta & \alpha^2 & \nu\beta^2 & 2\alpha\beta & 2\nu\alpha\beta^2 & 2\alpha^2\beta \end{bmatrix} \tag{C.15d}$$

$$[F_4] = \begin{bmatrix} 4\alpha^2 & 0 & 2\alpha\beta & 2\nu\alpha^2 & 0 & 4\nu\alpha\beta \\ & 2(1-\nu)\alpha^2 & (1-\nu)\alpha^2 & (1-\nu)\alpha\beta & 2(1-\nu)\alpha\beta & 0 \\ & & \beta^2 + \frac{1}{2}(1-\nu)\alpha^2 & \frac{1}{2}(1+\nu)\alpha\beta & (1-\nu)\alpha\beta & 2\nu\beta^2 \\ & \mathbb{S} & & \frac{1}{2}(1-\nu)\beta^2 + \alpha^2 & (1-\nu)\beta^2 & 2\alpha\beta \\ & & & & 2(1-\nu)\beta^2 & 0 \\ & & & & & 4\beta^2 \end{bmatrix}$$

(C.15e)

$$[F_5] = \begin{bmatrix} 4\alpha^2\beta & 2\nu\alpha^3 & 2\alpha\beta^2 & 4\nu\alpha^2\beta & 4\alpha^2\beta^2 & 4\nu\alpha^3\beta \\ (1-\nu)\alpha^3 & 2(1-\nu)\alpha^2\beta & 2(1-\nu)\alpha^2\beta & (1-\nu)\alpha\beta^2 & 2(1-\nu)\alpha^3\beta & 2(1-\nu)\alpha^2\beta^2 \\ 2\alpha\beta^2 + \frac{1}{2}(1-\nu)\alpha^3 & \alpha^2\beta & \beta^3 + (1-\nu)\alpha^2\beta & \frac{1}{2}(1-3\nu)\alpha\beta^2 & 2\alpha\beta^3 + (1-\nu)\alpha^3\beta & (1+\nu)\alpha^2\beta^2 \\ \frac{1}{2}(1-3\nu)\alpha^2\beta & (1-\nu)\alpha\beta^2 + \alpha^3 & \alpha\beta^2 & \frac{1}{2}(1-\nu)\beta^3 + 2\alpha^2\beta & (1+\nu)\alpha^2\beta^2 & (1-\nu)\alpha\beta^3 + 2\alpha^3\beta \\ (1-\nu)\alpha^2\beta & 2(1-\nu)\alpha\beta^2 & 2(1-\nu)\alpha\beta^2 & (1-\nu)\beta^3 & 2(1-\nu)\alpha^2\beta^2 & 2(1-\nu)\alpha\beta^3 \\ 2\nu\alpha\beta^2 & 2\alpha^2\beta & 2\nu\beta^3 & 4\alpha\beta^2 & 4\nu\alpha\beta^3 & 4\alpha^2\beta^2 \end{bmatrix}$$

(C.15f)

$$[F_6] = \begin{bmatrix} 4\alpha^2\beta^2 + \frac{1}{2}(1-\nu)\alpha^4 & (1+\nu)\alpha^3\beta & 2\alpha\beta^3 + (1-\nu)\alpha^3\beta & \frac{1}{2}(1+7\nu)\alpha^2\beta^2 & 4\alpha^2\beta^3 + (1-\nu)\alpha^4\beta & (1+3\nu)\alpha^3\beta^2 \\ & 2(1-\nu)\alpha^2\beta^2 + \alpha^4 & (2-\nu)\alpha^2\beta^2 & (1-\nu)\alpha\beta^3 + 2\alpha^3\beta & 2\alpha^3\beta^2 & 2(1-\nu)\alpha^2\beta^3 + 2\alpha^4\beta \\ & & \beta^4 + 2(1-\nu)\alpha^2\beta^2 & (1+\nu)\alpha\beta^3 & 2\alpha\beta^4 + 2(1-\nu)\alpha^3\beta^2 & 2\alpha^2\beta^3 \\ & & & \frac{1}{2}(1-\nu)\beta^4 + 4\alpha^2\beta^2 & (1+3\nu)\alpha^2\beta^3 & (1-\nu)\alpha\beta^4 + 4\alpha^3\beta^2 \\ & \mathrm{S} & & & 4\alpha^2\beta^4 + 2(1-\nu)\alpha^4\beta^2 & 2(1+\nu)\alpha^3\beta^3 \\ & & & & & 2(1-\nu)\alpha^2\beta^4 + 4\alpha^4\beta^2 \end{bmatrix}$$

(C.15g)

The effects of approximating the integral described by equation C.14 can be studied by looking closely at members contained within the matrix [F]. A one-point integration scheme assumes the integrand is constant over each subdomain. The constants value is computed at the mesh origin (i.e. $\alpha = \beta = 0.0$). Table C.1 shows some typical expressions contained within the matrix [F] and illustrates the difference between evaluating the integrand exactly and approximating it numerically with a one-point scheme.

| Position | Function $f(\alpha,\beta)$ | $\int_{-0.5}^{0.5} f(\alpha,\beta)\, d\alpha\, d\beta$ | 1-point Integration |
|---|---|---|---|
| F[3,15] | $\beta^2$ | $\frac{1}{12}$ | 0 |
| F[4,13] | $\frac{1}{2}(1-\nu)\alpha^2$ | $\frac{1}{24}(1-\nu)$ | 0 |
| F[7,7] | $4\alpha^2$ | $\frac{1}{3}$ | 0 |
| F[8,18] | $2(1-\nu)\alpha^2\beta^2$ | $\frac{1}{72}(1-\nu)$ | 0 |
| F[9,9] | $\beta^2 + \frac{1}{2}(1-\nu)\alpha^2$ | $\frac{1}{24}(3-\nu)$ | 0 |
| F[15,15] | $\beta^4 + 2(1-\nu)\alpha^2\beta^2$ | $\frac{1}{720}(19-10\nu)$ | 0 |

Table C.1: Numerical Approximation of the Stiffness Matrix Coefficients.

Clearly, table C.1 shows the effect of using a low-order integration scheme. The largest value appearing within the stiffness matrix is unity. Using a one-point scheme approximates a large number of the stiffness matrix coefficients with a zero value yet there exact values are comparable to unity. Numerically, the example presented in section 4.1.3 highlights the combined effect of making such an assumption.

## Appendix D

## THE APPLICATION OF THE CFD METHOD TO THE ANALYSIS OF BEAM PROBLEMS WITH FIXED EDGES

Neglecting the effects of axial deformation, the stress and strain equations governing the behaviour of one dimensional beam problems may be written in the form.

$$\varepsilon_x = \frac{1}{h^2}\frac{d^2w}{d\alpha^2}z \tag{D.1}$$

$$\sigma_x = \frac{E}{h^2}\frac{d^2w}{d\alpha^2}z \tag{D.2}$$

where w denotes the beams normal displacement and h is the spacing (assumed constant) between adjacent nodes.

Applying the principle of virtual displacements in discretized form, the stiffness matrix and load vector associated with each mesh may be computed from the equations

$$[K]_i = \frac{E\,I}{h^3}\int_0^1 \frac{d^2w}{d\alpha^2}\frac{d^2w}{d\alpha^2}\,d\alpha \tag{D.3}$$

$$[P]_i = q\,h\int_0^1 w\,d\alpha \tag{D.4}$$

where E denotes Young's Modulus, I the moment of inertia and q the uniformly distributed load acting on the beam.

For the four node mesh illustrated in figure D.1, the displacement, w, may be approximated by the function

$$w = \begin{bmatrix} \frac{1}{6}(-2\alpha + 3\alpha^2 - \alpha^3) \\ \frac{1}{2}(2 - \alpha - 2\alpha^2 + \alpha^3) \\ \frac{1}{2}(2\alpha + \alpha^2 - \alpha^3) \\ \frac{1}{6}(-\alpha + \alpha^3) \end{bmatrix}^t \begin{bmatrix} w_1 \\ w_2 \\ w_3 \\ w_4 \end{bmatrix} \tag{D.5}$$

where $w_1, w_2, \ldots, w_4$ are the unknown displacement components at each node. Differentiating the above expression with respect to the local coordinate $\alpha$ and substituting into equations D.3 and D.4 allows the stiffness and force coefficients for an internal mesh to be evaluated as

$$[K]_i = \frac{E\,I}{h^3} \begin{bmatrix} \frac{1}{3} & -\frac{1}{2} & 0 & \frac{1}{6} \\ -\frac{1}{2} & 1 & -\frac{1}{2} & 0 \\ 0 & -\frac{1}{2} & 1 & -\frac{1}{2} \\ \frac{1}{6} & -\frac{1}{2} & 0 & \frac{1}{3} \end{bmatrix} \tag{D.6}$$

$$[P]_i = qh \begin{bmatrix} -\frac{1}{24} \\ \frac{13}{24} \\ \frac{13}{24} \\ -\frac{1}{24} \end{bmatrix} \tag{D.7}$$

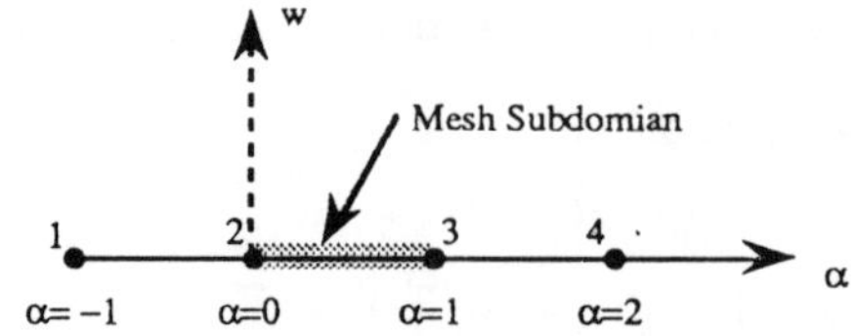


Figure D1: Four Node Mesh Configuration.

Similarly, by incorporating the fixed edge boundary conditions (i.e $\partial w/\partial\alpha = 0$ at $\alpha = 0$ and $\alpha = 1$) into the displacement function the following stiffness matrices and load vectors can be derived.

Boundary Scheme 1

Left Boundary Constraint

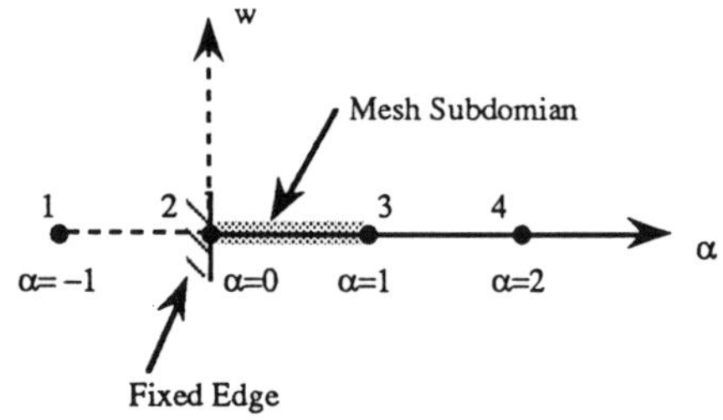


Right Boundary Constraint

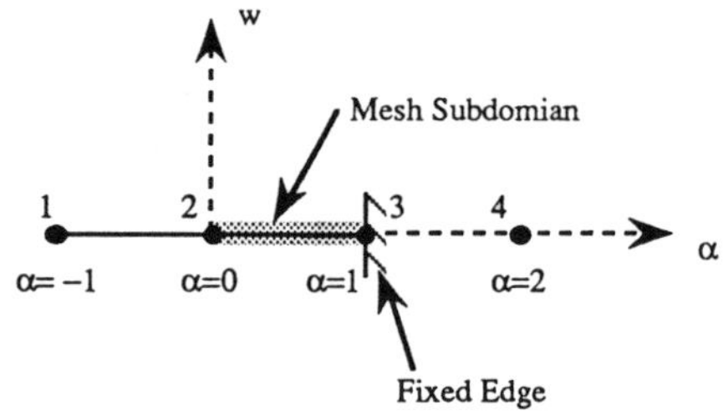


Constraint Equations:

$$\frac{\partial w}{\partial \alpha}\Big|_{\alpha=0} = 0 \qquad\qquad \frac{\partial w}{\partial \alpha}\Big|_{\alpha=1} = 0$$

$$w_1 = -\frac{3}{2}w_2 + 3w_3 - \frac{1}{2}w_4 \qquad\qquad w_4 = -\frac{1}{2}w_1 + 3w_2 - \frac{3}{2}w_3$$

(D.8)

Modified Displacement Approximations:

$$w = \begin{bmatrix} 0 \\ (1 - \frac{7}{4}\alpha^2 + \frac{3}{4}\alpha^3) \\ 2\alpha^2 - \alpha^3 \\ \frac{1}{4}(-\alpha^2 + \alpha^3) \end{bmatrix}^t \begin{bmatrix} w_1 \\ w_2 \\ w_3 \\ w_4 \end{bmatrix} \qquad w = \begin{bmatrix} -\frac{1}{4}\alpha + \frac{1}{2}\alpha^2 - \frac{1}{4}\alpha^3 \\ 1 - \alpha - \alpha^2 + \alpha^3 \\ \frac{5}{4}\alpha + \frac{1}{2}\alpha^2 - \frac{3}{4}\alpha^3 \\ 0 \end{bmatrix}^t \begin{bmatrix} w_1 \\ w_2 \\ w_3 \\ w_4 \end{bmatrix}$$

(D.9)

Stiffness Matrices:

$$[K]_i = \frac{E\,I}{h^3}\begin{bmatrix} 0 & 0 & 0 & 0 \\ 0 & \frac{13}{4} & -\frac{7}{2} & \frac{1}{4} \\ 0 & -\frac{7}{2} & 4 & -\frac{1}{2} \\ 0 & \frac{1}{4} & -\frac{1}{2} & \frac{1}{4} \end{bmatrix} \qquad [K]_i = \frac{E\,I}{h^3}\begin{bmatrix} \frac{1}{4} & -\frac{1}{2} & \frac{1}{4} & 0 \\ -\frac{1}{2} & 4 & -\frac{7}{2} & 0 \\ \frac{1}{4} & -\frac{7}{2} & \frac{13}{4} & 0 \\ 0 & 0 & 0 & 0 \end{bmatrix}$$

(D.10)

Load Vectors:

$$[P]_i = qh\begin{bmatrix} 0 \\ \frac{29}{48} \\ \frac{5}{12} \\ -\frac{1}{48} \end{bmatrix} \qquad [P]_i = qh\begin{bmatrix} -\frac{1}{48} \\ \frac{5}{12} \\ \frac{29}{48} \\ 0 \end{bmatrix}$$

(D.11)

In subsequent examples, modelling a fixed edge boundary condition by applying the constraint $\partial w/\partial\alpha = 0$ to the displacement function will be referred to as boundary scheme 1. As illustrated in chapter 4 and in the numerical results presented in this chapter, the above system of equations do not correctly model a fixed edge. With increasing mesh refinement, the deflected shape and bending stresses deteriorate in accuracy. In particular, at nodal points adjacent to the boundary an apparent discontinuity in bending stress exists (see figures D.3 and D.4).

As an alternative to directly setting the derivative $\partial w / \partial \alpha = 0$, the fixed edge was modeled by approximating the normal slope using a lower order displacement function. Thus, the boundary constraint equations are formed using a second order approximation. The stiffness matrices and load vectors are then evaluated using the full third order displacement function as a basis. In this appendix, the formulation derived below is referred to as boundary scheme 2.

Boundary Scheme 2

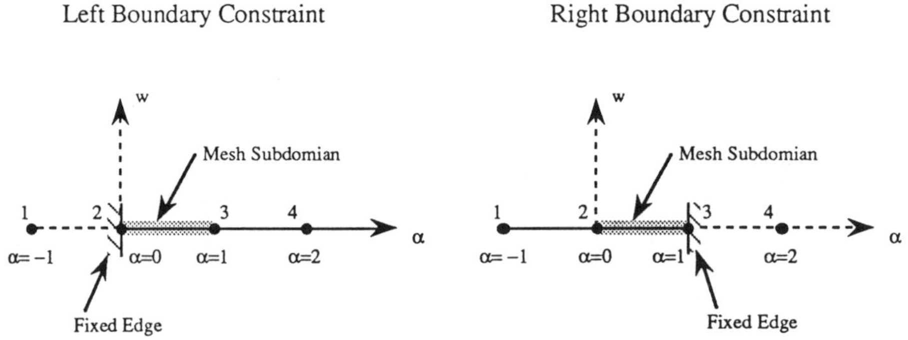


Constraint Equations:

$$\frac{\partial w}{\partial \alpha}\Big|_{\alpha=0} = 0 \qquad\qquad \frac{\partial w}{\partial \alpha}\Big|_{\alpha=1} = 0$$

$$w_1 = w_3 \qquad\qquad w_4 = w_2$$

(D.12)

Modified Displacement Approximations:

$$w = \begin{bmatrix} 0 \\ \frac{1}{2}(2 - \alpha - 2\alpha^2 + \alpha^3) \\ \frac{1}{3}(2\alpha + 3\alpha^2 - 2\alpha^3) \\ \frac{1}{6}(-\alpha + \alpha^3) \end{bmatrix}^t \begin{bmatrix} w_1 \\ w_2 \\ w_3 \\ w_4 \end{bmatrix} \qquad w = \begin{bmatrix} \frac{1}{6}(-2\alpha + 3\alpha^2 - \alpha^3) \\ \frac{1}{3}(3 - 2\alpha - 3\alpha^2 + 2\alpha^3) \\ \frac{1}{2}(2\alpha + \alpha^2 - \alpha^3) \\ 0 \end{bmatrix}^t \begin{bmatrix} w_1 \\ w_2 \\ w_3 \\ w_4 \end{bmatrix}$$

(D.13)

Stiffness Matrices:

$$[K]_i = \frac{E\ I}{h^3} \begin{bmatrix} 0 & 0 & 0 & 0 \\ 0 & 1 & -1 & 0 \\ 0 & -1 & \frac{4}{3} & -\frac{1}{3} \\ 0 & 0 & -\frac{1}{3} & \frac{1}{3} \end{bmatrix} \qquad [K]_i = \frac{E\ I}{h^3} \begin{bmatrix} \frac{1}{3} & -\frac{1}{3} & 0 & 0 \\ -\frac{1}{3} & \frac{4}{3} & -1 & 0 \\ 0 & -1 & 1 & 0 \\ 0 & 0 & 0 & 0 \end{bmatrix}$$

(D.14)

Load Vectors:

$$[P]_i = qh \begin{bmatrix} 0 \\ \frac{13}{24} \\ \frac{1}{2} \\ -\frac{1}{24} \end{bmatrix} \qquad [P]_i = qh \begin{bmatrix} -\frac{1}{24} \\ \frac{1}{2} \\ \frac{13}{24} \\ 0 \end{bmatrix}$$

(D.11)

A beam fixed at both ends and subjected to a uniformly distributed load was analysed using the formulations derived. The assumed dimensions and geometric properties are illustrated in figure D2. The deflections and bending moments at the centre and edge of the beam are presented in table D1 for various mesh discretizations and compared to the exact solutions.

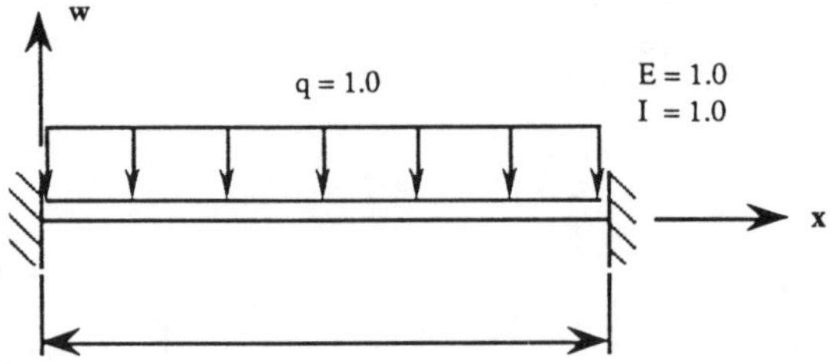


Figure D2: Beam with Fixed Boundary Conditions at Both Ends.

| | Boundary Scheme 1 | | | Boundary Scheme 2 | | |
|---|---|---|---|---|---|---|
| Mesh | w ( 0.5) | $M_x$ (0.0) | $M_x$ (0.5) | w ( 0.5) | $M_x$ (0.0) | $M_x$ (0.5) |
| 4 | 0.003624 | 0.079206 | -0.061861 | 0.005859 | 0.140625 | -0.046875 |
| 8 | 0.002972 | 0.054509 | -0.046660 | 0.003348 | 0.115513 | -0.044201 |
| 16 | 0.002806 | 0.039136 | -0.043615 | 0.002779 | 0.100399 | -0.042642 |
| 32 | 0.002717 | 0.030905 | -0.042663 | 0.002646 | 0.092110 | -0.041911 |
| 64 | 0.002664 | 0.026659 | -0.042174 | 0.002615 | 0.087783 | -0.041728 |
| 128 | 0.002635 | 0.024504 | -0.041923 | 0.002607 | 0.085573 | -0.041682 |
| 256 | 0.002620 | 0.023419 | -0.041796 | 0.002604 | 0.084457 | -0.041670 |
| 512 | 0.002612 | 0.022875 | -0.041731 | 0.002604 | 0.083896 | -0.041668 |
| 1024 | 0.002608 | 0.022602 | -0.041700 | 0.002604 | 0.083615 | -0.041667 |
| Exact | 0.002604 | 0.083333 | -0.041667 | 0.002604 | 0.083333 | -0.041667 |

Table D.1: Deflections and Bending Moments for a Beam Fixed at Both Ends.

Clearly, table D.1 indicates the problems associated with approximating a fixed edge with boundary scheme 1. Despite good correlation with the exact solution at mid-span, the bending moment at the support is converging to the wrong value. In contrast, the second approach (scheme 2) has produced results which compare favourably with the analytical solution.

As highlighted in figures D.3 and D.4 the use of scheme 1 introduces discontinuities in the bending moment diagram adjacent to the support. These effects are local and appear to only marginally effect the beams behaviour at mid-span.

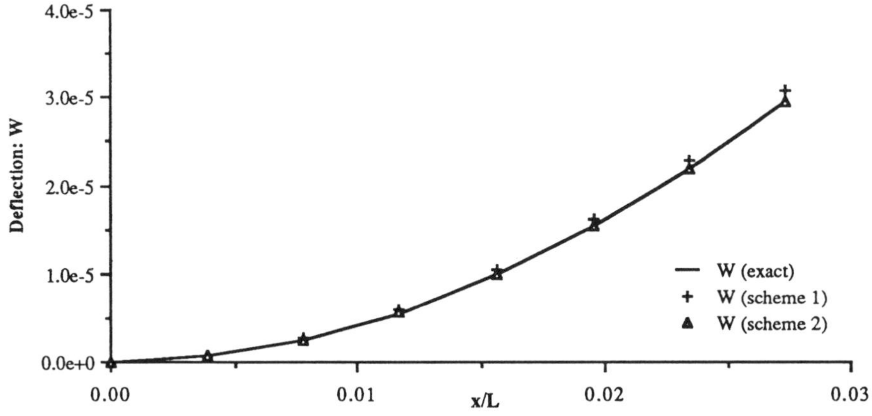


Figure D.3: Deflected Shape Adjacent to the Beams Edge (257 nodes).

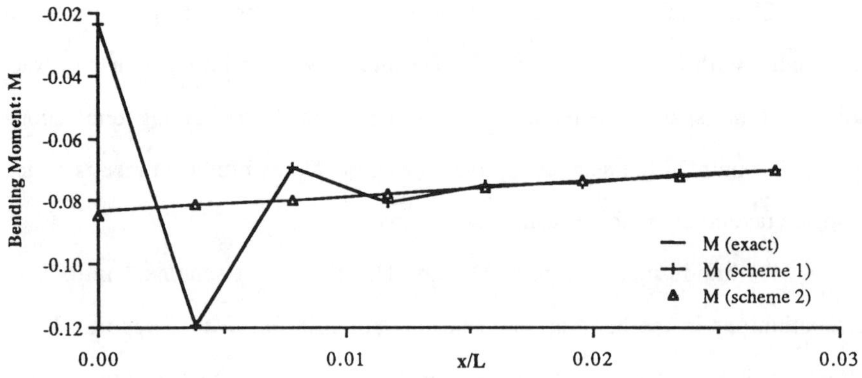


Figure D.4: Bending Stresses Adjacent to the Beams Edge (257 nodes).

To gain an understanding into why such a problem occurs, the variation in slope within a boundary mesh is considered. Figure D.5 illustrates the results obtained from a 257 node simulation and compares them with the analytical solution. Notably, two zero slope conditions exist within the mesh when the constraint associated with scheme 1 is applied. In contrast, the analytical solution has only one zero slope condition within the same domain. Numerically, this difference has the effect of introducing a 'wave' into the displacement function between the external fictitious node and the boundary node. Consequently, a local disturbance as illustrated in figure D.4 will be evident as the function fails to sufficiently approximate the beams deformation.

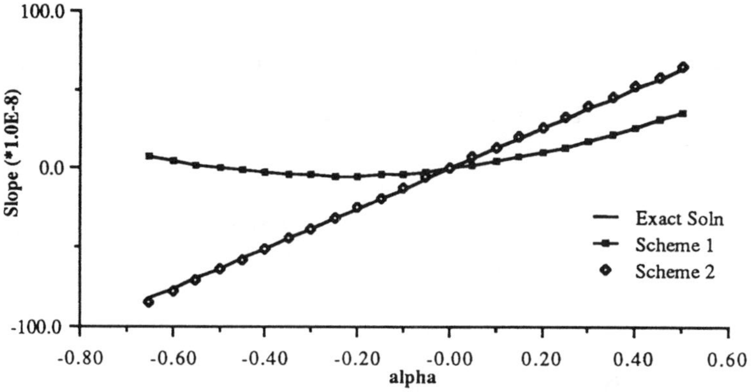


Figure D.5: Variation of Slope Within an Edge Mesh.

# Appendix E

# TRANSFORMATION OF THE GENERALIZED EIGENVALUE PROBLEM TO STANDARD FORM

For a shell structure undergoing free vibration, the governing equations may be written in the general form

$$[K][\phi] - \lambda [M] [\phi] = [0] \qquad \text{(E.1)}$$

where [K] and [M] are the stiffness and mass matrices respectively.

The purpose of this appendix is to detail the stages required in reducing the above system of equations to *standard form*.

i.e.

$$[\mathcal{K}] [\psi] - \lambda[\psi] = [0] \qquad \text{(E.2)}$$

This form enables the eigenvalue problem to be solved using such techniques as the Jacobian iteration method [86].

If the mass matrix [M] is symmetric and positive definite, then using Cholesky factorization, it can be decomposed into the form

$$[M] = [L] [L]^t \qquad \text{(E.3)}$$

where [L] is a unique lower triangular matrix with positive diagonal elements. Substituting equation (E.3) into (E.1) and pre-multiplying both sides by $[L]^{-1}$ yields

$$([L]^{-1} [K] - \lambda [L]^t)[\phi] = [0] \qquad \text{(E.4)}$$

Defining the matrix

$$[\psi] \;=\; [L]^t\,[\phi] \qquad \text{(E.5)}$$

and substituting it into equation (E.4) results in the expression

$$([L]^{-1}\,[K]\,[L^t]^{-1} \;-\; \lambda\,[L^t]\,[L^t]^{-1})[\psi] \;=\; [0] \qquad \text{(E.6)}$$

which may be written in the standard form

$$([\mathcal{K}] - \lambda\,[I])[\psi] \;=\; [0] \qquad \text{(E.7)}$$

where $$[\mathcal{K}] \;=\; [L]^{-1}[K]\,[L^t]^{-1} \qquad \text{(E.8)}$$

Solving the above set of equations yields the system's eigenvalues and by making use of equation (E.5) their associated eigenvectors.

# Appendix F

## NUMERICAL RESULTS

In this appendix, numerical results relating to the dynamic characteristics of the shell structures discussed in chapter 5 are presented in detail. The computed eigenvalues and eigenvectors were generated using a computer program developed under a Unix environment on an IBM RT (model 6151) workstation. The computer was configured with 12 megabytes of random access memory, a 720×512 monochrome graphics display, a twin button mouse and uses the 032 reduced instruction set central processing unit.

The cross-sections and perspective views presented in the numerical examples are screen dumps taken directly from a pre and post processing program written specifically for analysing the numerical results. The user interface part of this program is based on the interactive mouse, menu, and window environment commonly found on many computing systems. Both the analysis and processing programs were written entirely in the C programming language using standard C libraries and the low level graphics routines of GSL in the AIX operating system.

The numerical results presented in this appendix contain information primarily related to a structure's eigenvalues and eigenvectors. Typically, details associated with the first six modes of vibration are presented. For each eigenvector, normalized displacement components are tabulated along selected cross-sections at positions relative to the structures origin. Where the nodal positions have not corresponded to the required locations, the displacement values are interpolated using a second-order approximation.

The process of normalizing eigenvectors involves selecting the maximum tabulated displacement component associated with one of the cross-sections as a basis. The remaining displacements are expressed relative to this value. Similarly, displacement positions are defined relative to an appropriate structural dimension or in terms of angular divisions.

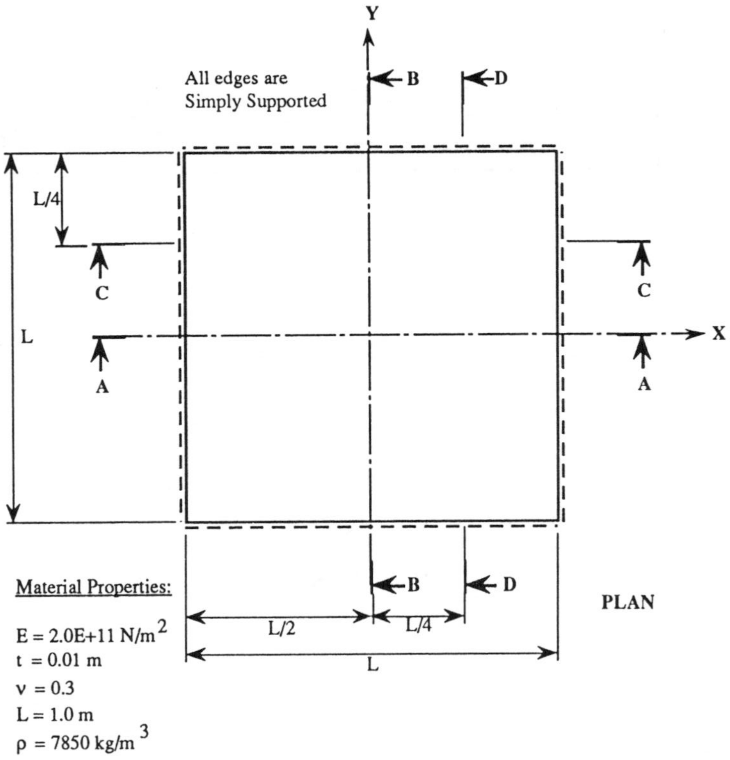


Figure F.1a: Simply Supported Plate.

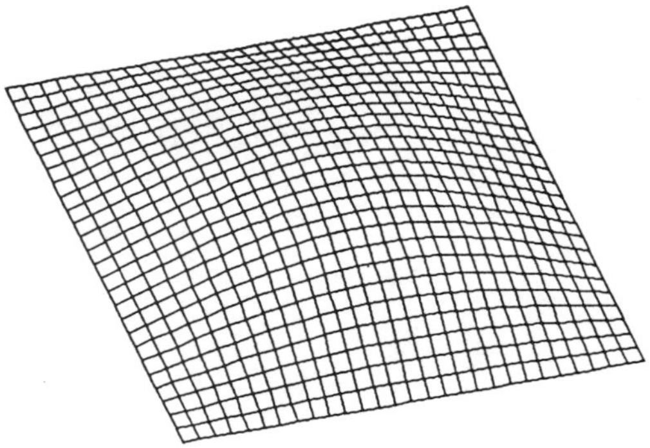

Perspective

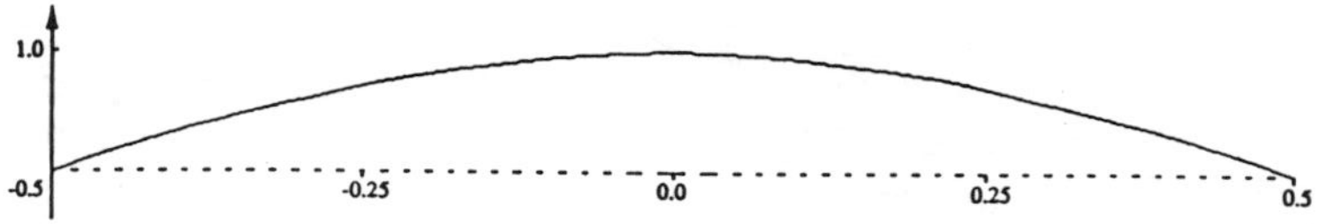


Section A-A

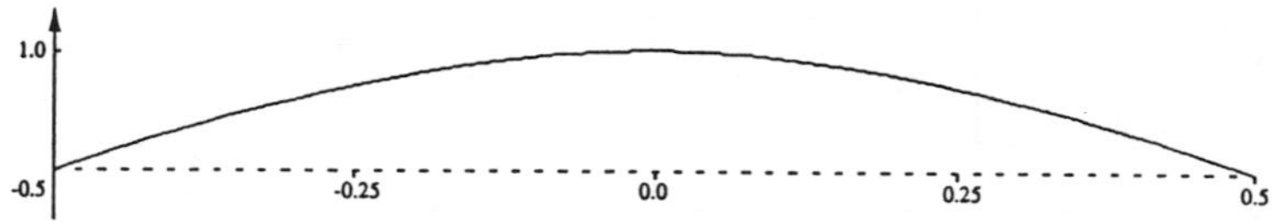


Section B-B

**Normalized Mode Characteristics:**

1st Mode: $\lambda = 19.71$

Normal Displacement Posn (x,y,z): (0.0,0.0,0.0)

| Position | | -0.5 | -0.375 | -0.25 | -0.125 | 0.0 | 0.125 | 0.25 | 0.375 | 0.5 |
|---|---|---|---|---|---|---|---|---|---|---|
| A-A | Uz | 0.000 | 0.383 | 0.707 | 0.924 | 1.000 | 0.924 | 0.707 | 0.383 | 0.000 |
| | Exact | 0.000 | 0.383 | 0.707 | 0.924 | 1.000 | 0.924 | 0.707 | 0.383 | 0.000 |

| Position | | -0.5 | -0.375 | -0.25 | -0.125 | 0.0 | 0.125 | 0.25 | 0.375 | 0.5 |
|---|---|---|---|---|---|---|---|---|---|---|
| B-B | Uz | 0.000 | 0.383 | 0.707 | 0.924 | 1.000 | 0.924 | 0.707 | 0.383 | 0.000 |
| | Exact | 0.000 | 0.383 | 0.707 | 0.924 | 1.000 | 0.924 | 0.707 | 0.383 | 0.000 |

Figure F.1b : Simply Supported Plate: 1st Mode.

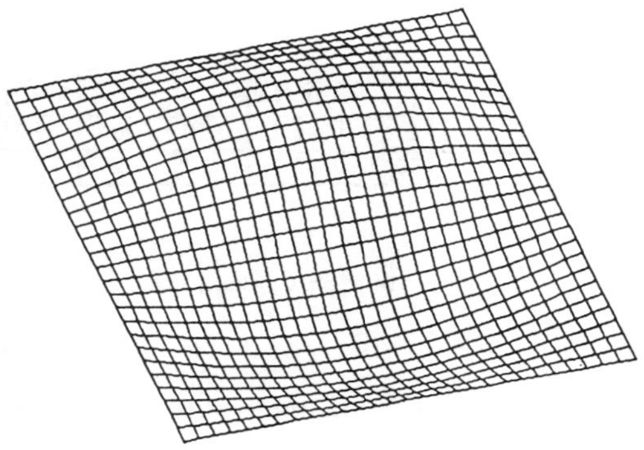

**Perspective**

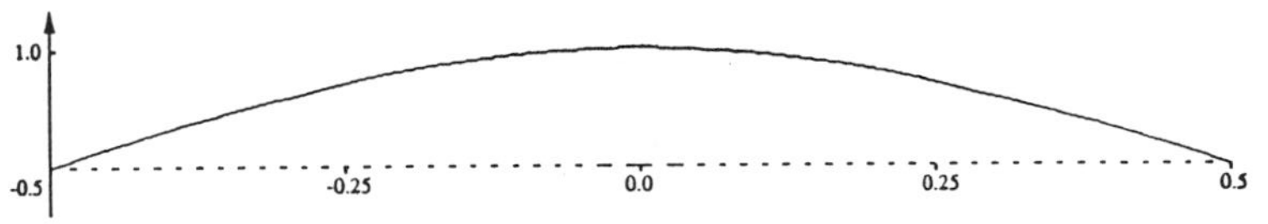


**Section C-C**

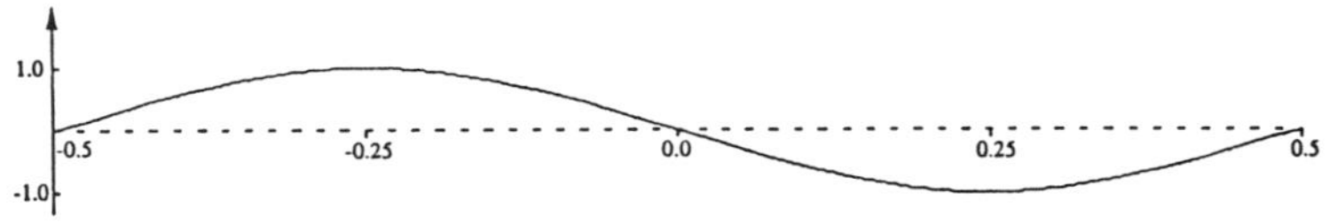


**Section B-B**

## Normalized Mode Characteristics:

2nd Mode: $\lambda = 49.00$

Normal Displacement Posn (x,y,z): (0.0,0.25,0.0)

| Position | | -0.5 | -0.375 | -0.25 | -0.125 | 0.0 | 0.125 | 0.25 | 0.375 | 0.5 |
|---|---|---|---|---|---|---|---|---|---|---|
| C-C | Uz | 0.000 | 0.383 | 0.707 | 0.924 | 1.000 | 0.924 | 0.707 | 0.383 | 0.000 |
| | Exact | 0.000 | 0.383 | 0.707 | 0.924 | 1.000 | 0.924 | 0.707 | 0.383 | 0.000 |

| Position | | -0.5 | -0.375 | -0.25 | -0.125 | 0.0 | 0.125 | 0.25 | 0.375 | 0.5 |
|---|---|---|---|---|---|---|---|---|---|---|
| B-B | Uz | 0.000 | 0.707 | 1.000 | 0.707 | 0.000 | -0.707 | -1.000 | -0.707 | 0.000 |
| | Exact | 0.000 | 0.707 | 1.000 | 0.707 | 0.000 | -0.707 | -1.000 | -0.707 | 0.000 |

Figure F.1c : Simply Supported Plate: 2nd Mode.

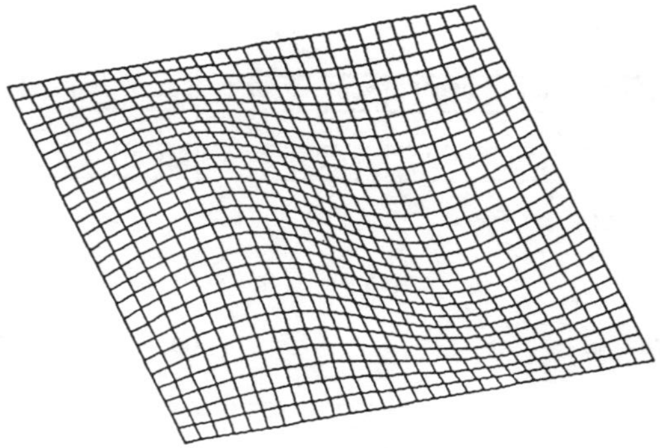

Perspective

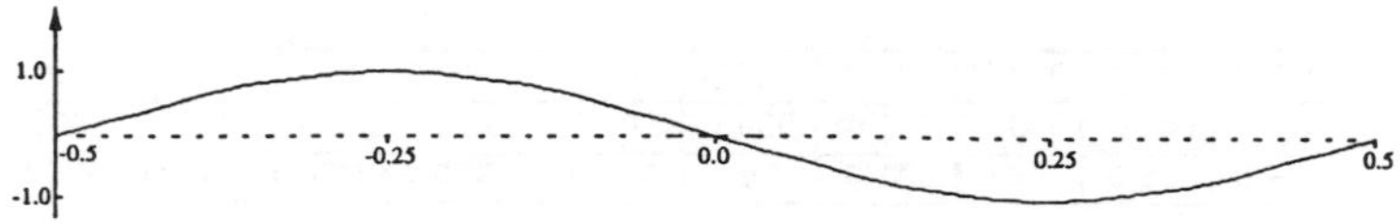


Section A-A

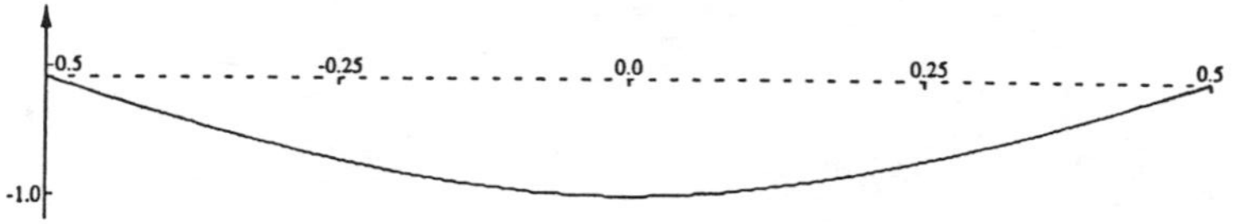


Section D-D

## Normalized Mode Characteristics:

3rd Mode: $\lambda = 49.04$

Normal Displacement Posn (x,y,z): (0.25,0.0,0.0)

| Position | | -0.5 | -0.375 | -0.25 | -0.125 | 0.0 | 0.125 | 0.25 | 0.375 | 0.5 |
|---|---|---|---|---|---|---|---|---|---|---|
| A-A | Uz | 0.000 | 0.707 | 1.000 | 0.707 | 0.000 | -0.707 | -1.000 | -0.707 | 0.000 |
| | Exact | 0.000 | 0.707 | 1.000 | 0.707 | 0.000 | -0.707 | -1.000 | -0.707 | 0.000 |

| Position | | -0.5 | -0.375 | -0.25 | -0.125 | 0.0 | 0.125 | 0.25 | 0.375 | 0.5 |
|---|---|---|---|---|---|---|---|---|---|---|
| D-D | Uz | 0.000 | -0.383 | -0.707 | -0.924 | -1.000 | -0.924 | -0.707 | -0.383 | 0.000 |
| | Exact | 0.000 | -0.383 | -0.707 | -0.924 | -1.000 | -0.924 | -0.707 | -0.383 | 0.000 |

Figure F.1d : Simply Supported Plate: 3rd Mode.

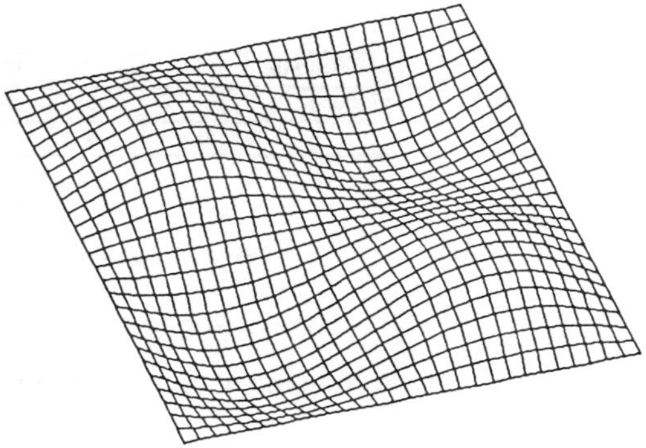

Perspective

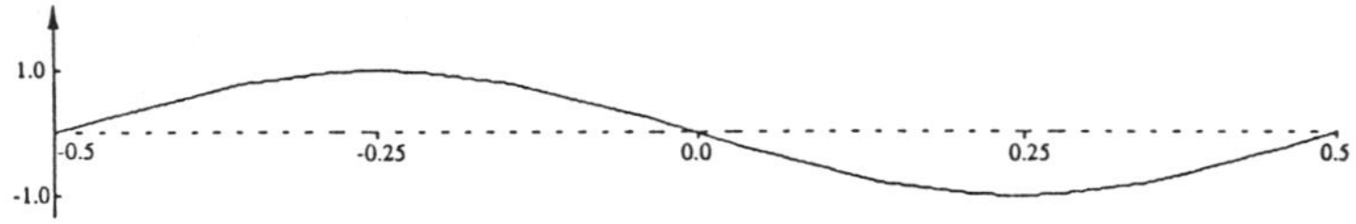


Section C-C

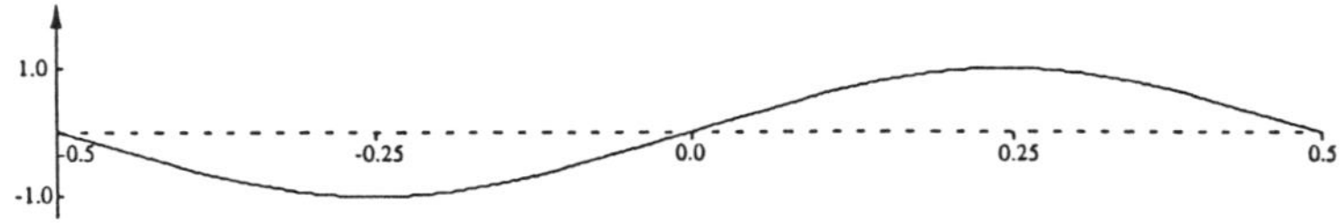


Section D-D

**Normalized Mode Characteristics:**

4th Mode: $\lambda = 78.47$

Normal Displacement Posn (x,y,z): (-0.25,0.25,0.0)

| Position | | -0.5 | -0.375 | -0.25 | -0.125 | 0.0 | 0.125 | 0.25 | 0.375 | 0.5 |
|---|---|---|---|---|---|---|---|---|---|---|
| C-C | Uz | 0.000 | 0.707 | 1.000 | 0.707 | 0.000 | -0.707 | -1.000 | -0.707 | 0.000 |
| | Exact | 0.000 | 0.707 | 1.000 | 0.707 | 0.000 | -0.707 | -1.000 | -0.707 | 0.000 |

| Position | | -0.5 | -0.375 | -0.25 | -0.125 | 0.0 | 0.125 | 0.25 | 0.375 | 0.5 |
|---|---|---|---|---|---|---|---|---|---|---|
| D-D | Uz | 0.000 | -0.707 | -1.000 | -0.707 | 0.000 | 0.707 | 1.000 | 0.707 | 0.000 |
| | Exact | 0.000 | -0.707 | -1.000 | -0.707 | 0.000 | 0.707 | 1.000 | 0.707 | 0.000 |

Figure F.1e : Simply Supported Plate: 4th Mode.

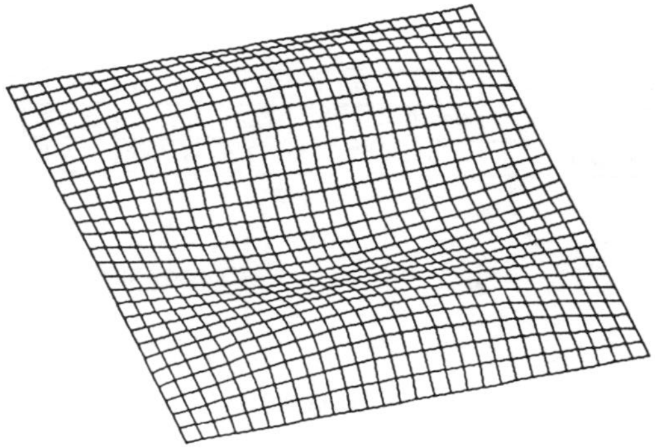

Perspective

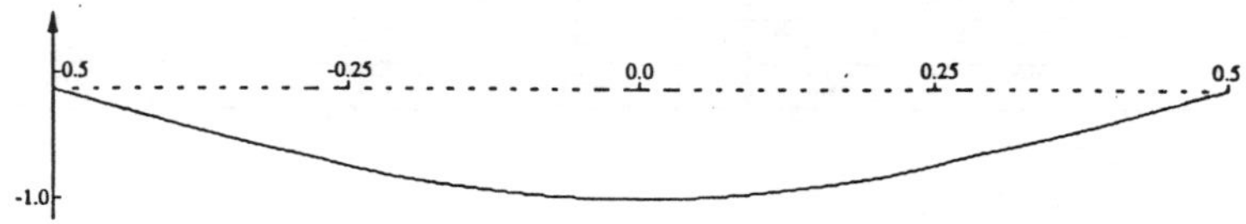


Section A-A

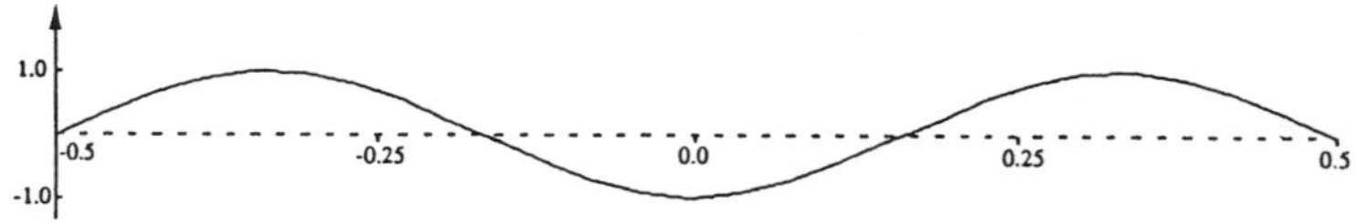


Section B-B

**Normalized Mode Characteristics:**

5th Mode: $\lambda$ = 96.88

Normal Displacement Posn (x,y,z): (0.0,0.0,0.0)

| Position | | -0.5 | -0.375 | -0.25 | -0.125 | 0.0 | 0.125 | 0.25 | 0.375 | 0.5 |
|---|---|---|---|---|---|---|---|---|---|---|
| A-A | Uz | 0.000 | -0.383 | -0.707 | -0.924 | -1.000 | -0.924 | -0.707 | -0.383 | 0.000 |
| | Exact | 0.000 | -0.383 | -0.707 | -0.924 | -1.000 | -0.924 | -0.707 | -0.383 | 0.000 |

| Position | | -0.5 | -0.375 | -0.25 | -0.125 | 0.0 | 0.125 | 0.25 | 0.375 | 0.5 |
|---|---|---|---|---|---|---|---|---|---|---|
| B-B | Uz | 0.000 | 0.923 | 0.709 | -0.384 | -1.000 | -0.384 | 0.709 | 0.923 | 0.000 |
| | Exact | 0.000 | 0.924 | 0.707 | -0.383 | -1.000 | -0.383 | 0.707 | 0.924 | 0.000 |

Figure F.1f : Simply Supported Plate: 5th Mode.

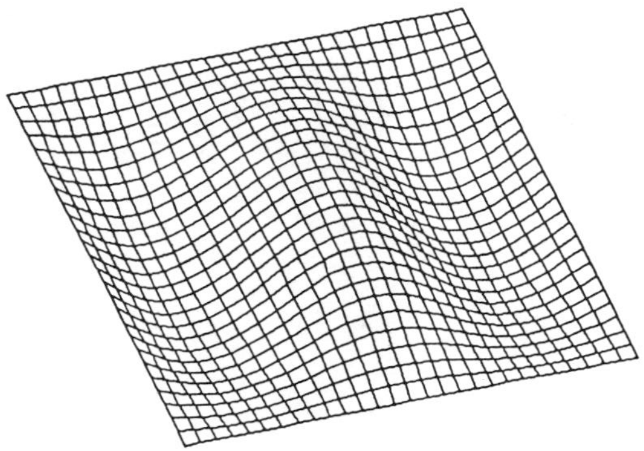

Perspective

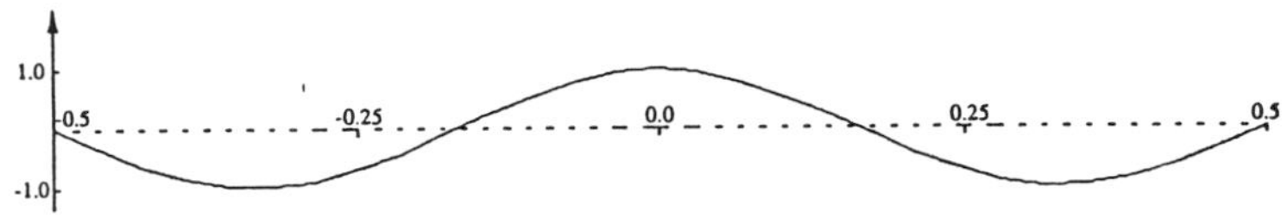


Section A-A

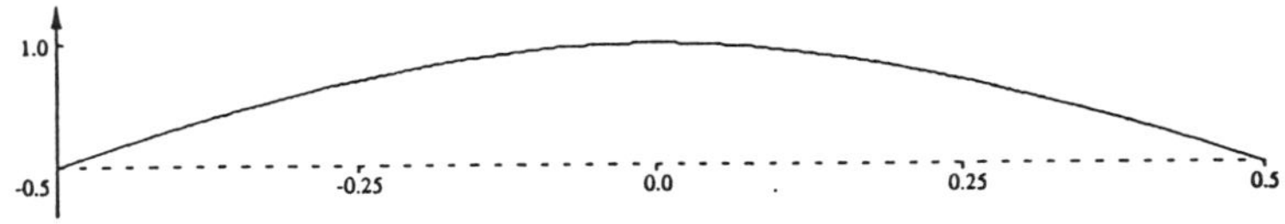


Section B-B

**Normalized Mode Characteristics:**

6th Mode: $\lambda = 97.12$

Normal Displacement Posn (x,y,z): (0.0,0.0,0.0)

| Position | | -0.5 | -0.375 | -0.25 | -0.125 | 0.0 | 0.125 | 0.25 | 0.375 | 0.5 |
|---|---|---|---|---|---|---|---|---|---|---|
| A-A | Uz | 0.000 | -0.923 | -0.707 | 0.385 | 1.000 | 0.385 | -0.707 | -0.923 | 0.000 |
| | Exact | 0.000 | -0.924 | -0.707 | 0.383 | 1.000 | 0.383 | -0.707 | -0.924 | 0.000 |

| Position | | -0.5 | -0.375 | -0.25 | -0.125 | 0.0 | 0.125 | 0.25 | 0.375 | 0.5 |
|---|---|---|---|---|---|---|---|---|---|---|
| B-B | Uz | 0.000 | 0.383 | 0.707 | 0.924 | 1.000 | 0.924 | 0.707 | 0.383 | 0.000 |
| | Exact | 0.000 | 0.383 | 0.707 | 0.924 | 1.000 | 0.924 | 0.707 | 0.383 | 0.000 |

Figure F.1g : Simply Supported Plate: 6th Mode.

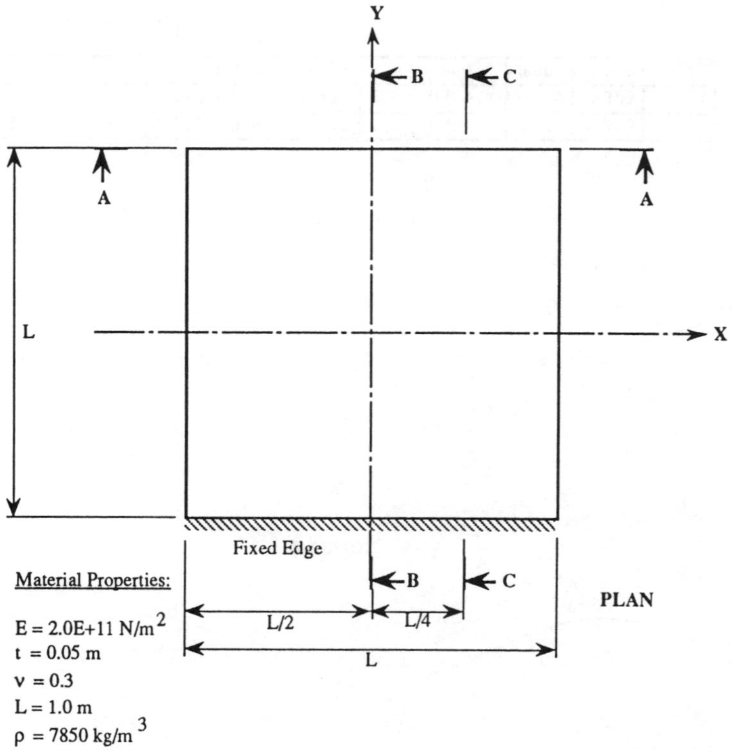


Figure F.2a: Cantilever Plate.

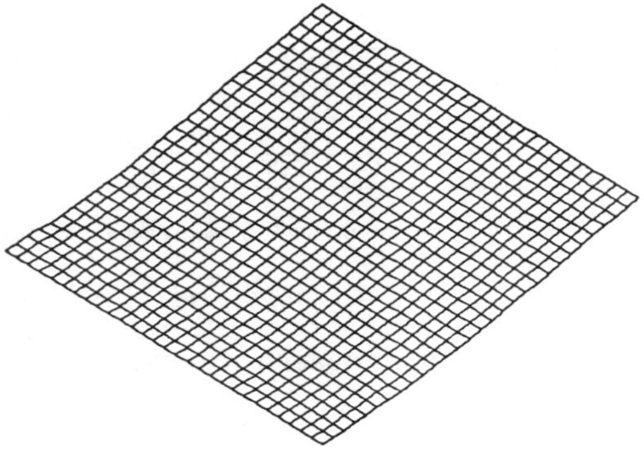

Perspective

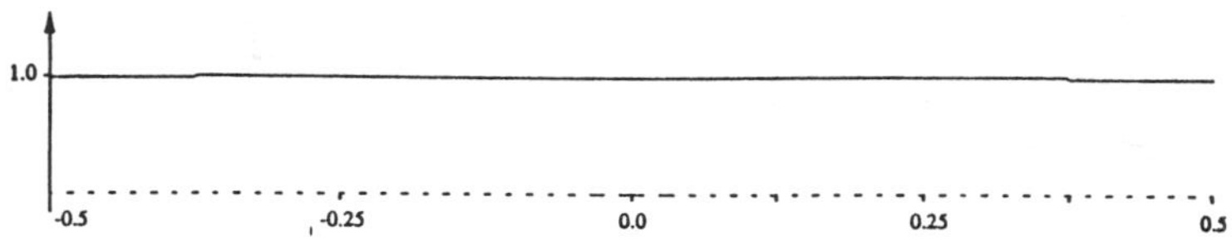


Section A-A

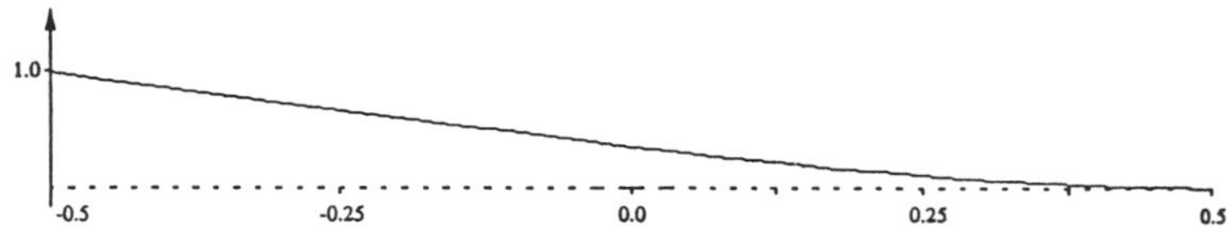


Section B-B

## Normalized Mode Characteristics:

1st Mode: $\omega$ = 264.6 rad/sec

Normal Displacement Posn (x,y): (0.0,0.5)

| Position | | -0.5 | -0.375 | -0.25 | -0.125 | 0.0 | 0.125 | 0.25 | 0.375 | 0.5 |
|---|---|---|---|---|---|---|---|---|---|---|
| A-A | Ux | 0.000 | 0.000 | 0.000 | 0.000 | 0.000 | 0.000 | 0.000 | 0.000 | 0.000 |
| | Uy | 0.000 | 0.000 | 0.000 | 0.000 | 0.000 | 0.000 | 0.000 | 0.000 | 0.000 |
| | Uz | 0.983 | 0.989 | 0.995 | 0.999 | 1.000 | 0.999 | 0.995 | 0.989 | 0.983 |

| Position | | -0.5 | -0.375 | -0.25 | -0.125 | 0.0 | 0.125 | 0.25 | 0.375 | 0.5 |
|---|---|---|---|---|---|---|---|---|---|---|
| B-B | Ux | 0.000 | 0.000 | 0.000 | 0.000 | 0.000 | 0.000 | 0.000 | 0.000 | 0.000 |
| | Uy | 0.000 | 0.000 | 0.000 | 0.000 | 0.000 | 0.000 | 0.000 | 0.000 | 0.000 |
| | Uz | 0.000 | 0.027 | 0.099 | 0.207 | 0.341 | 0.494 | 0.658 | 0.828 | 1.000 |

Figure F.2b : Cantilever Plate: 1st Mode.

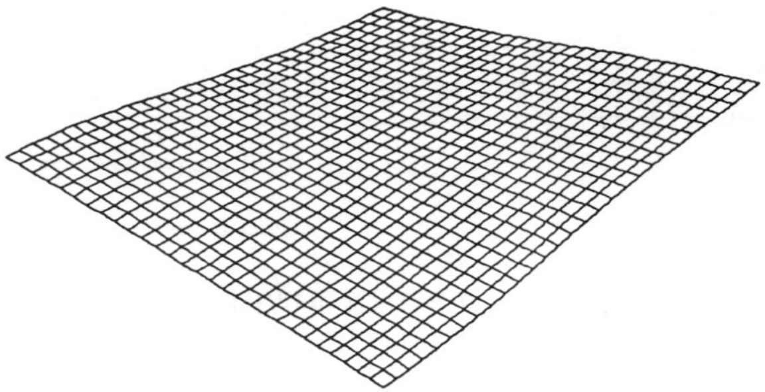

Perspective

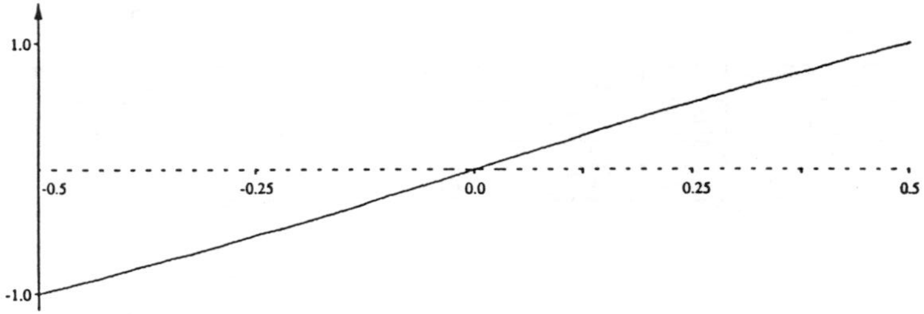


Section A-A

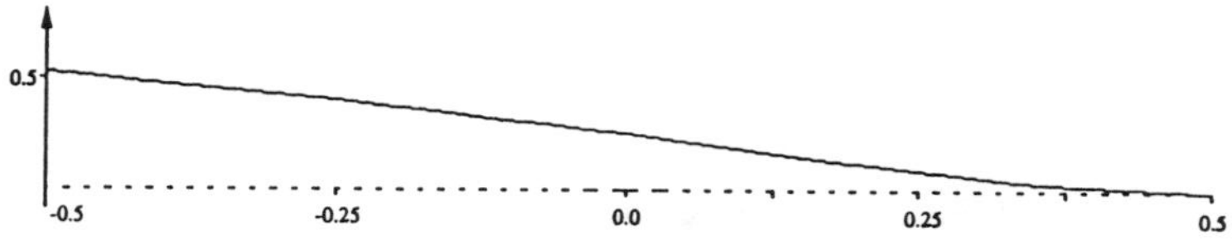


Section C-C

## Normalized Mode Characteristics:

2nd Mode: $\omega$ = 647.1 rad/sec

Normal Displacement Posn (x,y): (0.5,0.5)

| Position | | -0.5 | -0.375 | -0.25 | -0.125 | 0.0 | 0.125 | 0.25 | 0.375 | 0.5 |
|---|---|---|---|---|---|---|---|---|---|---|
| A-A | Ux | 0.000 | 0.000 | 0.000 | 0.000 | 0.000 | 0.000 | 0.000 | 0.000 | 0.000 |
| | Uy | 0.000 | 0.000 | 0.000 | 0.000 | 0.000 | 0.000 | 0.000 | 0.000 | 0.000 |
| | Uz | -1.000 | -0.774 | -0.531 | -0.271 | 0.000 | 0.271 | 0.531 | 0.774 | 1.000 |

| Position | | -0.5 | -0.375 | -0.25 | -0.125 | 0.0 | 0.125 | 0.25 | 0.375 | 0.5 |
|---|---|---|---|---|---|---|---|---|---|---|
| C-C | Ux | 0.000 | 0.000 | 0.000 | 0.000 | 0.000 | 0.000 | 0.000 | 0.000 | 0.000 |
| | Uy | 0.000 | 0.000 | 0.000 | 0.000 | 0.000 | 0.000 | 0.000 | 0.000 | 0.000 |
| | Uz | 0.000 | 0.026 | 0.088 | 0.168 | 0.252 | 0.333 | 0.406 | 0.471 | 0.531 |

Figure F.2c : Cantilever Plate: 2nd Mode.

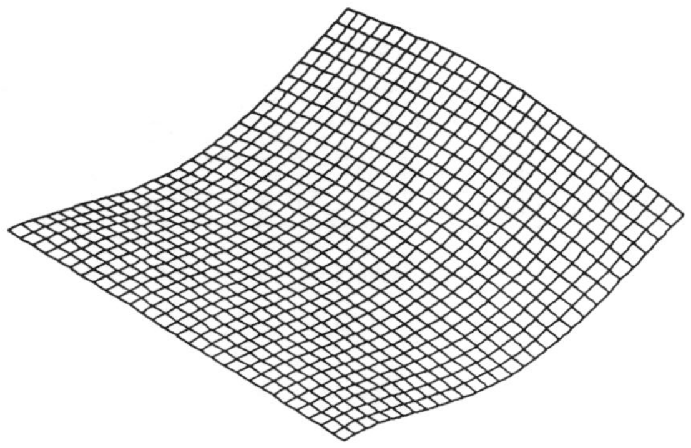

Perspective

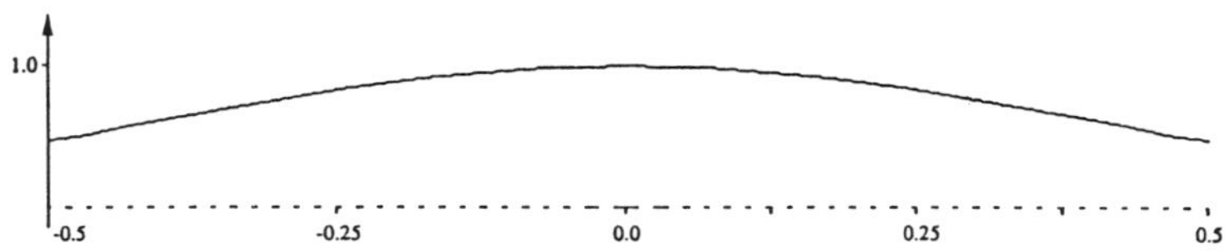


Section A-A

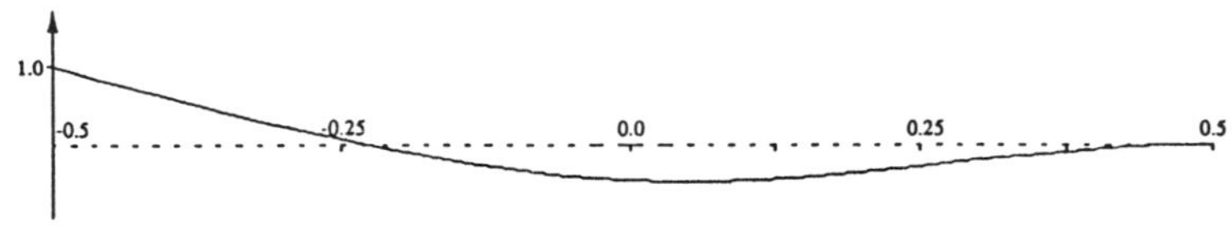


Section B-B

## Normalized Mode Characteristics:

3rd Mode: $\omega$ = 1609. rad/sec

Normal Displacement Posn (x,y): (0.0,0.5)

| Position | | -0.5 | -0.375 | -0.25 | -0.125 | 0.0 | 0.125 | 0.25 | 0.375 | 0.5 |
|---|---|---|---|---|---|---|---|---|---|---|
| A-A | Ux | 0.000 | 0.000 | 0.000 | 0.000 | 0.000 | 0.000 | 0.000 | 0.000 | 0.000 |
| | Uy | 0.000 | 0.000 | 0.000 | 0.000 | 0.000 | 0.000 | 0.000 | 0.000 | 0.000 |
| | Uz | 0.465 | 0.662 | 0.835 | 0.957 | 1.000 | 0.957 | 0.835 | 0.662 | 0.465 |

| Position | | -0.5 | -0.375 | -0.25 | -0.125 | 0.0 | 0.125 | 0.25 | 0.375 | 0.5 |
|---|---|---|---|---|---|---|---|---|---|---|
| B-B | Ux | 0.000 | 0.000 | 0.000 | 0.000 | 0.000 | 0.000 | 0.000 | 0.000 | 0.000 |
| | Uy | 0.000 | 0.000 | 0.000 | 0.000 | 0.000 | 0.000 | 0.000 | 0.000 | 0.000 |
| | Uz | 0.000 | -0.090 | -0.271 | -0.417 | -0.433 | -0.271 | 0.059 | 0.503 | 1.000 |

Figure F.2d : Cantilever Plate: 3rd Mode.

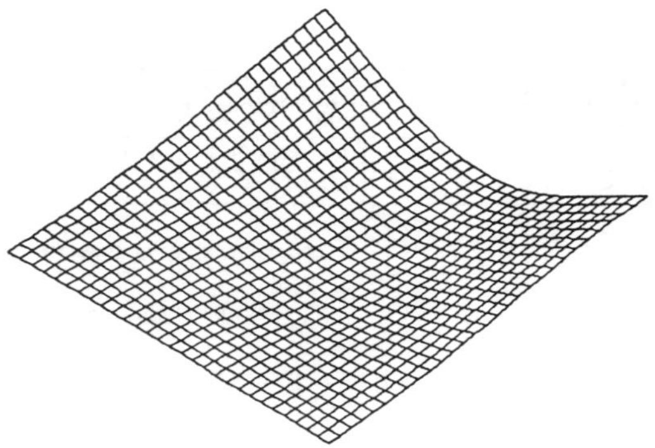

Perspective

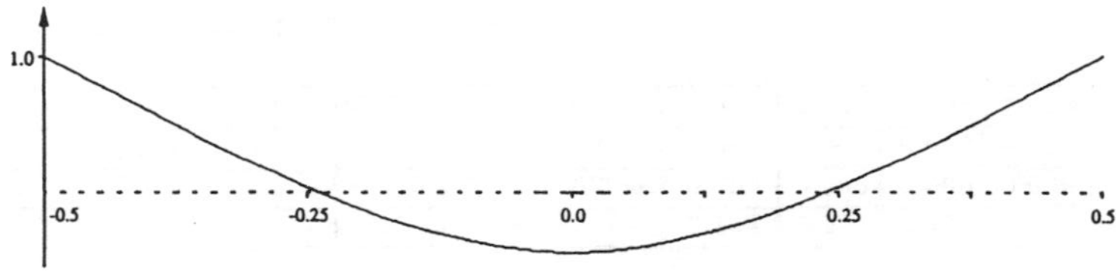


Section A-A

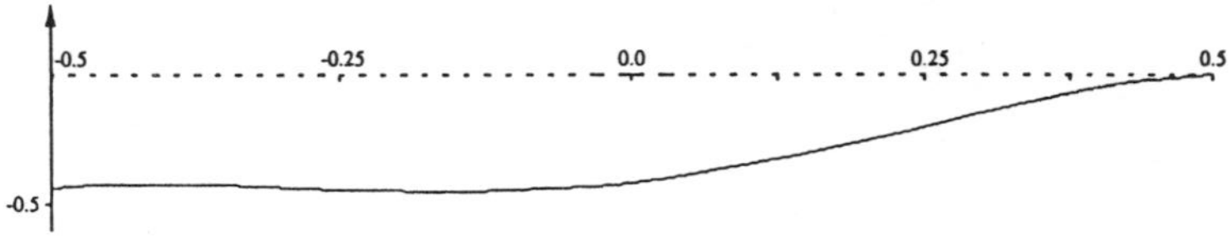


Section B-B

## Normalized Mode Characteristics:

4th Mode: $\omega$ = 2059. rad/sec

Normal Displacement Posn (x,y): (0.5,0.5)

| Position | | -0.5 | -0.375 | -0.25 | -0.125 | 0.0 | 0.125 | 0.25 | 0.375 | 0.5 |
|---|---|---|---|---|---|---|---|---|---|---|
| A-A | Ux | 0.000 | 0.000 | 0.000 | 0.000 | 0.000 | 0.000 | 0.000 | 0.000 | 0.000 |
| | Uy | 0.000 | 0.000 | 0.000 | 0.000 | 0.000 | 0.000 | 0.000 | 0.000 | 0.000 |
| | Uz | 1.000 | 0.511 | 0.040 | -0.310 | -0.441 | -0.310 | 0.040 | 0.511 | 1.000 |

| Position | | -0.5 | -0.375 | -0.25 | -0.125 | 0.0 | 0.125 | 0.25 | 0.375 | 0.5 |
|---|---|---|---|---|---|---|---|---|---|---|
| B-B | Ux | 0.000 | 0.000 | 0.000 | 0.000 | 0.000 | 0.000 | 0.000 | 0.000 | 0.000 |
| | Uy | 0.000 | 0.000 | 0.000 | 0.000 | 0.000 | 0.000 | 0.000 | 0.000 | 0.000 |
| | Uz | 0.000 | -0.070 | -0.203 | -0.332 | -0.419 | -0.453 | -0.448 | -0.431 | -0.441 |

Figure F.2e : Cantilever Plate: 4th Mode.

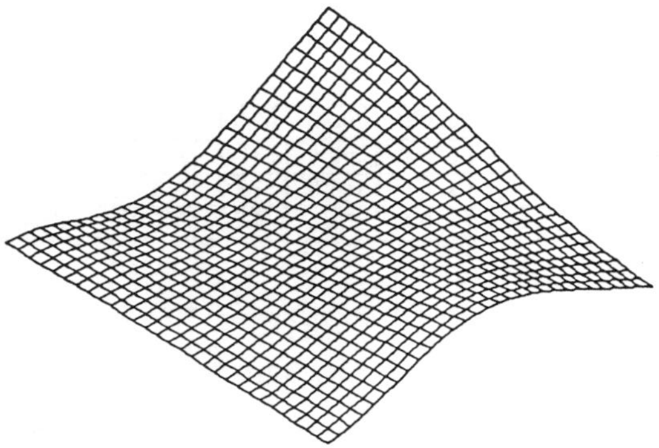

Perspective

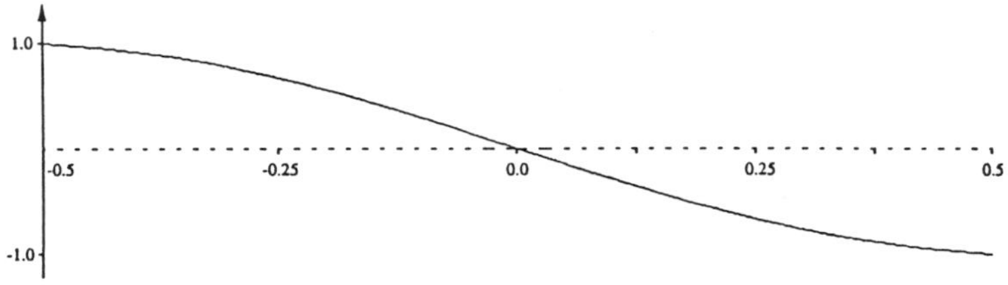


Section A-A

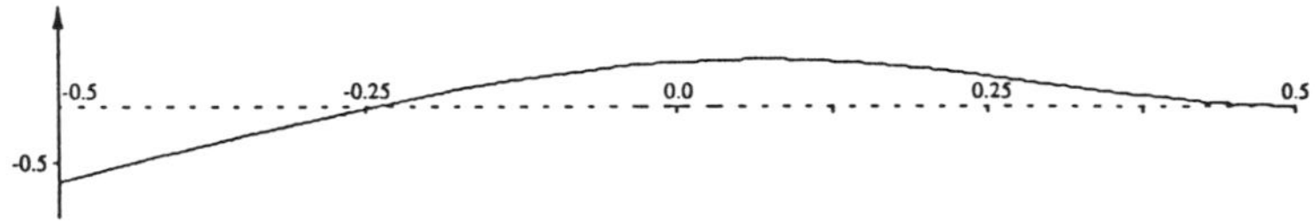


Section C-C

**Normalized Mode Characteristics:**

5th Mode: $\omega$ = 2340. rad/sec

Normal Displacement Posn (x,y): (0.5,0.5)

| Position | | -0.5 | -0.375 | -0.25 | -0.125 | 0.0 | 0.125 | 0.25 | 0.375 | 0.5 |
|---|---|---|---|---|---|---|---|---|---|---|
| A-A | Ux | 0.000 | 0.000 | 0.000 | 0.000 | 0.000 | 0.000 | 0.000 | 0.000 | 0.000 |
| | Uy | 0.000 | 0.000 | 0.000 | 0.000 | 0.000 | 0.000 | 0.000 | 0.000 | 0.000 |
| | Uz | 1.000 | 0.881 | 0.670 | 0.364 | 0.000 | -0.364 | -0.670 | -0.881 | -1.000 |

| Position | | -0.5 | -0.375 | -0.25 | -0.125 | 0.0 | 0.125 | 0.25 | 0.375 | 0.5 |
|---|---|---|---|---|---|---|---|---|---|---|
| C-C | Ux | 0.000 | 0.000 | 0.000 | 0.000 | 0.000 | 0.000 | 0.000 | 0.000 | 0.000 |
| | Uy | 0.000 | 0.000 | 0.000 | 0.000 | 0.000 | 0.000 | 0.000 | 0.000 | 0.000 |
| | Uz | 0.000 | 0.095 | 0.267 | 0.391 | 0.389 | 0.242 | -0.021 | -0.339 | -0.670 |

Figure F.2f : Cantilever Plate: 5th Mode.

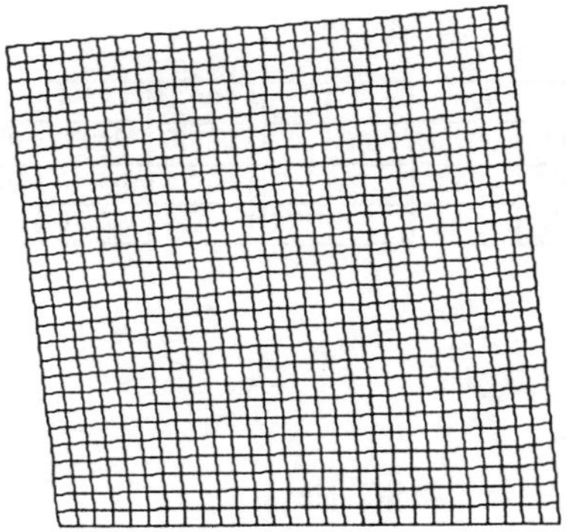

**Plan**

## Normalized Mode Characteristics:

6th Mode: $\omega$ = 3324. rad/sec

Normal Displacement Posn (x,y,z): (0.5,0.5)

| Position | | -0.5 | -0.375 | -0.25 | -0.125 | 0.0 | 0.125 | 0.25 | 0.375 | 0.5 |
|---|---|---|---|---|---|---|---|---|---|---|
| A-A | Ux | -1.000 | -0.996 | -0.987 | -0.978 | -0.975 | -0.978 | -0.987 | -0.996 | -1.000 |
| | Uy | -0.408 | -0.295 | -0.189 | -0.092 | 0.000 | 0.092 | 0.189 | 0.295 | 0.408 |
| | Uz | 0.000 | 0.000 | 0.000 | 0.000 | 0.000 | 0.000 | 0.000 | 0.000 | 0.000 |

| Position | | -0.5 | -0.375 | -0.25 | -0.125 | 0.0 | 0.125 | 0.25 | 0.375 | 0.5 |
|---|---|---|---|---|---|---|---|---|---|---|
| B-B | Ux | 0.000 | -0.080 | -0.190 | -0.321 | -0.462 | -0.606 | -0.744 | -0.870 | -0.975 |
| | Uy | 0.000 | 0.000 | 0.000 | 0.000 | 0.000 | 0.000 | 0.000 | 0.000 | 0.000 |
| | Uz | 0.000 | 0.000 | 0.000 | 0.000 | 0.000 | 0.000 | 0.000 | 0.000 | 0.000 |

Figure F.2g : Cantilever Plate: 6th Mode.

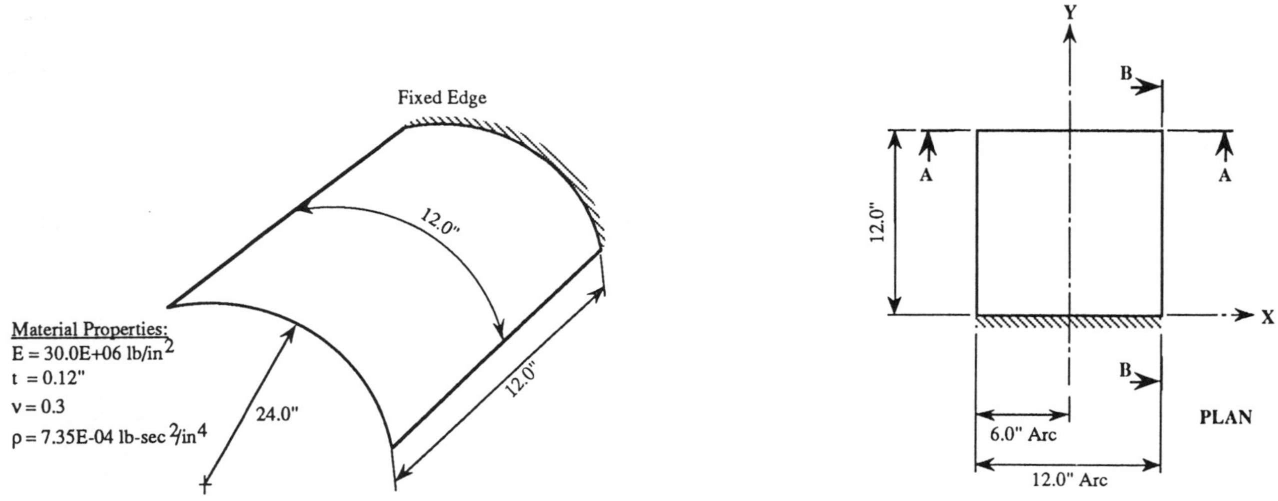


Figure F3.a: Curved Fan Blade.

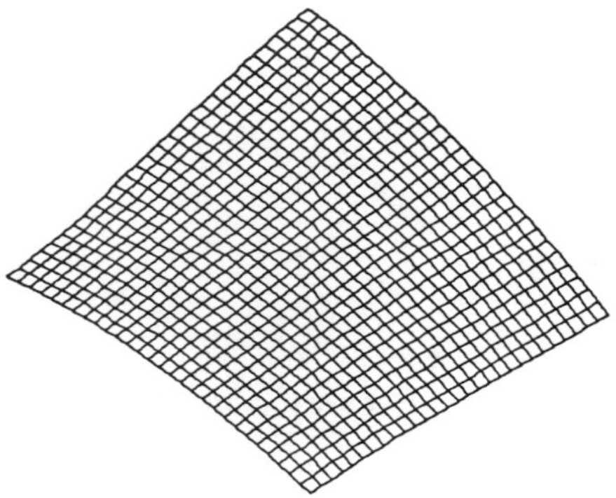

Perspective

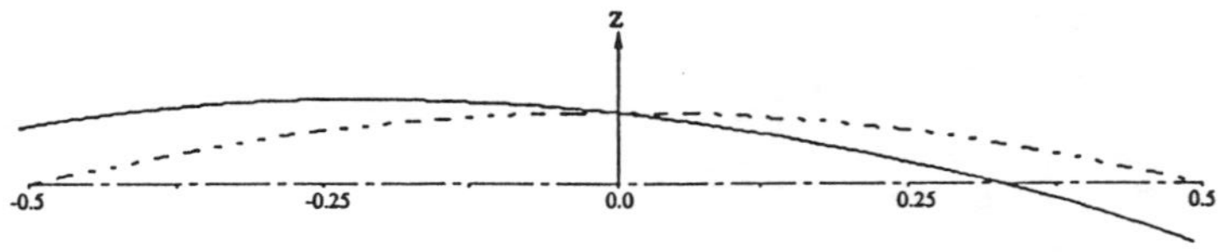


Section A-A

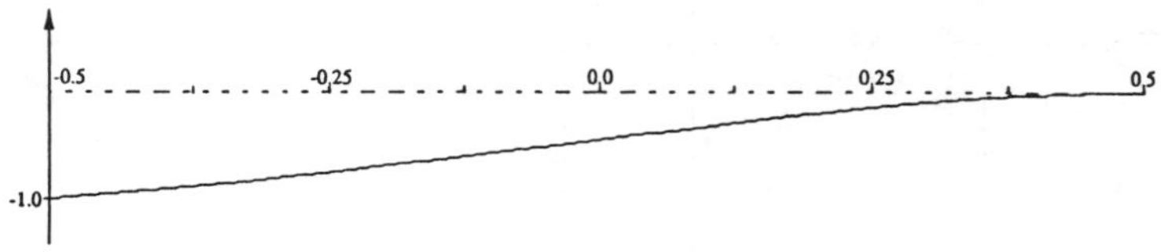


Section B-B

**Normalized Mode Characteristics:**

1st Mode: f = 86.2 hz

Normal Displacement Posn (x,y): (-5.937695,12.0)

| Position | | -0.5 | -0.375 | -0.25 | -0.125 | 0.0 | 0.125 | 0.25 | 0.375 | 0.5 |
|---|---|---|---|---|---|---|---|---|---|---|
| A-A | Ux | -0.148 | -0.099 | -0.061 | -0.037 | -0.028 | -0.037 | -0.061 | -0.099 | -0.148 |
| | Uy | 0.010 | -0.001 | -0.005 | -0.004 | 0.000 | 0.004 | 0.005 | 0.001 | -0.010 |
| | Uz | 1.000 | 0.775 | 0.531 | 0.270 | 0.000 | -0.270 | -0.531 | -0.775 | -1.000 |

| Position | | -0.5 | -0.375 | -0.25 | -0.125 | 0.0 | 0.125 | 0.25 | 0.375 | 0.5 |
|---|---|---|---|---|---|---|---|---|---|---|
| B-B | Ux | 0.000 | -0.007 | -0.024 | -0.046 | -0.070 | -0.093 | -0.115 | -0.133 | -0.148 |
| | Uy | 0.000 | -0.004 | -0.007 | -0.009 | -0.010 | -0.010 | -0.010 | -0.010 | -0.010 |
| | Uz | 0.000 | -0.047 | -0.155 | -0.295 | -0.449 | -0.605 | -0.752 | -0.884 | -1.000 |

Figure F3.b: Curved Fan Blade: 1st Mode.

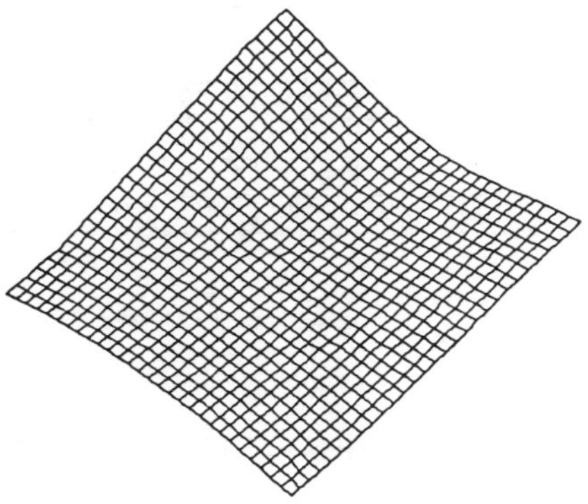

Perspective

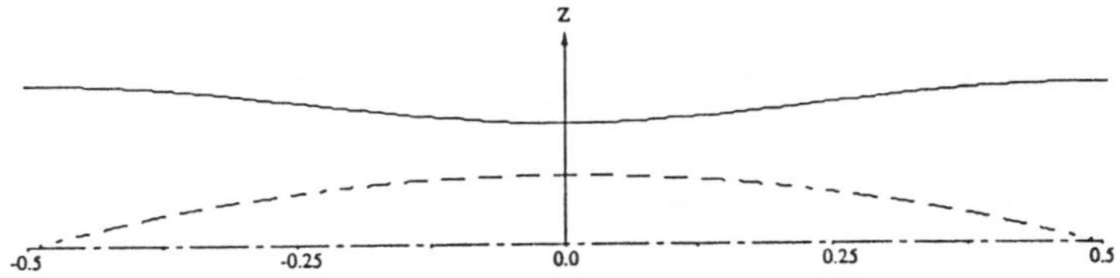


Section A-A

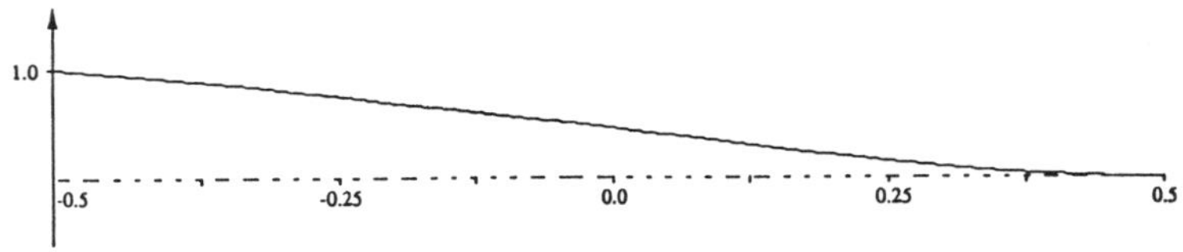


Section B-B

**Normalized Mode Characteristics:**

2nd Mode: f = 138.5 hz

Normal Displacement Posn (x,y): (-5.937695,12.0)

| Position | | -0.5 | -0.375 | -0.25 | -0.125 | 0.0 | 0.125 | 0.25 | 0.375 | 0.5 |
|---|---|---|---|---|---|---|---|---|---|---|
| A-A | Ux | -0.102 | -0.053 | -0.019 | -0.003 | 0.000 | 0.003 | 0.019 | 0.053 | 0.102 |
| | Uy | 0.024 | 0.006 | -0.004 | -0.009 | -0.010 | -0.009 | -0.004 | 0.006 | 0.024 |
| | Uz | 1.000 | 0.779 | 0.562 | 0.397 | 0.335 | 0.397 | 0.562 | 0.779 | 1.000 |

| Position | | -0.5 | -0.375 | -0.25 | -0.125 | 0.0 | 0.125 | 0.25 | 0.375 | 0.5 |
|---|---|---|---|---|---|---|---|---|---|---|
| B-B | Ux | 0.000 | 0.004 | 0.017 | 0.034 | 0.051 | 0.068 | 0.082 | 0.094 | 0.102 |
| | Uy | 0.000 | 0.008 | 0.015 | 0.019 | 0.022 | 0.023 | 0.024 | 0.024 | 0.024 |
| | Uz | 0.000 | 0.046 | 0.155 | 0.295 | 0.450 | 0.606 | 0.753 | 0.885 | 1.000 |

Figure F.3c : Curved Fan Blade: 2nd Mode.

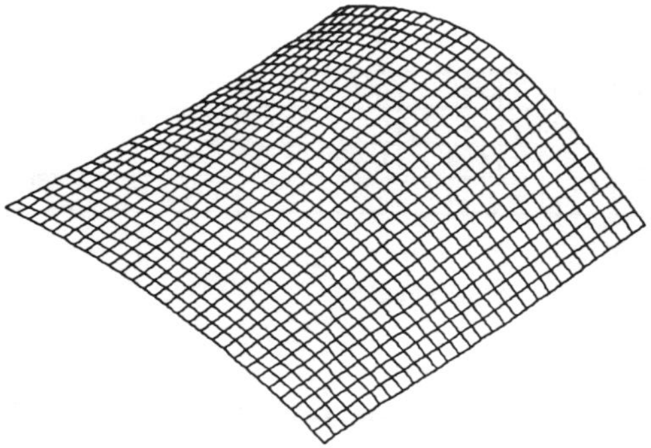

Perspective

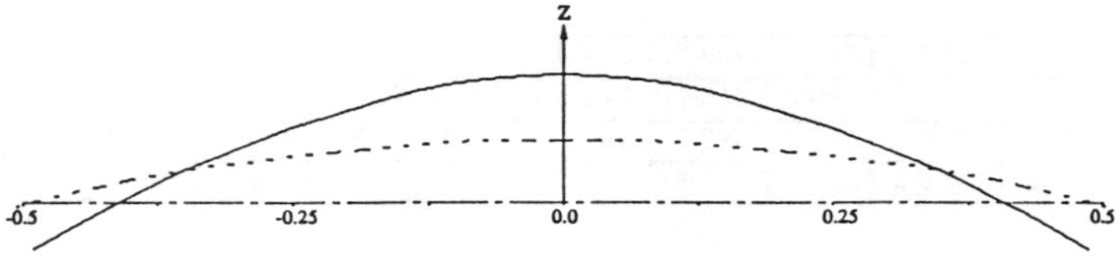


Section A-A

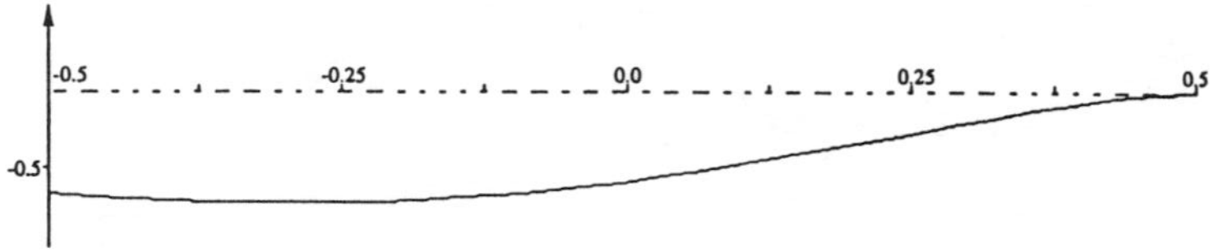


Section B-B

## Normalized Mode Characteristics:

3rd Mode: f = 247.0 hz

Normal Displacement Posn (x,y): (0.0,12.0)

| Position | | -0.5 | -0.375 | -0.25 | -0.125 | 0.0 | 0.125 | 0.25 | 0.375 | 0.5 |
|---|---|---|---|---|---|---|---|---|---|---|
| A-A | Ux | 0.253 | 0.125 | 0.040 | 0.004 | 0.000 | -0.004 | -0.040 | -0.125 | -0.253 |
| | Uy | 0.030 | 0.011 | -0.004 | -0.014 | -0.017 | -0.014 | -0.004 | 0.011 | 0.030 |
| | Uz | -0.679 | -0.098 | 0.450 | 0.852 | 1.000 | 0.852 | 0.450 | -0.098 | -0.679 |

| Position | | -0.5 | -0.375 | -0.25 | -0.125 | 0.0 | 0.125 | 0.25 | 0.375 | 0.5 |
|---|---|---|---|---|---|---|---|---|---|---|
| B-B | Ux | 0.000 | -0.027 | -0.072 | -0.120 | -0.163 | -0.199 | -0.226 | -0.244 | -0.253 |
| | Uy | 0.000 | 0.006 | 0.011 | 0.017 | 0.022 | 0.026 | 0.028 | 0.030 | 0.030 |
| | Uz | 0.000 | -0.102 | -0.280 | -0.458 | -0.601 | -0.694 | -0.734 | -0.725 | -0.679 |

Figure F.3d : Curved Fan Blade: 3rd Mode.

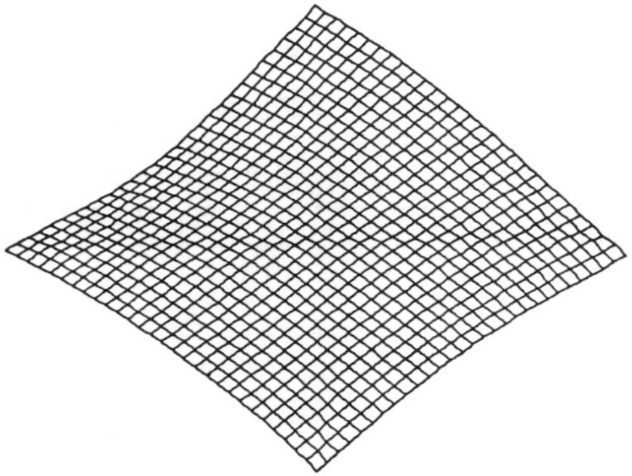

Perspective

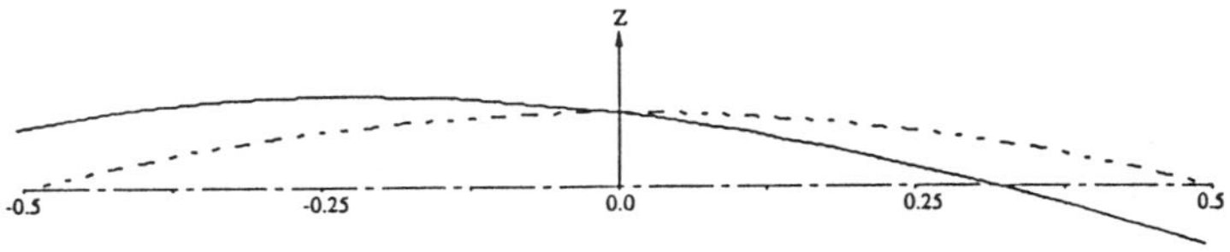


Section A-A

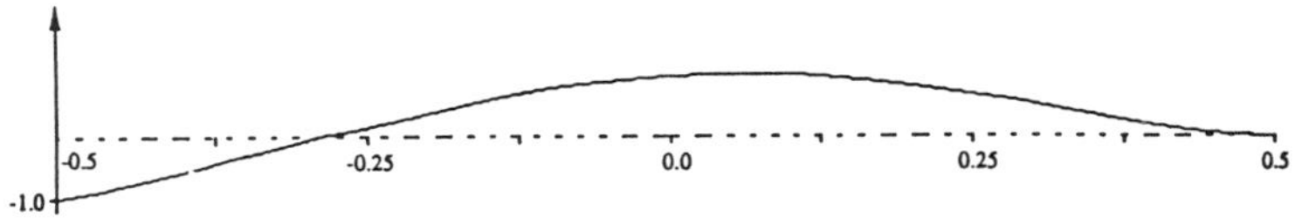


Section B-B

## Normalized Mode Characteristics:

4th Mode: f = 341.5 hz

Normal Displacement Posn (x,y): (-5.937695,12.0)

| Position | | -0.5 | -0.375 | -0.25 | -0.125 | 0.0 | 0.125 | 0.25 | 0.375 | 0.5 |
|---|---|---|---|---|---|---|---|---|---|---|
| A-A | Ux | -0.139 | -0.105 | -0.064 | -0.034 | -0.023 | -0.034 | -0.064 | -0.105 | -0.139 |
| | Uy | 0.035 | -0.003 | -0.014 | -0.010 | 0.000 | 0.010 | 0.014 | 0.003 | -0.035 |
| | Uz | 1.000 | 0.846 | 0.599 | 0.304 | 0.000 | -0.304 | -0.599 | -0.846 | -1.000 |

| Position | | -0.5 | -0.375 | -0.25 | -0.125 | 0.0 | 0.125 | 0.25 | 0.375 | 0.5 |
|---|---|---|---|---|---|---|---|---|---|---|
| B-B | Ux | 0.000 | 0.039 | 0.110 | 0.159 | 0.161 | 0.114 | 0.321 | -0.059 | -0.139 |
| | Uy | 0.000 | 0.011 | 0.011 | 0.003 | -0.010 | -0.023 | -0.031 | -0.034 | -0.035 |
| | Uz | 0.000 | 0.236 | 0.634 | 0.910 | 0.913 | 0.619 | 0.111 | -0.468 | -1.000 |

Figure F.3e : Curved Fan Blade: 4th Mode.

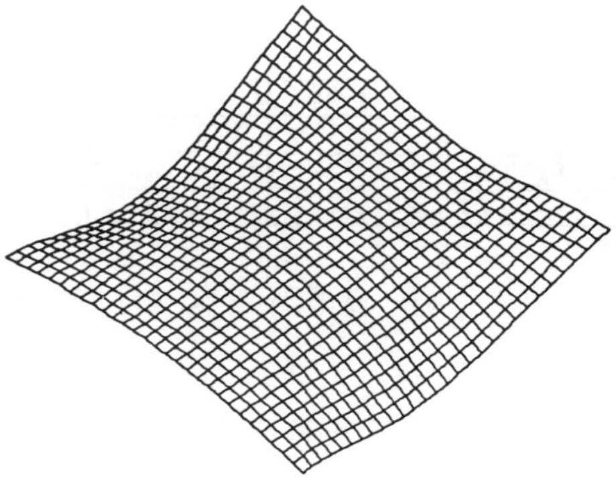

Perspective

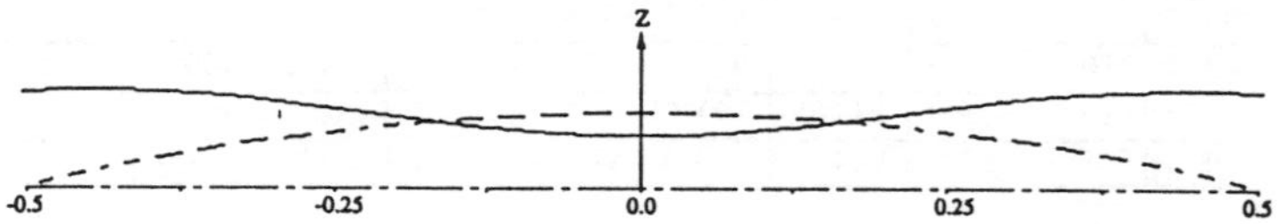


Section A-A

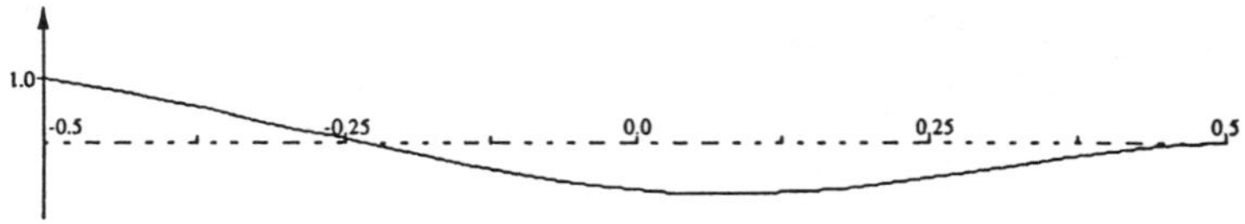


Section B-B

**Normalized Mode Characteristics:**

5th Mode: f = 385.5 hz

Normal Displacement Posn (x,y): (-5.937695,12.0)

| Position | | -0.5 | -0.375 | -0.25 | -0.125 | 0.0 | 0.125 | 0.25 | 0.375 | 0.5 |
|---|---|---|---|---|---|---|---|---|---|---|
| A-A | Ux | -0.186 | -0.109 | -0.043 | -0.008 | 0.000 | 0.008 | 0.043 | 0.109 | 0.186 |
| | Uy | 0.020 | -0.002 | -0.004 | 0.001 | 0.003 | 0.001 | -0.004 | -0.002 | 0.020 |
| | Uz | 1.000 | 0.650 | 0.232 | -0.115 | -0.249 | -0.115 | 0.232 | 0.650 | 1.000 |

| Position | | -0.5 | -0.375 | -0.25 | -0.125 | 0.0 | 0.125 | 0.25 | 0.375 | 0.5 |
|---|---|---|---|---|---|---|---|---|---|---|
| B-B | Ux | 0.000 | -0.035 | -0.093 | -0.127 | -0.117 | -0.060 | 0.026 | 0.115 | 0.186 |
| | Uy | 0.000 | -0.009 | -0.010 | -0.004 | 0.005 | 0.013 | 0.018 | 0.020 | 0.020 |
| | Uz | 0.000 | -0.202 | -0.528 | -0.731 | -0.686 | -0.385 | 0.084 | 0.581 | 1.000 |

Figure F.3f : Curved Fan Blade: 5th Mode.

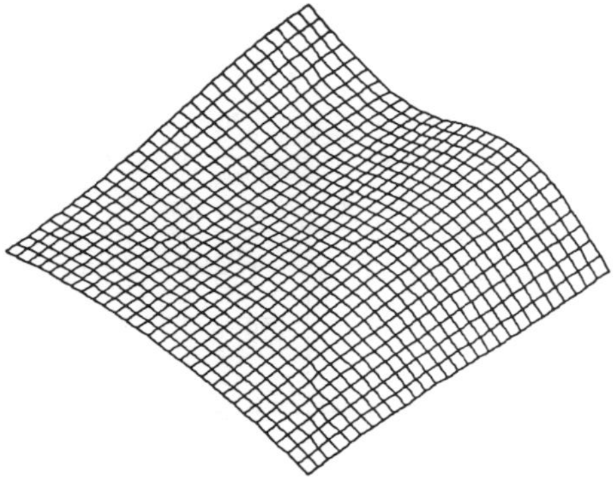

Perspective

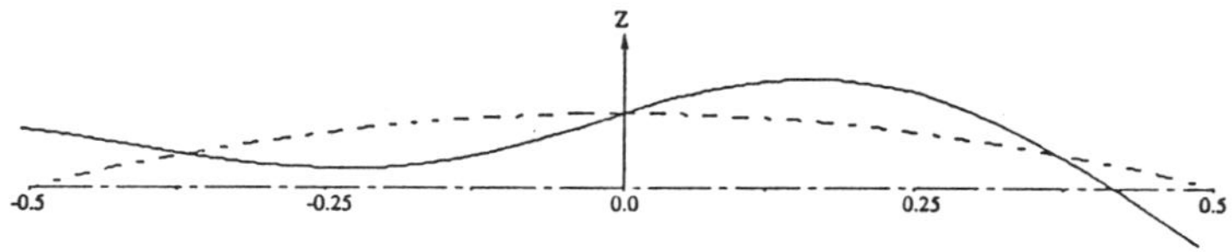


Section A-A

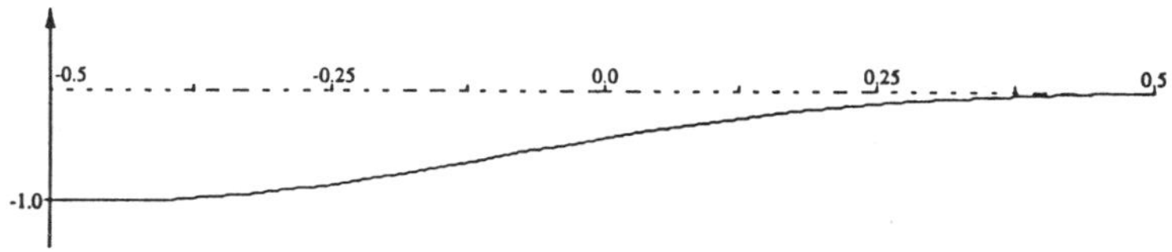


Section B-B

**Normalized Mode Characteristics:**

6th Mode: f = 525.0 hz

Normal Displacement Posn (x,y): (-5.937695,12.0)

| Position | | -0.5 | -0.375 | -0.25 | -0.125 | 0.0 | 0.125 | 0.25 | 0.375 | 0.5 |
|---|---|---|---|---|---|---|---|---|---|---|
| A-A | Ux | -0.261 | -0.048 | 0.059 | 0.058 | 0.040 | 0.058 | 0.059 | -0.048 | -0.261 |
| | Uy | -0.005 | 0.000 | 0.003 | 0.004 | 0.000 | -0.004 | -0.003 | 0.000 | 0.005 |
| | Uz | 1.000 | 0.035 | -0.641 | -0.621 | 0.000 | 0.621 | 0.641 | -0.035 | -1.000 |

| Position | | -0.5 | -0.375 | -0.25 | -0.125 | 0.0 | 0.125 | 0.25 | 0.375 | 0.5 |
|---|---|---|---|---|---|---|---|---|---|---|
| B-B | Ux | 0.000 | -0.015 | -0.039 | -0.075 | -0.122 | -0.175 | -0.221 | -0.252 | -0.261 |
| | Uy | 0.000 | 0.004 | 0.006 | 0.005 | 0.003 | 0.002 | 0.003 | 0.004 | 0.005 |
| | Uz | 0.000 | -0.041 | -0.115 | -0.244 | -0.436 | -0.662 | -0.864 | -0.984 | -1.000 |

Figure F.3g : Curved Fan Blade: 6th Mode.

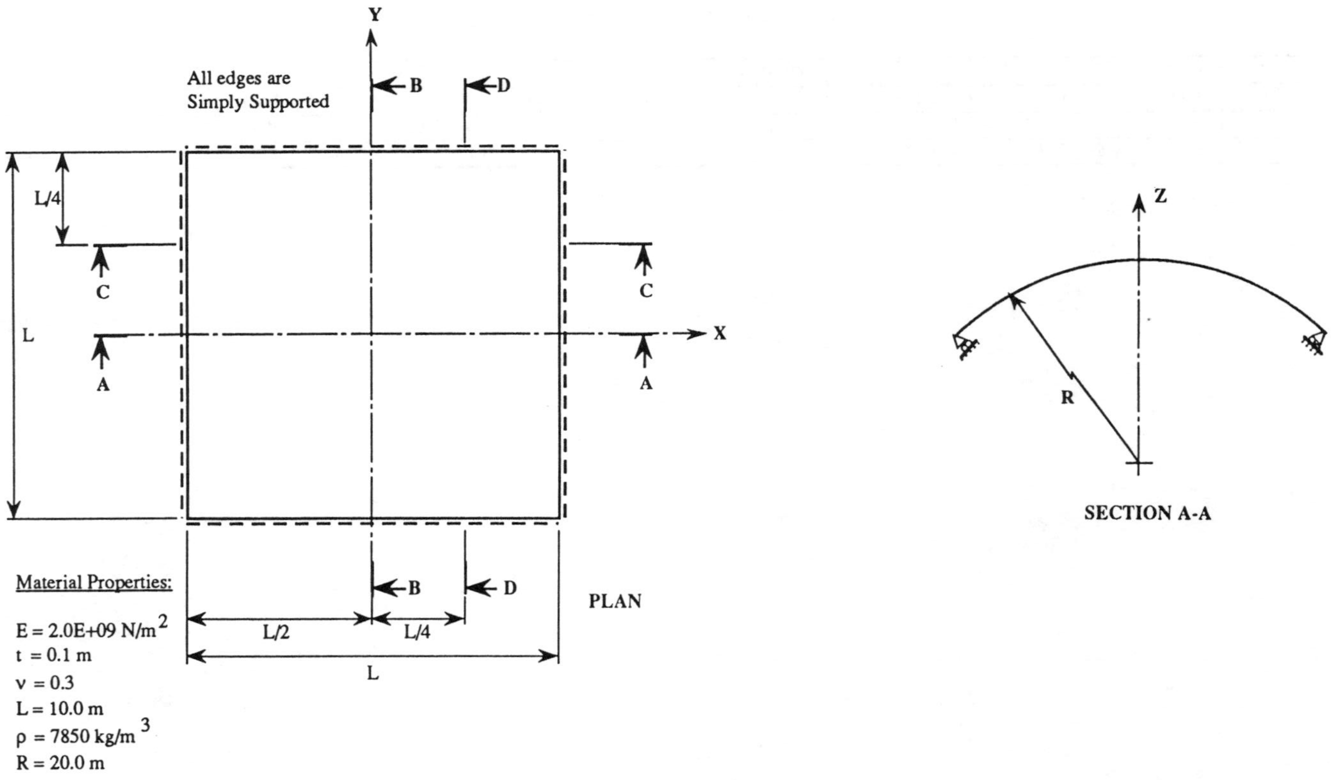


Figure F4.a: Spherical Shell.

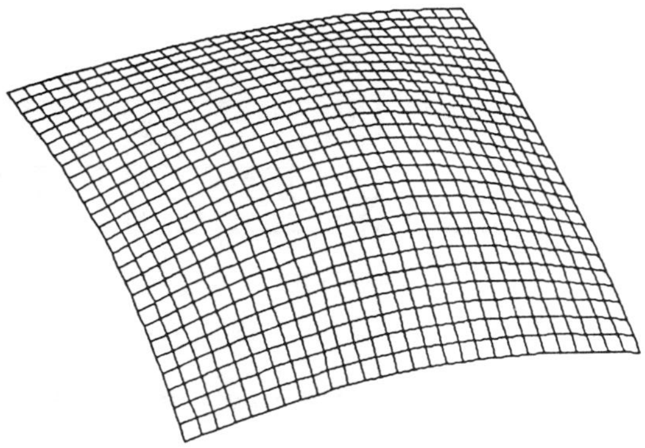

**Perspective**

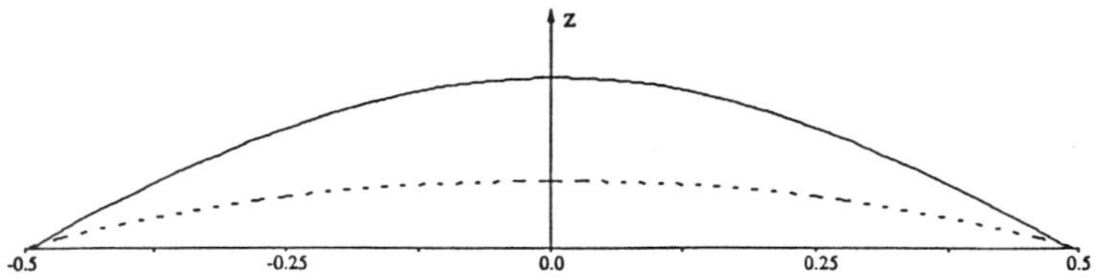


**Section A-A**

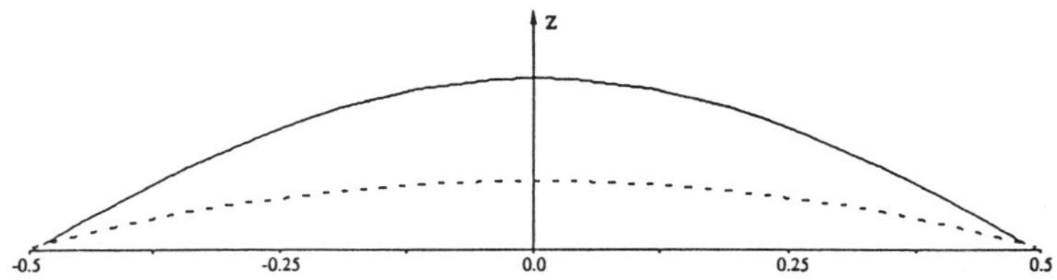


**Section B-B**

## Normalized Mode Characteristics:

1st Mode: $\omega$ = 25.3 rad/sec

Normal Displacement Posn (x,y): (0.0,0.0)

| Position | | -0.5 | -0.375 | -0.25 | -0.125 | 0.0 | 0.125 | 0.25 | 0.375 | 0.5 |
|---|---|---|---|---|---|---|---|---|---|---|
| A-A | Ux | 0.095 | 0.025 | -0.013 | -0.017 | 0.000 | 0.017 | 0.013 | -0.025 | -0.095 |
| | Uy | 0.000 | 0.000 | 0.000 | 0.000 | 0.000 | 0.000 | 0.000 | 0.000 | 0.000 |
| | Uz | 0.024 | 0.356 | 0.676 | 0.913 | 1.000 | 0.913 | 0.676 | 0.356 | 0.024 |

| Position | | -0.5 | -0.375 | -0.25 | -0.125 | 0.0 | 0.125 | 0.25 | 0.375 | 0.5 |
|---|---|---|---|---|---|---|---|---|---|---|
| B-B | Ux | 0.000 | 0.000 | 0.000 | 0.000 | 0.000 | 0.000 | 0.000 | 0.000 | 0.000 |
| | Uy | 0.095 | 0.025 | -0.013 | -0.017 | 0.000 | 0.017 | 0.013 | -0.025 | -0.095 |
| | Uz | 0.024 | 0.356 | 0.676 | 0.913 | 1.000 | 0.913 | 0.676 | 0.356 | 0.024 |

Figure F4.b : Spherical Shell: 1st Mode.

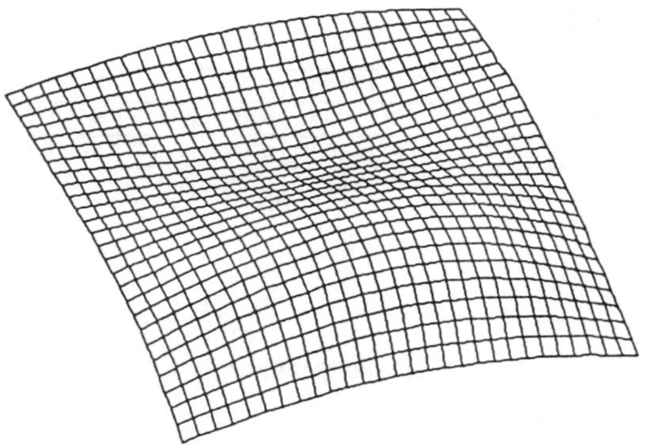

Perspective

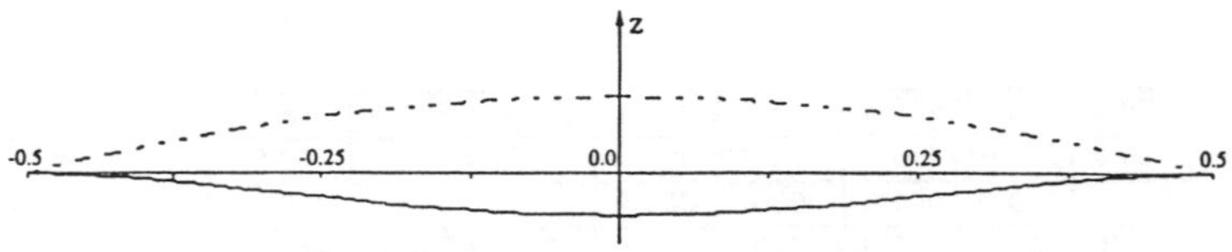


Section C-C

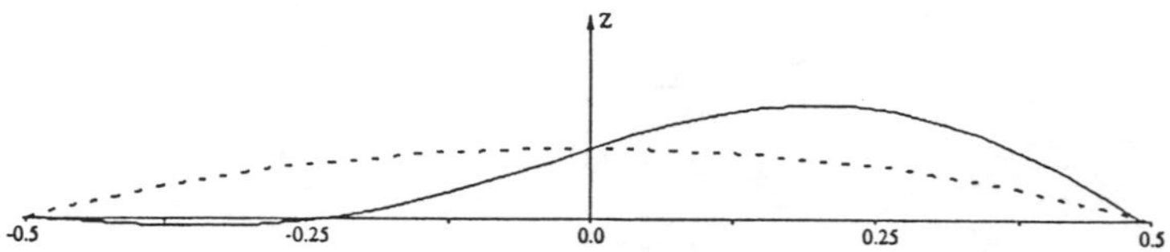


Section B-B

**Normalized Mode Characteristics:**

2nd Mode: $\omega$ = 26.2 rad/sec

Normal Displacement Posn (x,y): (0.0,2.5)

| Position | | -0.5 | -0.375 | -0.25 | -0.125 | 0.0 | 0.125 | 0.25 | 0.375 | 0.5 |
|---|---|---|---|---|---|---|---|---|---|---|
| C-C | Ux | -0.040 | 0.033 | 0.059 | 0.042 | 0.000 | -0.042 | -0.059 | -0.033 | 0.040 |
| | Uy | -0.001 | -0.050 | -0.093 | -0.122 | -0.132 | -0.122 | -0.093 | -0.050 | -0.001 |
| | Uz | -0.010 | -0.378 | -0.701 | -0.922 | -1.000 | -0.922 | -0.701 | -0.378 | -0.010 |

| Position | | -0.5 | -0.375 | -0.25 | -0.125 | 0.0 | 0.125 | 0.25 | 0.375 | 0.5 |
|---|---|---|---|---|---|---|---|---|---|---|
| B-B | Ux | 0.000 | 0.000 | 0.000 | 0.000 | 0.000 | 0.000 | 0.000 | 0.000 | 0.000 |
| | Uy | 0.077 | -0.081 | -0.132 | -0.109 | -0.089 | -0.109 | -0.132 | -0.081 | 0.077 |
| | Uz | 0.020 | 0.710 | 1.000 | 0.707 | 0.000 | -0.707 | -1.000 | -0.710 | -0.020 |

Figure F.4c : Spherical Shell: 2nd Mode.

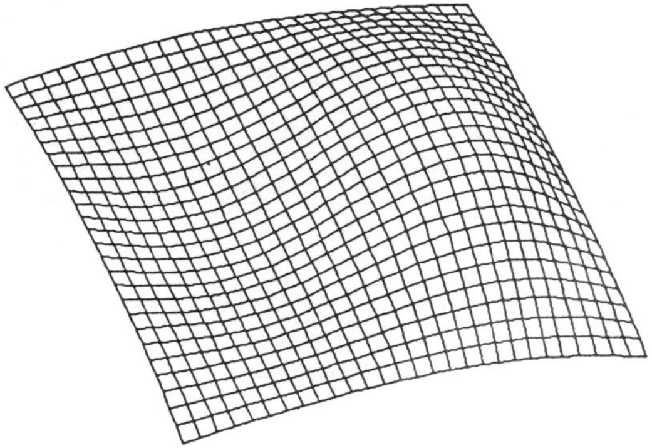

Perspective

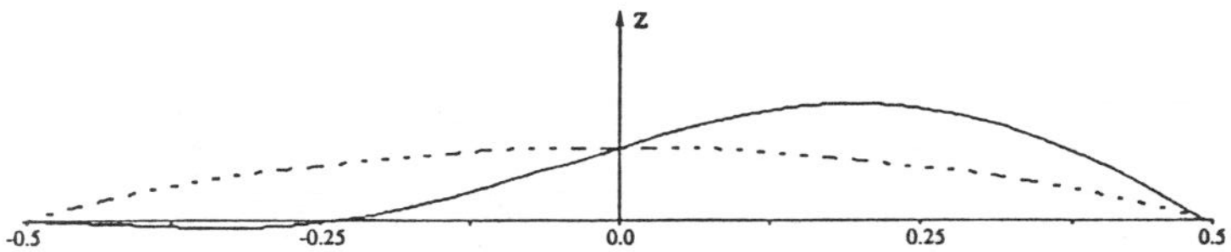


Section A-A

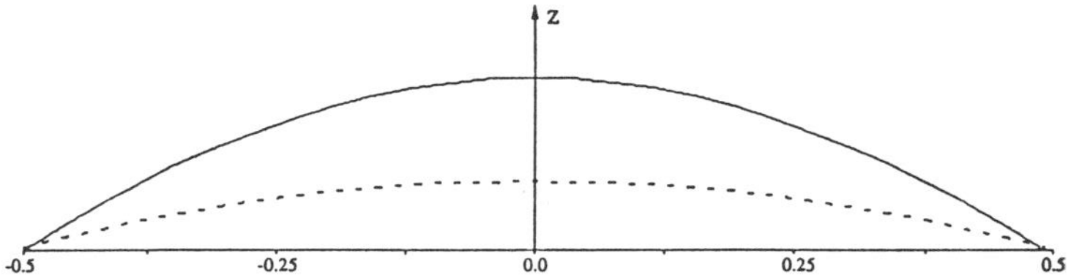


Section D-D

## Normalized Mode Characteristics:

3rd Mode: $\omega$ = 26.2 rad/sec

Normal Displacement Posn (x,y): (2.5,0.0)

| Position | | -0.5 | -0.375 | -0.25 | -0.125 | 0.0 | 0.125 | 0.25 | 0.375 | 0.5 |
|---|---|---|---|---|---|---|---|---|---|---|
| A-A | Ux | -0.077 | 0.081 | 0.132 | 0.109 | 0.089 | 0.109 | 0.132 | 0.081 | -0.077 |
| | Uy | 0.000 | 0.000 | 0.000 | 0.000 | 0.000 | 0.000 | 0.000 | 0.000 | 0.000 |
| | Uz | -0.020 | -0.710 | -1.000 | -0.707 | 0.000 | 0.707 | 1.000 | 0.710 | 0.020 |

| Position | | -0.5 | -0.375 | -0.25 | -0.125 | 0.0 | 0.125 | 0.25 | 0.375 | 0.5 |
|---|---|---|---|---|---|---|---|---|---|---|
| D-D | Ux | 0.001 | 0.050 | 0.092 | 0.121 | 0.132 | 0.121 | 0.092 | 0.050 | 0.001 |
| | Uy | 0.040 | -0.033 | -0.059 | -0.042 | 0.000 | 0.042 | 0.059 | 0.033 | -0.044 |
| | Uz | 0.010 | 0.378 | 0.701 | 0.922 | 1.000 | 0.922 | 0.701 | 0.378 | 0.010 |

Figure F.4d : Spherical Shell: 3rd Mode.

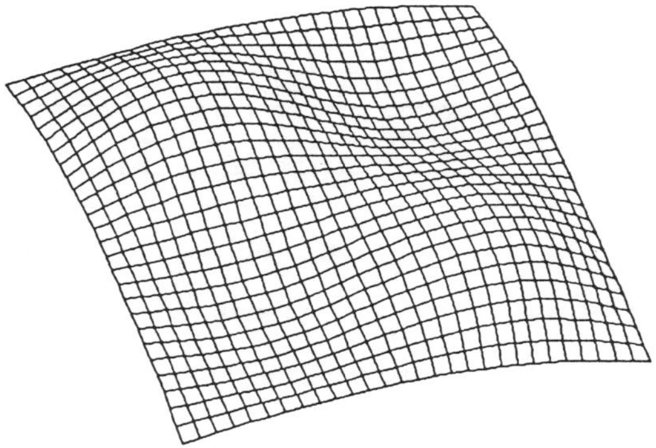

**Perspective**

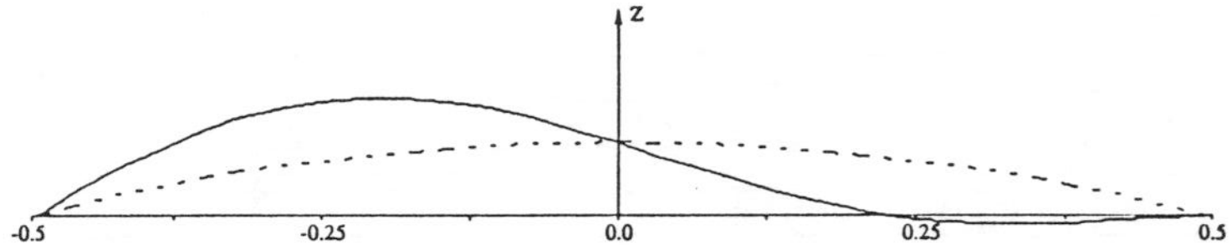


**Section C-C**

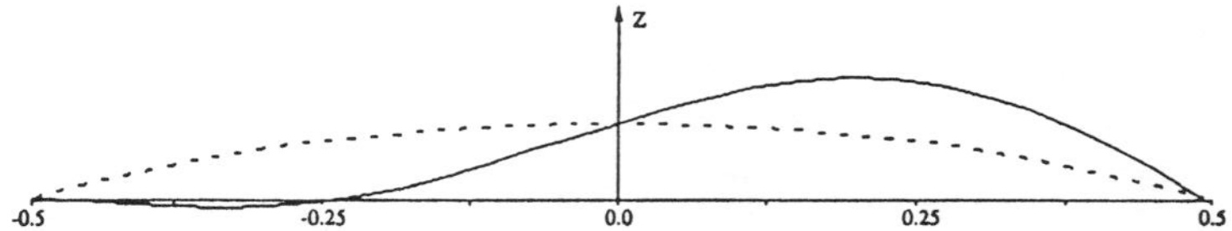


**Section D-D**

## Normalized Mode Characteristics:

4th Mode: $\omega$ = 27.8 rad/sec

Normal Displacement Posn (x,y): (-2.5,2.5)

| Position | | -0.5 | -0.375 | -0.25 | -0.125 | 0.0 | 0.125 | 0.25 | 0.375 | 0.5 |
|---|---|---|---|---|---|---|---|---|---|---|
| C-C | Ux | 0.051 | -0.099 | -0.128 | -0.083 | -0.054 | -0.083 | -0.128 | -0.099 | 0.051 |
| | Uy | 0.002 | 0.092 | 0.128 | 0.090 | 0.000 | -0.090 | -0.128 | -0.092 | -0.002 |
| | Uz | 0.013 | 0.710 | 1.000 | 0.705 | 0.000 | -0.705 | -1.000 | -0.710 | -0.013 |

| Position | | -0.5 | -0.375 | -0.25 | -0.125 | 0.0 | 0.125 | 0.25 | 0.375 | 0.5 |
|---|---|---|---|---|---|---|---|---|---|---|
| D-D | Ux | 0.002 | 0.092 | 0.128 | 0.090 | 0.000 | -0.090 | -0.128 | -0.092 | -0.002 |
| | Uy | 0.051 | -0.099 | -0.128 | -0.083 | -0.054 | -0.083 | -0.128 | -0.099 | 0.051 |
| | Uz | 0.013 | 0.710 | 1.000 | 0.705 | 0.000 | -0.705 | -1.000 | -0.710 | -0.013 |

Figure F.4e : Spherical Shell: 4th Mode.

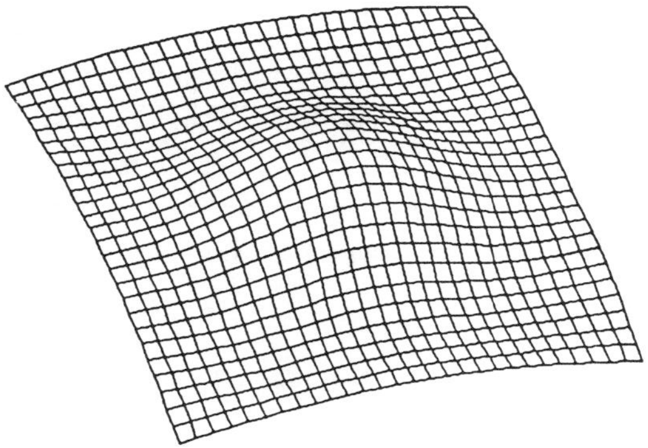

Perspective

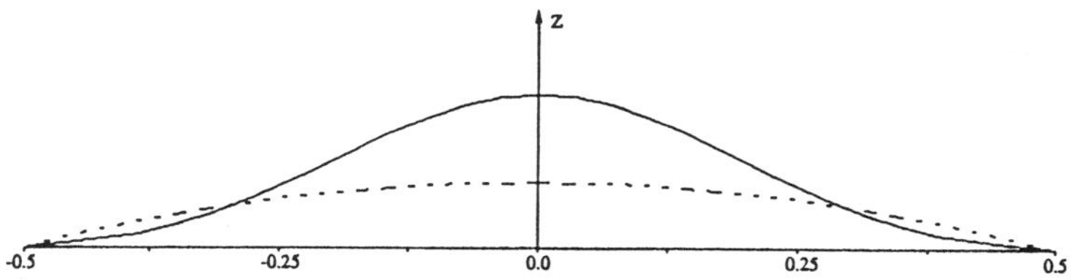


Section A-A

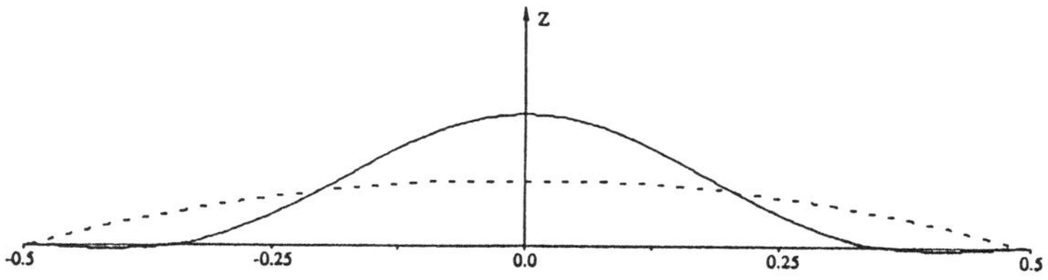


Section B-B

**Normalized Mode Characteristics:**

5th Mode: $\omega$ = 29.0 rad/sec

Normal Displacement Posn (x,y): (0.0,0.0)

| Position | | -0.5 | -0.375 | -0.25 | -0.125 | 0.0 | 0.125 | 0.25 | 0.375 | 0.5 |
|---|---|---|---|---|---|---|---|---|---|---|
| A-A | Ux | -0.014 | 0.024 | 0.004 | -0.019 | 0.000 | 0.019 | -0.004 | -0.024 | 0.014 |
| | Uy | 0.000 | 0.000 | 0.000 | 0.000 | 0.000 | 0.000 | 0.000 | 0.000 | 0.000 |
| | Uz | -0.004 | -0.124 | 0.165 | 0.719 | 1.000 | 0.719 | 0.165 | -0.124 | -0.004 |

| Position | | -0.5 | -0.375 | -0.25 | -0.125 | 0.0 | 0.125 | 0.25 | 0.375 | 0.5 |
|---|---|---|---|---|---|---|---|---|---|---|
| B-B | Ux | 0.000 | 0.000 | 0.000 | 0.000 | 0.000 | 0.000 | 0.000 | 0.000 | 0.000 |
| | Uy | -0.034 | 0.077 | 0.058 | 0.003 | 0.000 | -0.003 | -0.058 | -0.077 | 0.034 |
| | Uz | -0.009 | -0.455 | -0.191 | 0.584 | 1.000 | 0.584 | -0.191 | -0.455 | -0.009 |

Figure F.4f : Spherical Shell: 5th Mode.

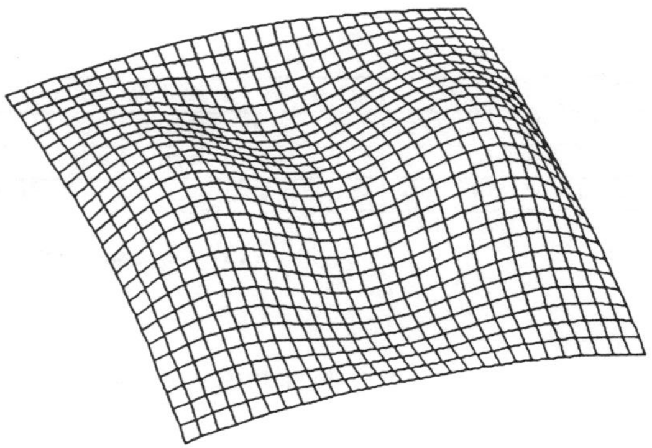

Perspective

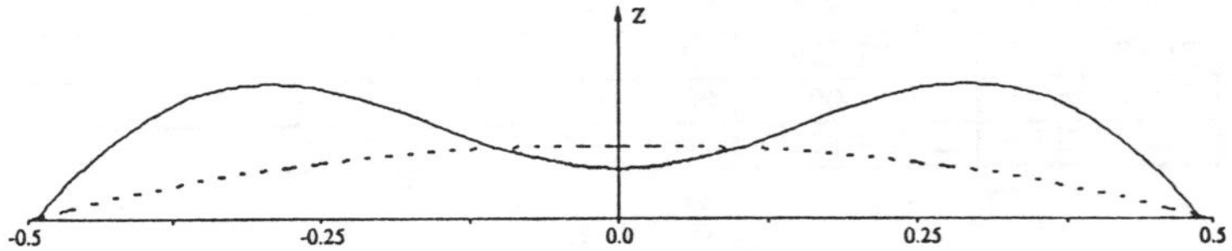


Section A-A

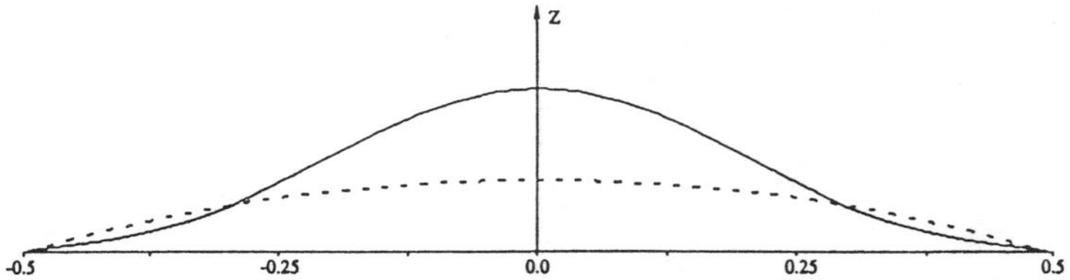


Section D-D

## Normalized Mode Characteristics:

6th Mode: $\omega$ = 29.1 rad/sec

Normal Displacement Posn (x,y): (-3.75,0.0)

| Position | | -0.5 | -0.375 | -0.25 | -0.125 | 0.0 | 0.125 | 0.25 | 0.375 | 0.5 |
|---|---|---|---|---|---|---|---|---|---|---|
| A-A | Ux | 0.062 | -0.162 | -0.155 | -0.056 | 0.000 | 0.056 | 0.155 | 0.162 | -0.062 |
| | Uy | 0.000 | 0.000 | 0.000 | 0.000 | 0.000 | 0.000 | 0.000 | 0.000 | 0.000 |
| | Uz | 0.016 | 1.000 | 0.973 | 0.151 | -0.331 | 0.151 | 0.973 | 1.000 | 0.016 |

| Position | | -0.5 | -0.375 | -0.25 | -0.125 | 0.0 | 0.125 | 0.25 | 0.375 | 0.5 |
|---|---|---|---|---|---|---|---|---|---|---|
| D-D | Ux | 0.000 | 0.009 | 0.056 | 0.123 | 0.155 | 0.123 | 0.056 | 0.009 | 0.000 |
| | Uy | -0.007 | 0.025 | 0.006 | -0.017 | 0.000 | 0.017 | -0.006 | -0.025 | 0.007 |
| | Uz | -0.002 | -0.105 | 0.170 | 0.702 | 0.973 | 0.702 | 0.170 | -0.105 | -0.002 |

Figure F.4g : Spherical Shell: 6th Mode.

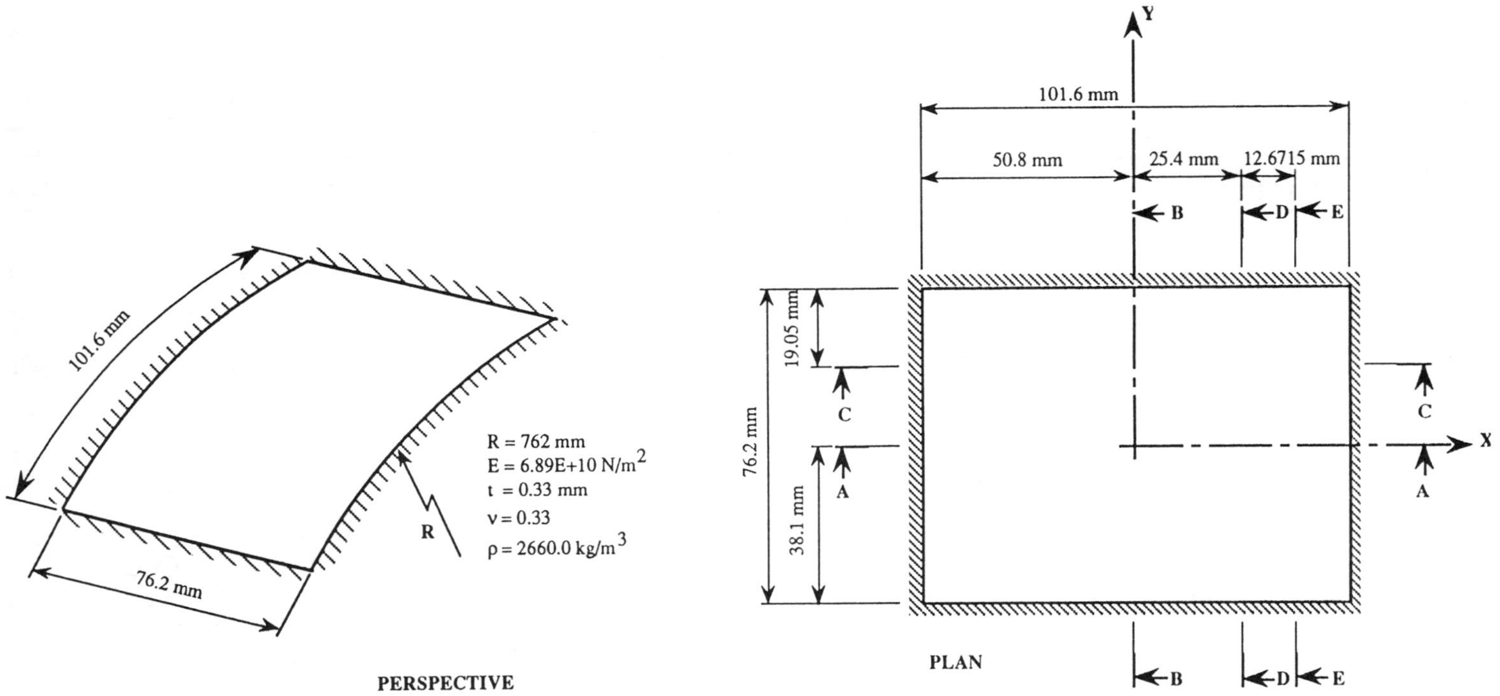


Figure F.5a: Cylindrical Panel.

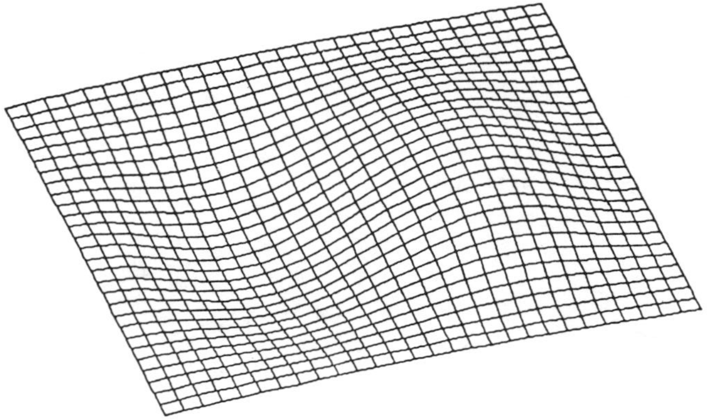

Perspective

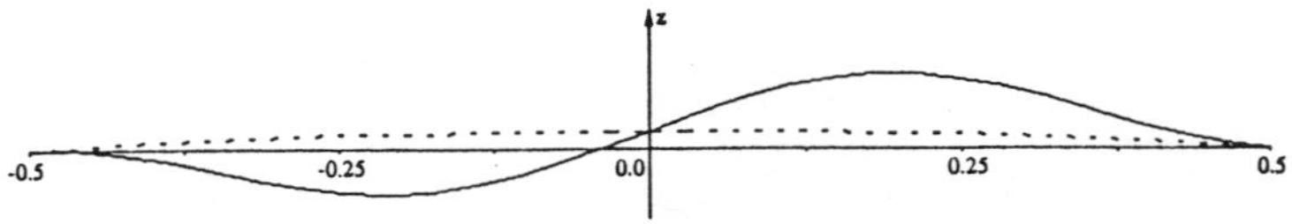


Section A-A

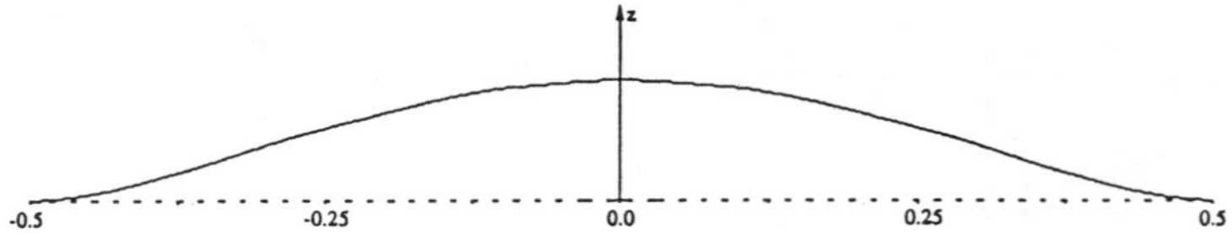


Section D-D

## Normalized Mode Characteristics:

1st Mode: f = 859.5 Hz

Normal Displacement Posn (x,y): (0.025381,0.0)

| Position | | -0.5 | -0.375 | -0.25 | -0.125 | 0.0 | 0.125 | 0.25 | 0.375 | 0.5 |
|---|---|---|---|---|---|---|---|---|---|---|
| A-A | Ux | 0.000 | 0.020 | 0.040 | 0.035 | 0.027 | 0.035 | 0.040 | 0.020 | 0.000 |
| | Uy | 0.000 | 0.000 | 0.000 | 0.000 | 0.000 | 0.000 | 0.000 | 0.000 | 0.000 |
| | Uz | 0.000 | -0.424 | -1.000 | -0.878 | 0.000 | 0.878 | 1.000 | 0.424 | 0.000 |

| Position | | -0.5 | -0.375 | -0.25 | -0.125 | 0.0 | 0.125 | 0.25 | 0.375 | 0.5 |
|---|---|---|---|---|---|---|---|---|---|---|
| D-D | Ux | 0.000 | 0.010 | 0.025 | 0.036 | 0.040 | 0.036 | 0.025 | 0.010 | 0.000 |
| | Uy | 0.000 | -0.001 | -0.001 | -0.001 | 0.000 | 0.001 | 0.001 | 0.001 | 0.000 |
| | Uz | 0.000 | 0.227 | 0.601 | 0.893 | 1.000 | 0.893 | 0.601 | 0.227 | 0.000 |

Figure F.5b : Cylindrical Panel: 1st Mode.

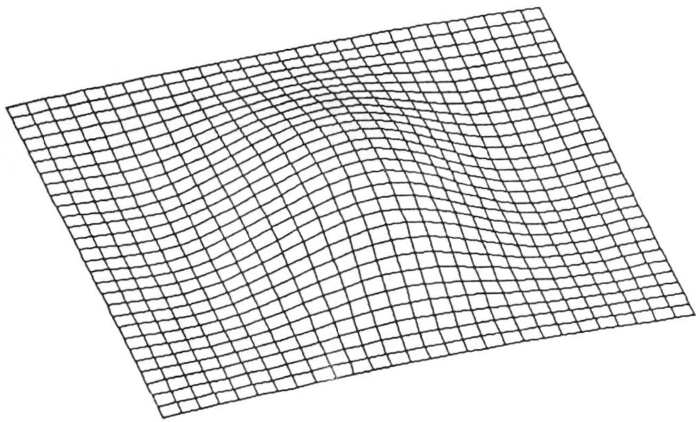

Perspective

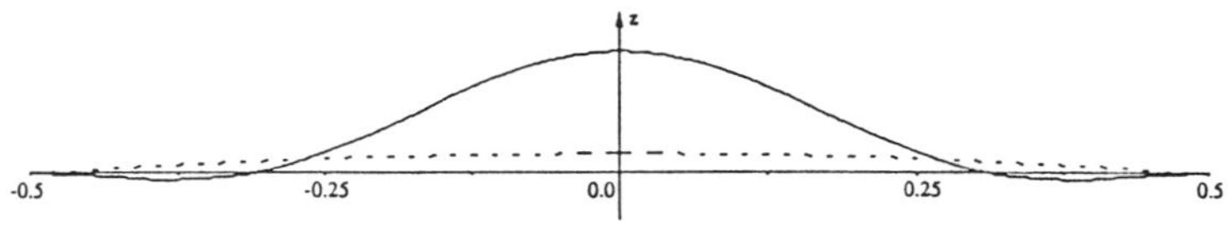


Section A-A

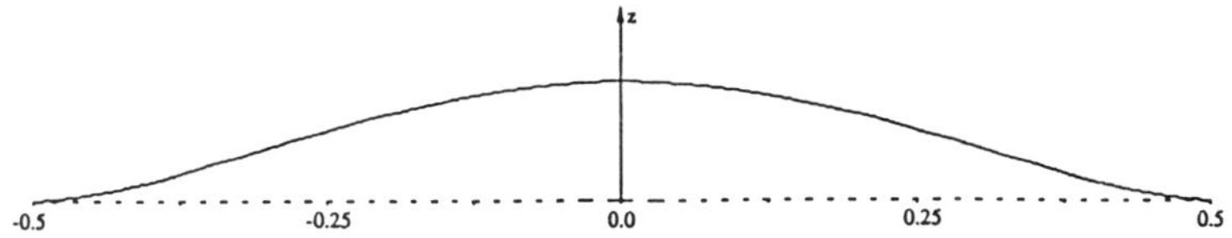


Section B-B

**Normalized Mode Characteristics:**

2nd Mode: f = 943.9 Hz

Normal Displacement Posn (x,y): (0.0,0.0)

| Position | | -0.5 | -0.375 | -0.25 | -0.125 | 0.0 | 0.125 | 0.25 | 0.375 | 0.5 |
|---|---|---|---|---|---|---|---|---|---|---|
| A-A | Ux | 0.000 | 0.012 | 0.008 | -0.002 | 0.000 | 0.002 | -0.008 | -0.012 | 0.000 |
| | Uy | 0.000 | 0.000 | 0.000 | 0.000 | 0.000 | 0.000 | 0.000 | 0.000 | 0.000 |
| | Uz | 0.000 | -0.148 | 0.072 | 0.673 | 1.000 | 0.673 | 0.072 | -0.148 | 0.000 |

| Position | | -0.5 | -0.375 | -0.25 | -0.125 | 0.0 | 0.125 | 0.25 | 0.375 | 0.5 |
|---|---|---|---|---|---|---|---|---|---|---|
| B-B | Ux | 0.000 | 0.000 | 0.000 | 0.000 | 0.000 | 0.000 | 0.000 | 0.000 | 0.000 |
| | Uy | 0.000 | -0.000 | -0.000 | -0.002 | 0.000 | 0.002 | 0.000 | 0.000 | 0.000 |
| | Uz | 0.000 | 0.217 | 0.587 | 0.887 | 1.000 | 0.887 | 0.587 | 0.217 | 0.000 |

Figure F.5c : Cylindrical Panel: 2nd Mode.

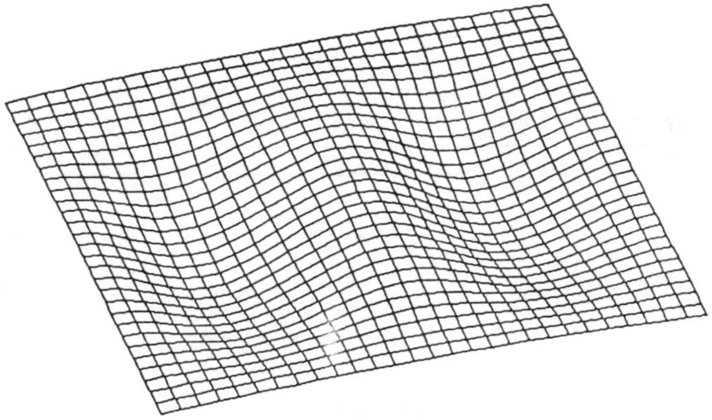

Perspective

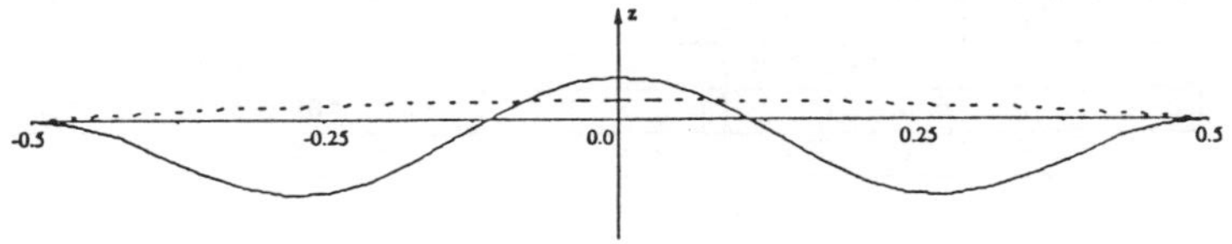


Section A-A

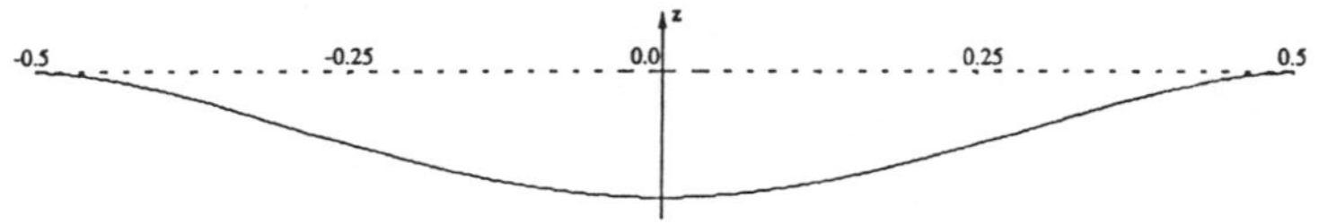


Section D-D

**Normalized Mode Characteristics:**

3rd Mode: f = 1262.3 Hz

Normal Displacement Posn (x,y): (-0.025381,0.0)

| Position | | -0.5 | -0.375 | -0.25 | -0.125 | 0.0 | 0.125 | 0.25 | 0.375 | 0.5 |
|---|---|---|---|---|---|---|---|---|---|---|
| A-A | Ux | 0.000 | 0.027 | 0.037 | 0.013 | 0.000 | -0.013 | -0.037 | -0.027 | 0.000 |
| | Uy | 0.000 | 0.000 | 0.000 | 0.000 | 0.000 | 0.000 | 0.000 | 0.000 | 0.000 |
| | Uz | 0.000 | -0.615 | -1.000 | -0.300 | 0.265 | -0.300 | -1.000 | -0.615 | 0.000 |

| Position | | -0.5 | -0.375 | -0.25 | -0.125 | 0.0 | 0.125 | 0.25 | 0.375 | 0.5 |
|---|---|---|---|---|---|---|---|---|---|---|
| D-D | Ux | 0.000 | -0.008 | -0.022 | -0.033 | -0.037 | -0.033 | -0.022 | -0.008 | 0.000 |
| | Uy | 0.000 | -0.000 | -0.000 | -0.000 | 0.000 | 0.000 | 0.000 | 0.000 | 0.000 |
| | Uz | 0.000 | -0.217 | -0.582 | -0.885 | -1.000 | -0.885 | -0.582 | -0.217 | 0.000 |

Figure F.5d : Cylindrical Panel: 3rd Mode.

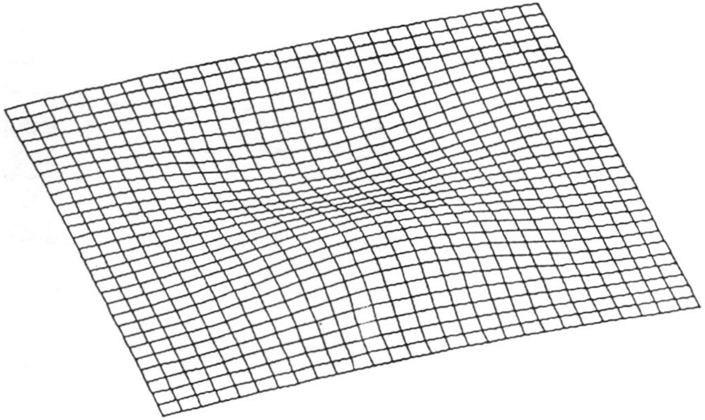

Perspective

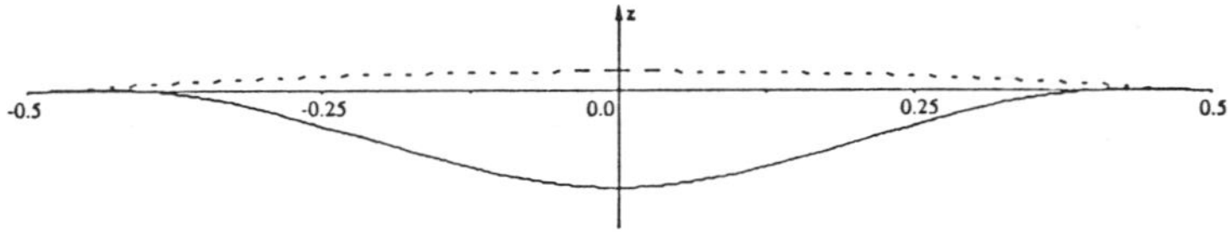


Section C-C

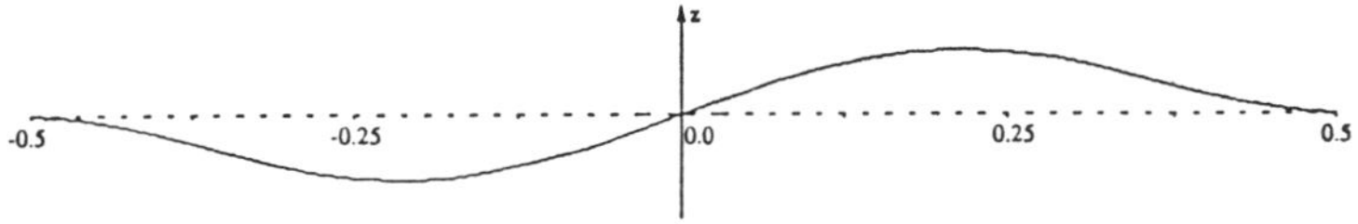


Section B-B

**Normalized Mode Characteristics:**

4th Mode: f = 1347.0 Hz

Normal Displacement Posn (x,y): (0.0,0.01905)

| Position | | -0.5 | -0.375 | -0.25 | -0.125 | 0.0 | 0.125 | 0.25 | 0.375 | 0.5 |
|---|---|---|---|---|---|---|---|---|---|---|
| C-C | Ux | 0.000 | 0.001 | 0.008 | 0.009 | 0.000 | -0.009 | -0.008 | -0.001 | 0.000 |
| | Uy | 0.000 | -0.001 | -0.001 | -0.001 | -0.001 | -0.001 | -0.001 | -0.001 | 0.000 |
| | Uz | 0.000 | -0.092 | -0.430 | -0.826 | -1.000 | -0.826 | -0.430 | -0.092 | 0.000 |

| Position | | -0.5 | -0.375 | -0.25 | -0.125 | 0.0 | 0.125 | 0.25 | 0.375 | 0.5 |
|---|---|---|---|---|---|---|---|---|---|---|
| B-B | Ux | 0.000 | 0.000 | 0.000 | 0.000 | 0.000 | 0.000 | 0.000 | 0.000 | 0.000 |
| | Uy | 0.000 | -0.000 | -0.001 | -0.003 | -0.003 | -0.003 | -0.001 | -0.000 | 0.000 |
| | Uz | 0.000 | 0.482 | 1.000 | 0.823 | 0.000 | -0.823 | -1.000 | -0.482 | 0.000 |

Figure F.5e : Cylindrical Panel: 4th Mode.

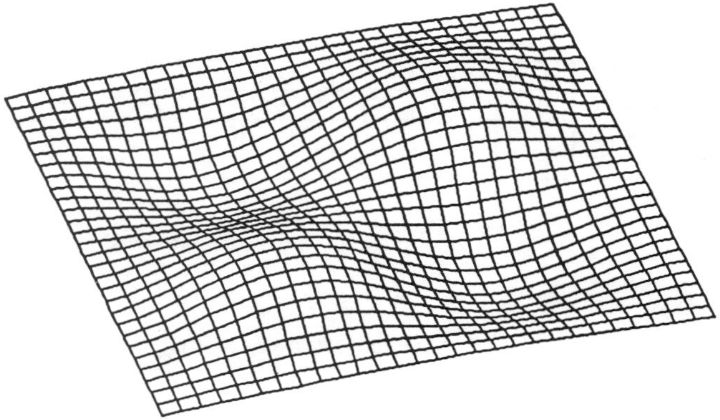

Perspective

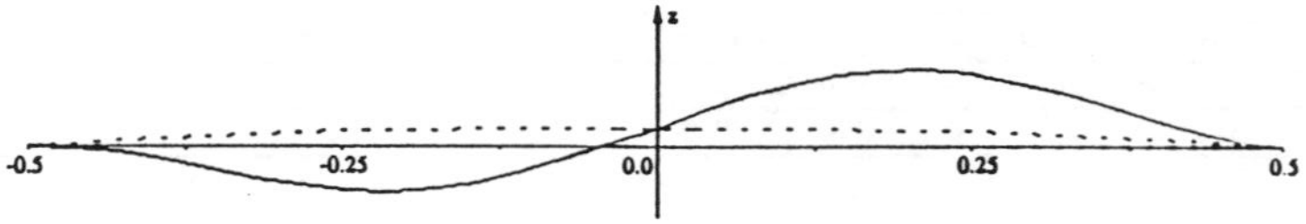


Section C-C

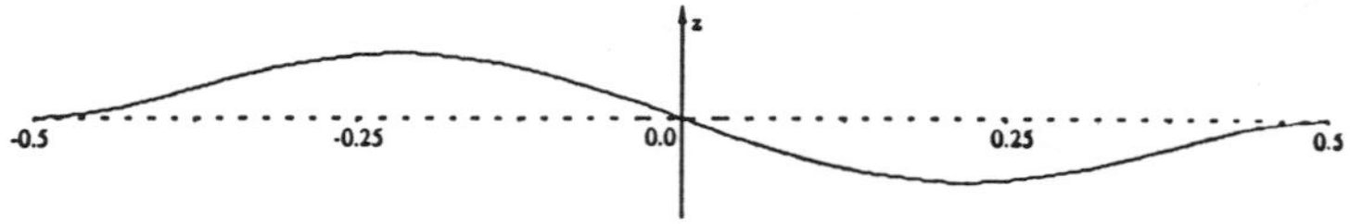


Section D-D

**Normalized Mode Characteristics:**

5th Mode: f = 1415.8 Hz

Normal Displacement Posn (x,y): (0.025381,0.01905)

| Position | | -0.5 | -0.375 | -0.25 | -0.125 | 0.0 | 0.125 | 0.25 | 0.375 | 0.5 |
|---|---|---|---|---|---|---|---|---|---|---|
| C-C | Ux | 0.000 | 0.019 | 0.034 | 0.025 | 0.016 | 0.025 | 0.034 | 0.019 | 0.000 |
| | Uy | 0.000 | -0.000 | 0.000 | 0.000 | 0.000 | -0.000 | -0.000 | 0.000 | 0.000 |
| | Uz | 0.000 | -0.456 | -1.000 | -0.849 | 0.000 | 0.849 | 1.000 | 0.456 | 0.000 |

| Position | | -0.5 | -0.375 | -0.25 | -0.125 | 0.0 | 0.125 | 0.25 | 0.375 | 0.5 |
|---|---|---|---|---|---|---|---|---|---|---|
| D-D | Ux | 0.000 | -0.018 | -0.034 | -0.028 | 0.000 | 0.028 | 0.034 | 0.018 | 0.000 |
| | Uy | 0.000 | -0.000 | -0.000 | -0.000 | -0.000 | -0.000 | -0.000 | -0.000 | 0.000 |
| | Uz | 0.000 | -0.496 | -1.000 | -0.811 | 0.000 | 0.811 | 1.000 | 0.496 | 0.000 |

Figure F.5f : Cylindrical Panel: 5th Mode.

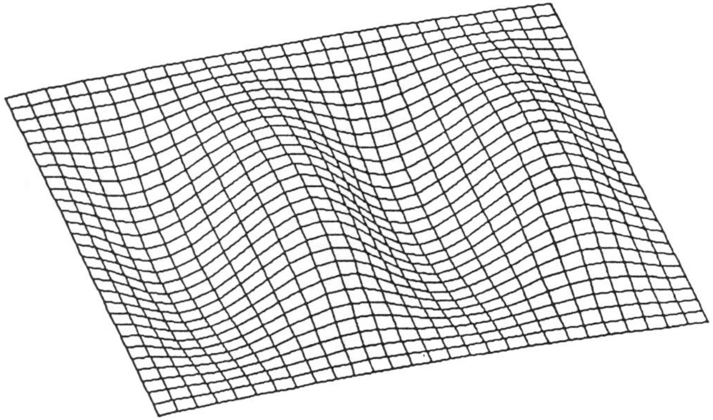

Perspective

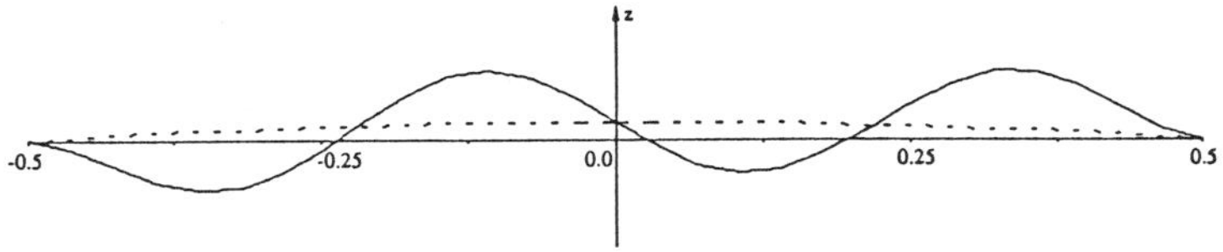


Section A-A

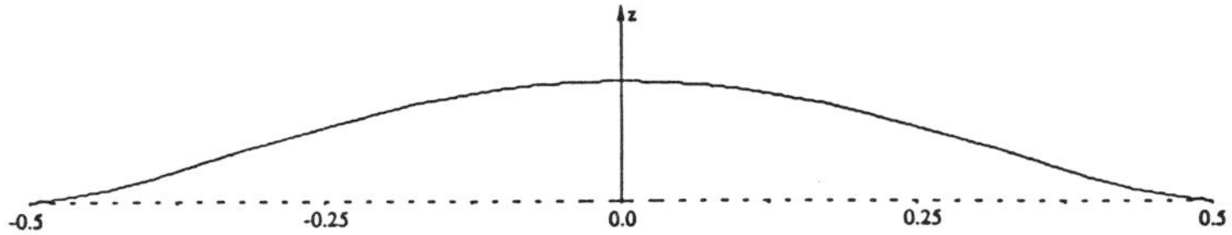


Section E-E

**Normalized Mode Characteristics:**

6th Mode: f = 1684.7 Hz

Normal Displacement Posn (x,y): (0.0380715,0.0)

| Position | | -0.5 | -0.375 | -0.25 | -0.125 | 0.0 | 0.125 | 0.25 | 0.375 | 0.5 |
|---|---|---|---|---|---|---|---|---|---|---|
| A-A | Ux | 0.000 | 0.054 | 0.031 | -0.005 | -0.001 | -0.005 | 0.031 | 0.054 | 0.000 |
| | Uy | 0.000 | 0.000 | 0.000 | 0.000 | 0.000 | 0.000 | 0.000 | 0.000 | 0.000 |
| | Uz | 0.000 | -0.991 | -0.409 | 0.927 | 0.000 | -0.927 | 0.409 | 0.991 | 0.000 |

| Position | | -0.5 | -0.375 | -0.25 | -0.125 | 0.0 | 0.125 | 0.25 | 0.375 | 0.5 |
|---|---|---|---|---|---|---|---|---|---|---|
| E-E | Ux | 0.000 | 0.014 | 0.034 | 0.049 | 0.055 | 0.049 | 0.034 | 0.014 | 0.000 |
| | Uy | 0.000 | -0.001 | -0.001 | -0.001 | 0.000 | 0.001 | 0.001 | 0.001 | 0.000 |
| | Uz | 0.000 | 0.252 | 0.622 | 0.899 | 1.000 | 0.899 | 0.622 | 0.252 | 0.000 |

Figure F.5g : Cylindrical Panel: 6th Mode.

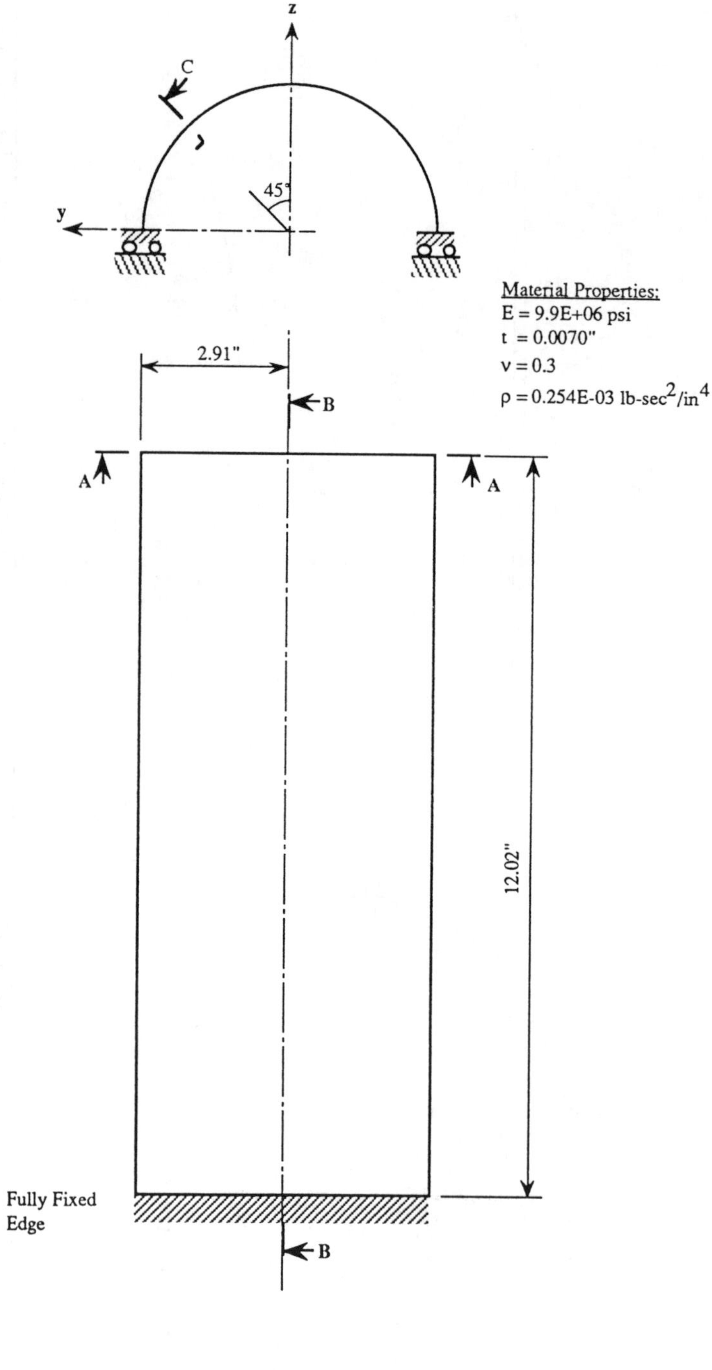


Figure F.6a: Cylindrical Tank.

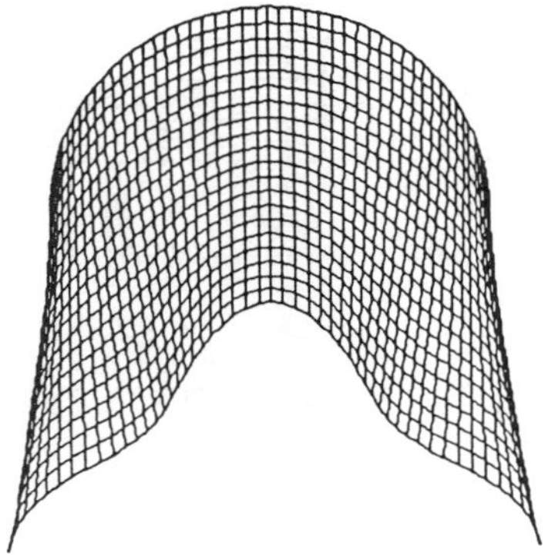

Perspective

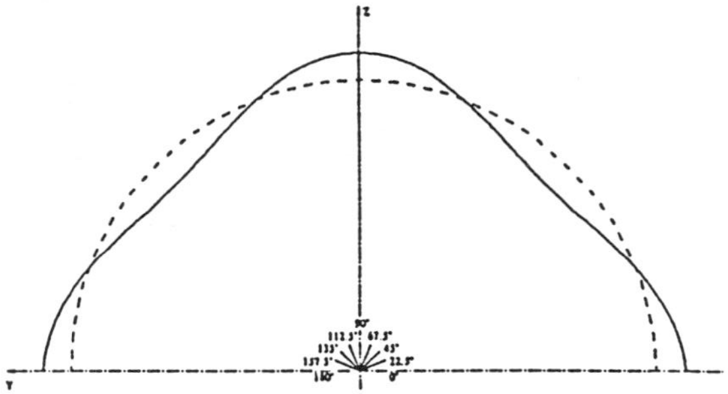


Section A-A

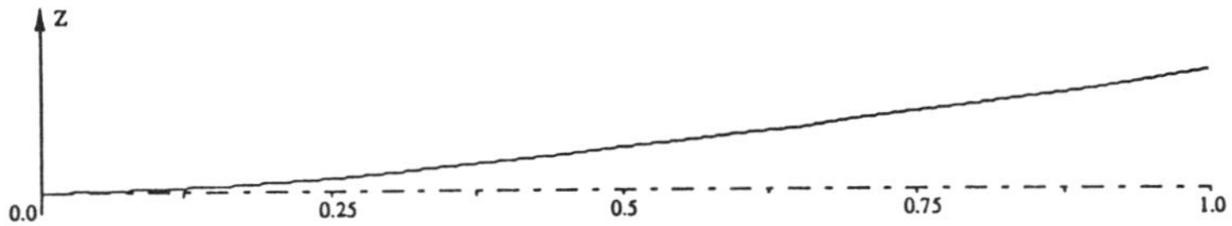


Section B-B

## Normalized Mode Characteristics:

1st Mode: f = 176.7 Hz

Normal Displacement Posn (x,y,z): (0.305308,-0.073914,0.0)

| Position | | 0° | 22.5° | 45° | 67.5° | 90° | 112.5° | 135° | 157.5° | 180° |
|---|---|---|---|---|---|---|---|---|---|---|
| A-A | Ux | -0.019 | 0.000 | 0.019 | 0.000 | 0.019 | 0.000 | 0.019 | 0.000 | -0.019 |
| | Uy | -1.000 | -0.082 | 0.665 | 0.214 | 0.000 | -0.214 | -0.665 | 0.082 | 1.000 |
| | Uz | 0.000 | -0.219 | -0.664 | 0.092 | 0.939 | 0.092 | -0.664 | -0.219 | 0.000 |

| Position | | 0.0 | 0.125 | 0.25 | 0.375 | 0.5 | 0.625 | 0.75 | 0.875 | 1.0 |
|---|---|---|---|---|---|---|---|---|---|---|
| B-B | Ux | 0.000 | -0.005 | -0.010 | -0.013 | -0.016 | -0.017 | -0.018 | -0.019 | -0.019 |
| | Uy | 0.000 | 0.000 | 0.000 | 0.000 | 0.000 | 0.000 | 0.000 | 0.000 | 0.000 |
| | Uz | 0.000 | 0.030 | 0.100 | 0.201 | 0.326 | 0.467 | 0.618 | 0.773 | 0.939 |

Figure F.6b : Cylindrical Tank: 1st Mode.

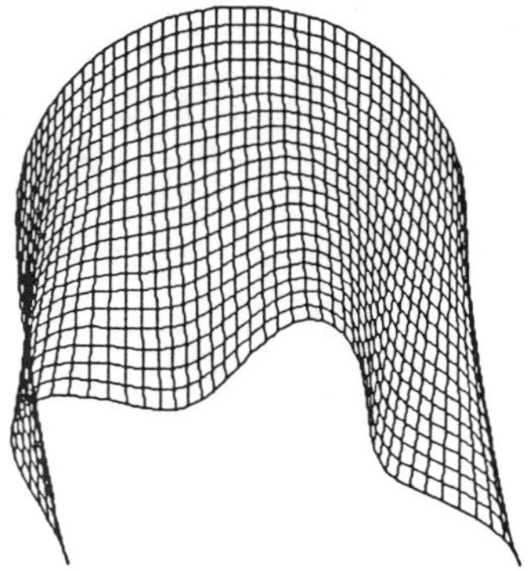

Perspective

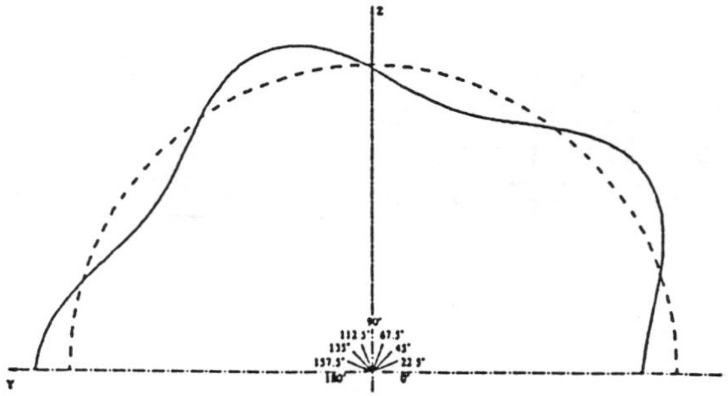


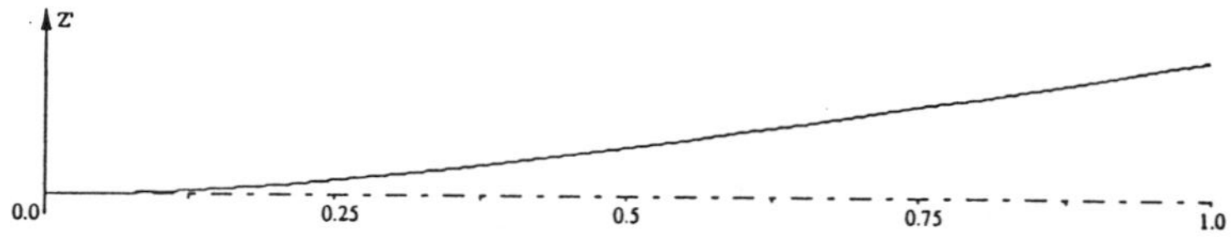


Section C-C

**Normalized Mode Characteristics:**

2nd Mode: f = 208.3 Hz

Normal Displacement Posn (x,y,z): (0.305308,-0.073914,0.0)

| Position | | 0° | 22.5° | 45° | 67.5° | 90° | 112.5° | 135° | 157.5° | 180° |
|---|---|---|---|---|---|---|---|---|---|---|
| A-A | Ux | 0.013 | -0.005 | -0.008 | 0.010 | 0.000 | -0.010 | 0.008 | 0.005 | -0.013 |
| | Uy | 1.000 | -0.266 | -0.541 | 0.254 | 0.176 | 0.254 | -0.541 | -0.266 | 1.000 |
| | Uz | 0.000 | 0.290 | 0.363 | -0.795 | 0.000 | 0.795 | -0.363 | -0.290 | 0.000 |

| Position | | 0.0 | 0.125 | 0.25 | 0.375 | 0.5 | 0.625 | 0.75 | 0.875 | 1.0 |
|---|---|---|---|---|---|---|---|---|---|---|
| C-C | Ux | 0.000 | -0.002 | -0.004 | -0.006 | -0.007 | -0.008 | -0.008 | -0.008 | -0.008 |
| | Uy | 0.000 | -0.016 | -0.056 | -0.115 | -0.188 | -0.270 | -0.356 | -0.444 | -0.541 |
| | Uz | 0.000 | 0.011 | 0.038 | 0.078 | 0.127 | 0.181 | 0.238 | 0.296 | 0.363 |

Figure F.6c : Cylindrical Tank: 2nd Mode.

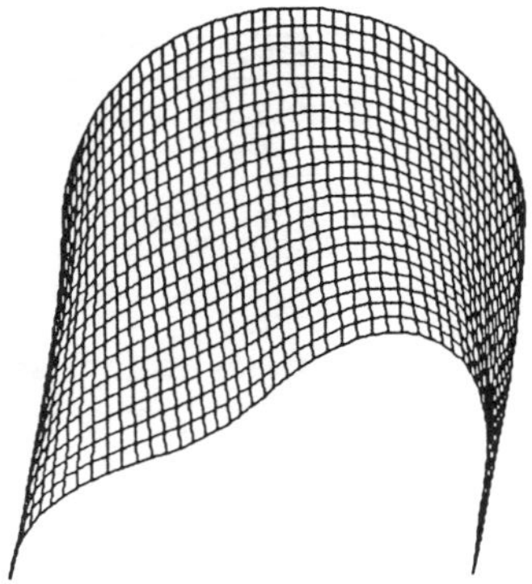

Perspective

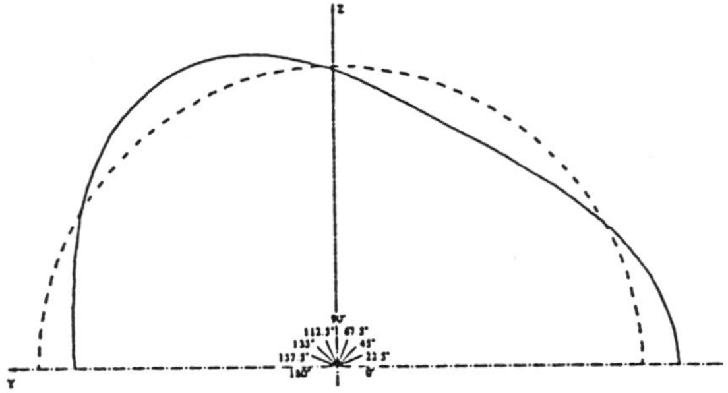


Section A-A

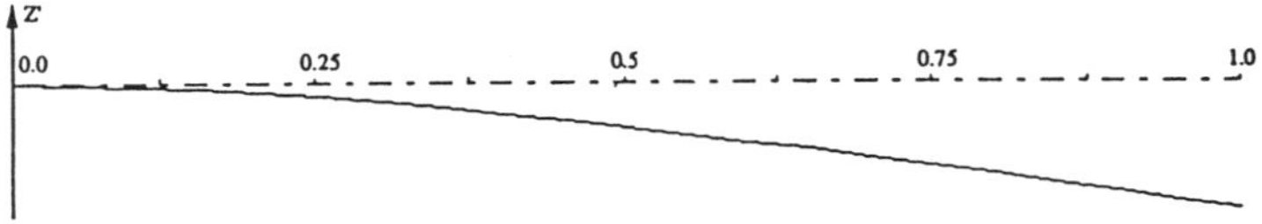


Section C-C

## Normalized Mode Characteristics:

3rd Mode: f = 234.0 Hz

Normal Displacement Posn (x,y,z): (0.305308,-0.073914,0.0)

| Position | | 0° | 22.5° | 45° | 67.5° | 90° | 112.5° | 135° | 157.5° | 180° |
|---|---|---|---|---|---|---|---|---|---|---|
| A-A | Ux | -0.030 | -0.011 | 0.021 | 0.028 | 0.000 | -0.028 | -0.021 | 0.011 | 0.030 |
| | Uy | -1.000 | -0.405 | 0.283 | 0.407 | 0.287 | 0.407 | 0.283 | -0.405 | -1.000 |
| | Uz | 0.000 | -0.116 | -0.570 | 0.700 | 0.000 | 0.700 | 0.570 | 0.116 | 0.000 |

| Position | | 0.0 | 0.125 | 0.25 | 0.375 | 0.5 | 0.625 | 0.75 | 0.875 | 1.0 |
|---|---|---|---|---|---|---|---|---|---|---|
| C-C | Ux | 0.000 | 0.006 | 0.011 | 0.015 | 0.018 | 0.020 | 0.021 | 0.021 | 0.021 |
| | Uy | 0.000 | 0.011 | 0.033 | 0.064 | 0.102 | 0.144 | 0.189 | 0.235 | 0.283 |
| | Uz | 0.000 | -0.021 | -0.065 | -0.127 | -0.204 | -0.290 | -0.381 | -0.474 | -0.570 |

Figure F.6d : Cylindrical Tank: 3rd Mode.

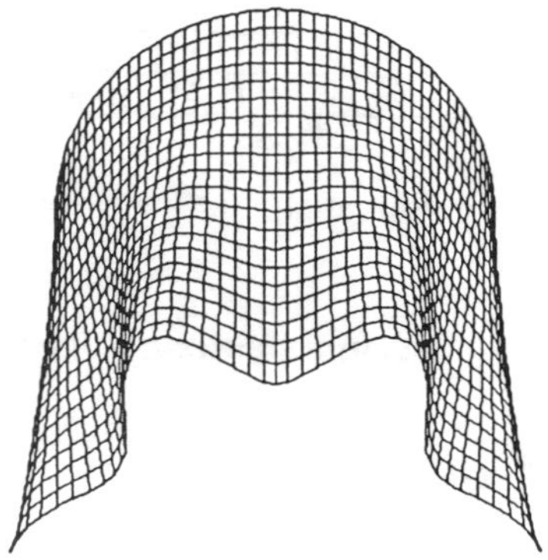

Perspective

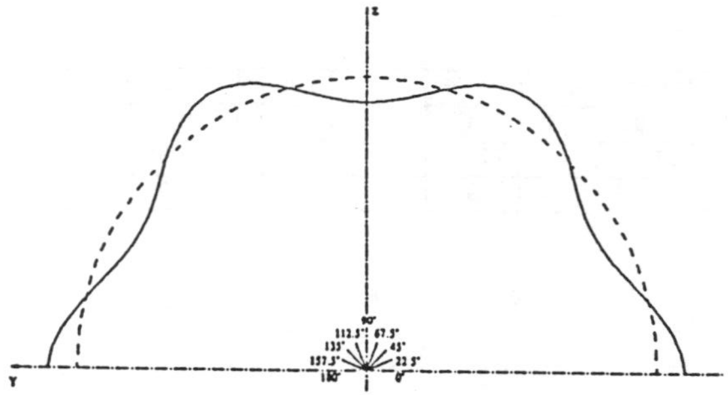


Section A-A

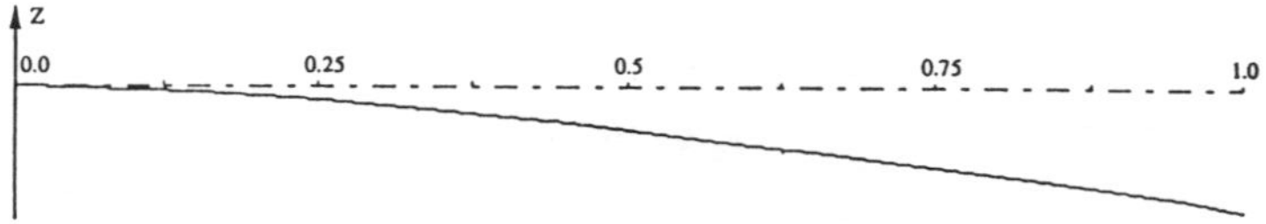


Section B-B

**Normalized Mode Characteristics:**

4th Mode: f = 285.5 Hz

Normal Displacement Posn (x,y,z): (0.305308,0.073914,0.0)

| Position | | 0° | 22.5° | 45° | 67.5° | 90° | 112.5° | 135° | 157.5° | 180° |
|---|---|---|---|---|---|---|---|---|---|---|
| A-A | Ux | -0.009 | 0.006 | 0.000 | -0.005 | 0.008 | -0.005 | 0.000 | 0.006 | -0.009 |
| | Uy | -1.000 | 0.553 | 0.091 | -0.331 | 0.000 | 0.331 | -0.091 | -0.553 | 1.000 |
| | Uz | 0.000 | -0.336 | 0.112 | 0.531 | 0.885 | 0.531 | 0.112 | -0.336 | 0.000 |

| Position | | 0.0 | 0.125 | 0.25 | 0.375 | 0.5 | 0.625 | 0.75 | 0.875 | 1.0 |
|---|---|---|---|---|---|---|---|---|---|---|
| B-B | Ux | 0.000 | 0.002 | 0.004 | 0.006 | 0.007 | 0.007 | 0.008 | 0.008 | 0.008 |
| | Uy | 0.000 | 0.000 | 0.000 | 0.000 | 0.000 | 0.000 | 0.000 | 0.000 | 0.000 |
| | Uz | 0.000 | -0.027 | -0.093 | -0.191 | -0.311 | -0.444 | -0.583 | -0.722 | -0.885 |

Figure F.6e : Cylindrical Tank: 4th Mode.

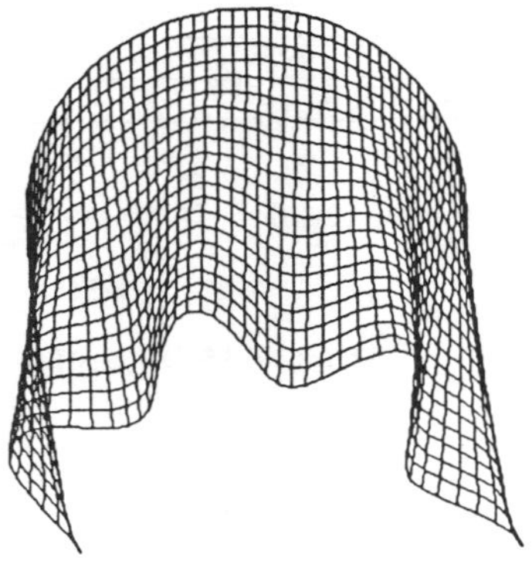

Perspective

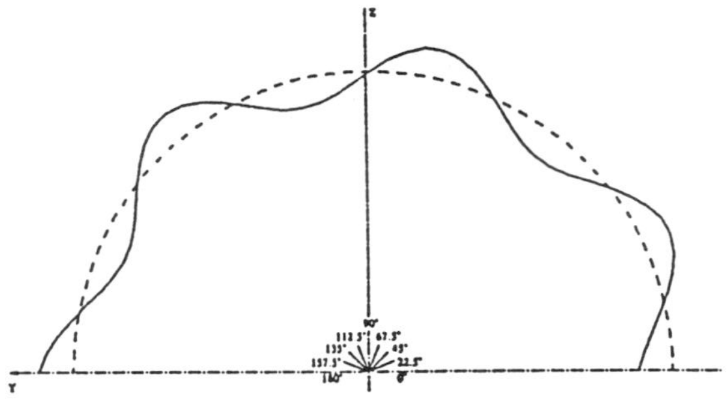


Section A-A

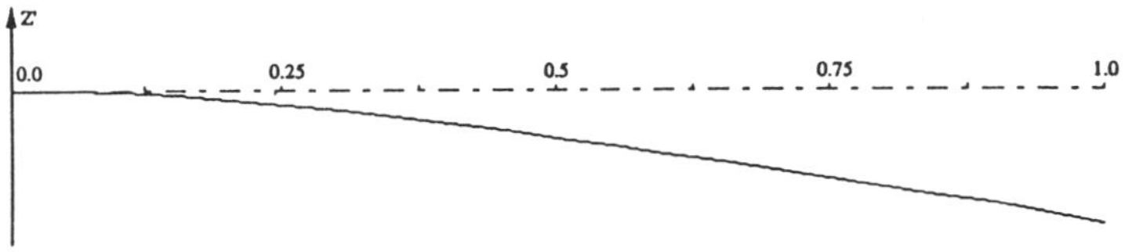


Section C-C

**Normalized Mode Characteristics:**

5th Mode: f = 388.5 Hz

Normal Displacement Posn (x,y,z): (0.305308,-0.073914,0.0)

| Position | | 0° | 22.5° | 45° | 67.5° | 90° | 112.5° | 135° | 157.5° | 180° |
|---|---|---|---|---|---|---|---|---|---|---|
| A-A | Ux | 0.007 | -0.005 | 0.004 | -0.002 | 0.000 | 0.002 | -0.004 | 0.005 | -0.007 |
| | Uy | 1.000 | -0.722 | 0.375 | -0.027 | -0.117 | -0.027 | 0.375 | -0.722 | 1.000 |
| | Uz | 0.000 | 0.343 | -0.491 | 0.349 | 0.000 | -0.349 | 0.491 | -0.343 | 0.000 |

| Position | | 0.0 | 0.125 | 0.25 | 0.375 | 0.5 | 0.625 | 0.75 | 0.875 | 1.0 |
|---|---|---|---|---|---|---|---|---|---|---|
| C-C | Ux | 0.000 | 0.001 | 0.002 | 0.003 | 0.003 | 0.004 | 0.004 | 0.004 | 0.004 |
| | Uy | 0.000 | 0.012 | 0.042 | 0.085 | 0.135 | 0.190 | 0.248 | 0.304 | 0.375 |
| | Uz | 0.000 | -0.016 | -0.055 | -0.111 | -0.179 | -0.252 | -0.328 | -0.403 | -0.491 |

Figure F.6f : Cylindrical Tank: 5th Mode.

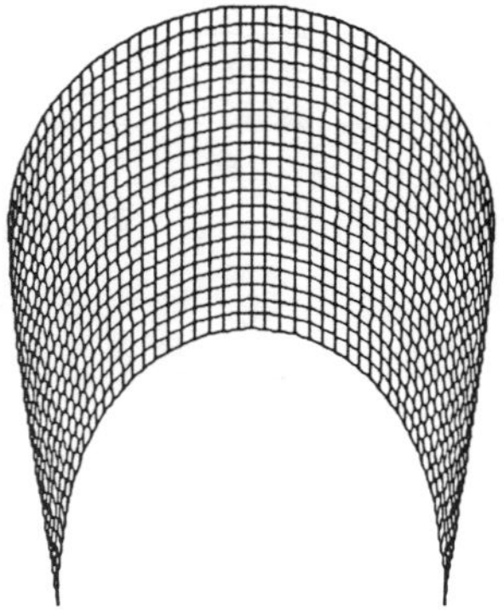

Perspective

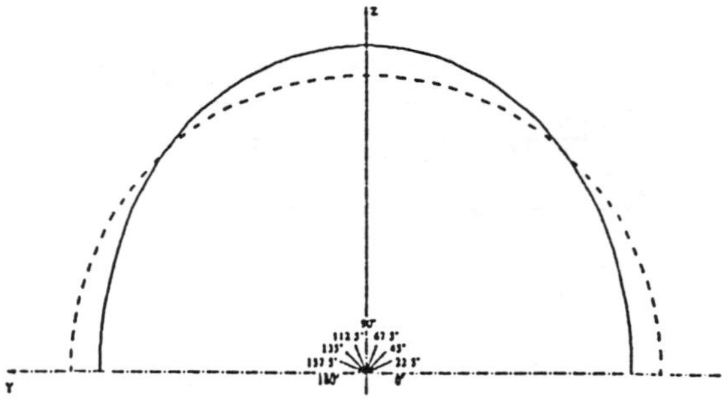


Section A-A

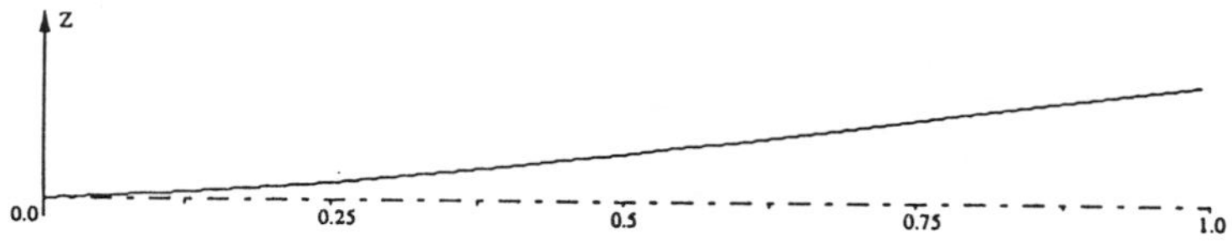


Section B-B

**Normalized Mode Characteristics:**

6th Mode: f = 455.9 Hz

Normal Displacement Posn (x,y,z): (0.305308,-0.073914,0.0)

| Position | | 0° | 22.5° | 45° | 67.5° | 90° | 112.5° | 135° | 157.5° | 180° |
|---|---|---|---|---|---|---|---|---|---|---|
| A-A | Ux | 0.073 | 0.052 | 0.000 | -0.052 | -0.073 | -0.052 | 0.000 | 0.052 | 0.073 |
| | Uy | 1.000 | 0.788 | 0.352 | 0.053 | 0.000 | 0.053 | 0.352 | 0.788 | 1.000 |
| | Uz | 0.000 | 0.053 | 0.351 | 0.789 | 0.991 | 0.789 | 0.351 | 0.053 | 0.000 |

| Position | | 0.0 | 0.125 | 0.25 | 0.375 | 0.5 | 0.625 | 0.75 | 0.875 | 1.0 |
|---|---|---|---|---|---|---|---|---|---|---|
| B-B | Ux | 0.000 | -0.021 | -0.038 | -0.052 | -0.062 | -0.068 | -0.071 | -0.073 | -0.073 |
| | Uy | 0.000 | 0.000 | 0.000 | 0.000 | 0.000 | 0.000 | 0.000 | 0.000 | 0.000 |
| | Uz | 0.000 | 0.053 | 0.136 | 0.248 | 0.382 | 0.529 | 0.683 | 0.837 | 0.991 |

Figure F.6g : Cylindrical Tank: 6th Mode.

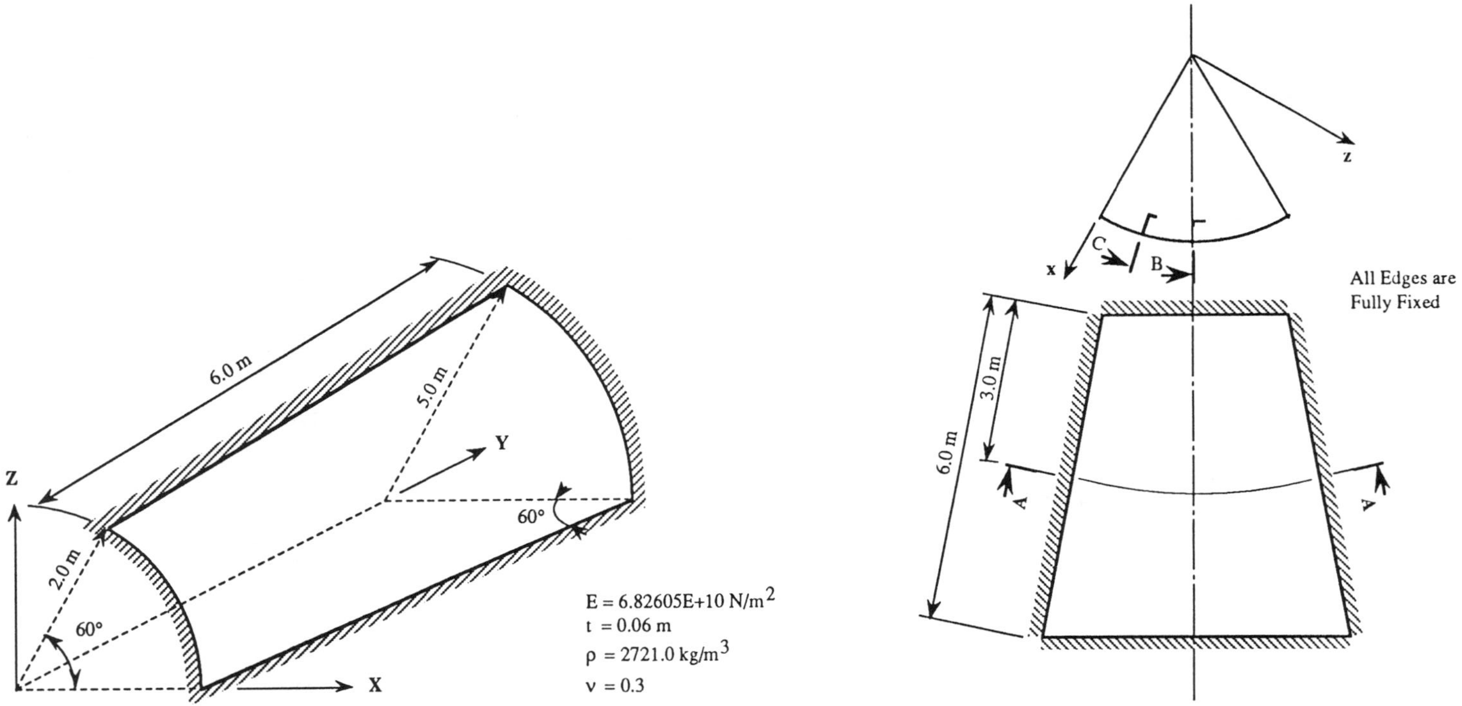


Figure F.7a : Conoidal Shell.

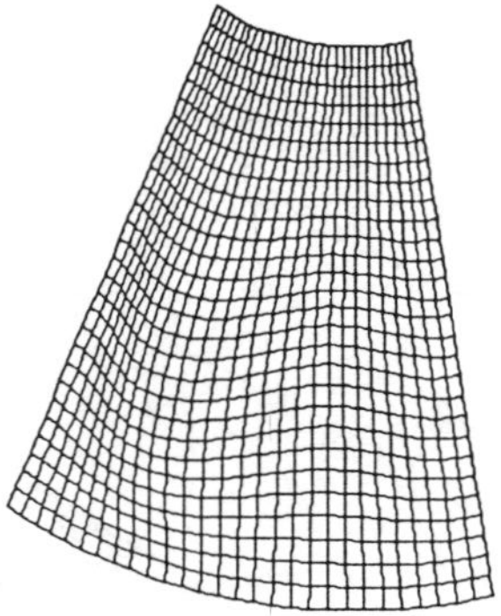

Perspective

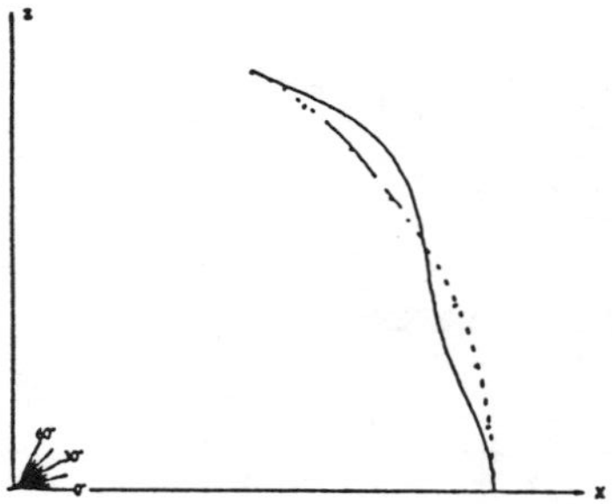

Section A-A

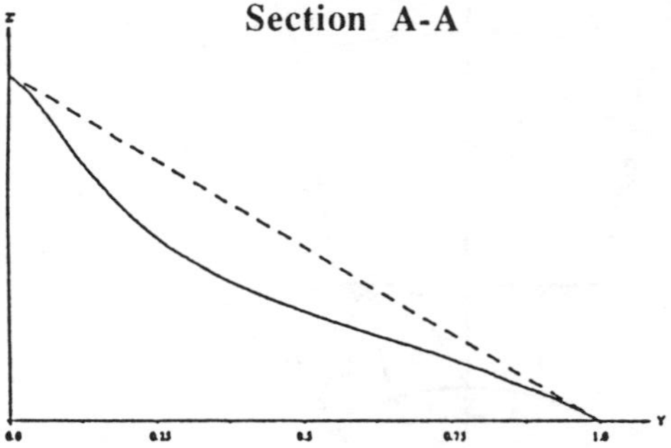

Section C-C

**Normalized Mode Characteristics:**

1st Mode: $\omega$ = 522.8 rad/sec

Normal Displacement Posn (x,y,z): (3.74296,1.94856,1.00292)

| Position | | 0° | 7.5° | 15° | 22.5° | 30° | 37.5° | 45° | 52.5° | 60° |
|---|---|---|---|---|---|---|---|---|---|---|
| A-A | Ux | 0.000 | -0.373 | -0.881 | -0.804 | -0.130 | 0.498 | 0.565 | 0.221 | 0.000 |
| | Uy | 0.000 | -0.213 | -0.512 | -0.454 | 0.000 | 0.454 | 0.512 | 0.213 | 0.000 |
| | Uz | 0.000 | -0.040 | -0.144 | -0.112 | 0.226 | 0.641 | 0.692 | 0.303 | 0.000 |

| Position | | 0.0 | 0.125 | 0.25 | 0.375 | 0.5 | 0.625 | 0.75 | 0.875 | 1.0 |
|---|---|---|---|---|---|---|---|---|---|---|
| C-C | Ux | 0.000 | -0.453 | -0.873 | -1.000 | -0.882 | -0.623 | -0.352 | -0.133 | 0.000 |
| | Uy | 0.000 | -0.246 | -0.489 | -0.570 | -0.512 | -0.373 | -0.217 | -0.087 | 0.000 |
| | Uz | 0.000 | -0.086 | -0.169 | -0.180 | -0.144 | -0.091 | -0.044 | -0.014 | 0.000 |

Figure F.7b : Conoidal Shell: 1st Mode.

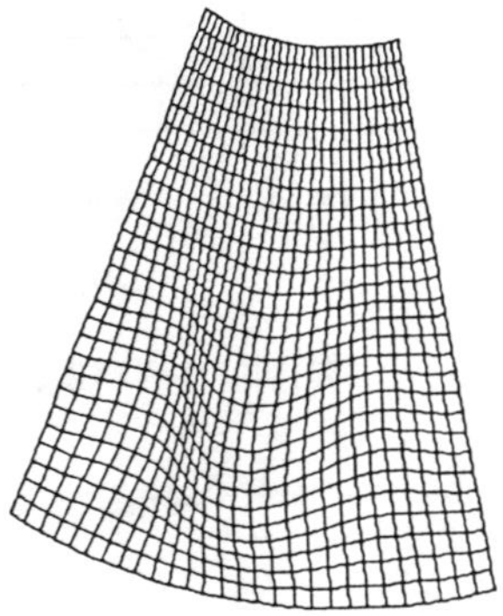

Perspective

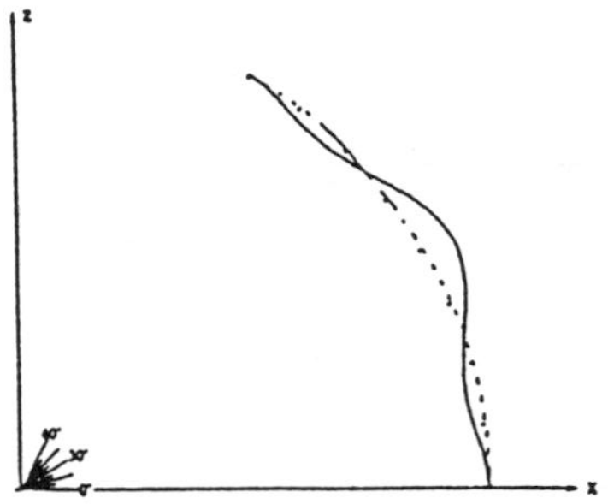

Section A-A

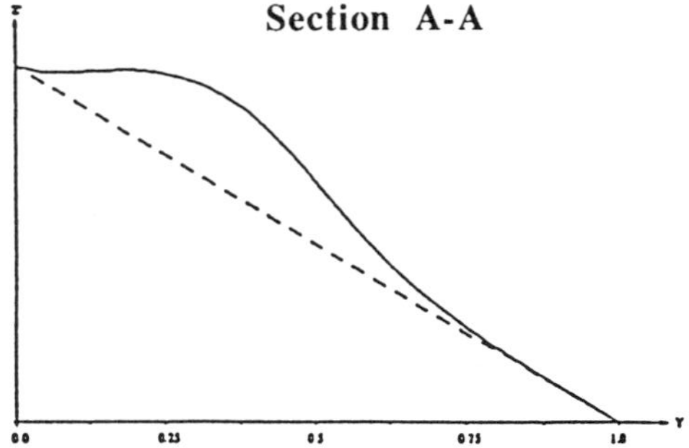


Section B-B

**Normalized Mode Characteristics:**

2nd Mode: $\omega$ = 627.6 rad/sec

Normal Displacement Posn (x,y,z): (3.68061,1.29904,2.125)

| Position | | 0° | 7.5° | 15° | 22.5° | 30° | 37.5° | 45° | 52.5° | 60° |
|---|---|---|---|---|---|---|---|---|---|---|
| A-A | Ux | 0.000 | -0.226 | -0.184 | 0.267 | 0.531 | 0.287 | -0.079 | -0.120 | 0.000 |
| | Uy | 0.000 | -0.134 | -0.101 | 0.193 | 0.373 | 0.193 | -0.101 | -0.134 | 0.000 |
| | Uz | 0.000 | -0.008 | 0.015 | 0.177 | 0.307 | 0.143 | -0.166 | -0.192 | 0.000 |

| Position | | 0.0 | 0.125 | 0.25 | 0.375 | 0.5 | 0.625 | 0.75 | 0.875 | 1.0 |
|---|---|---|---|---|---|---|---|---|---|---|
| B-B | Ux | 0.000 | 0.589 | 1.000 | 0.889 | 0.531 | 0.219 | 0.058 | 0.008 | 0.000 |
| | Uy | 0.000 | 0.378 | 0.658 | 0.601 | 0.373 | 0.164 | 0.050 | 0.010 | 0.000 |
| | Uz | 0.000 | 0.340 | 0.577 | 0.513 | 0.307 | 0.126 | 0.034 | 0.005 | 0.000 |

Figure F.7c : Conoidal Shell: 2nd Mode.

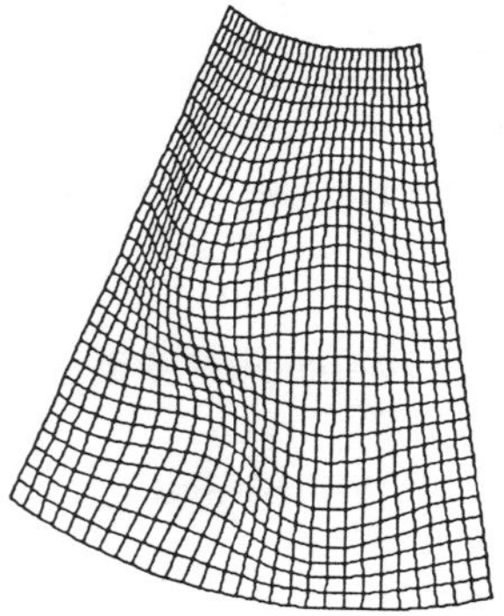

Perspective

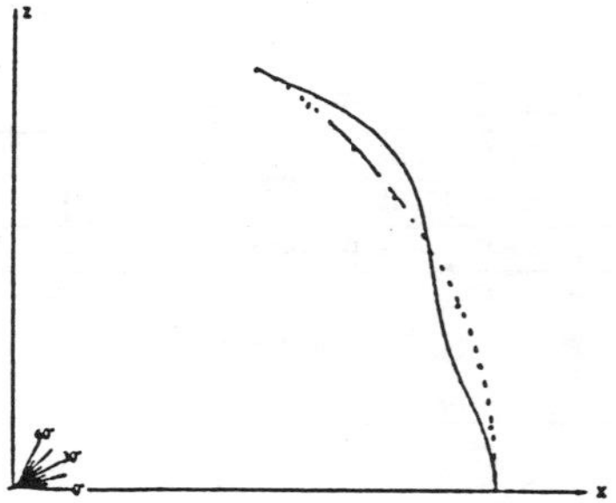

Section A-A

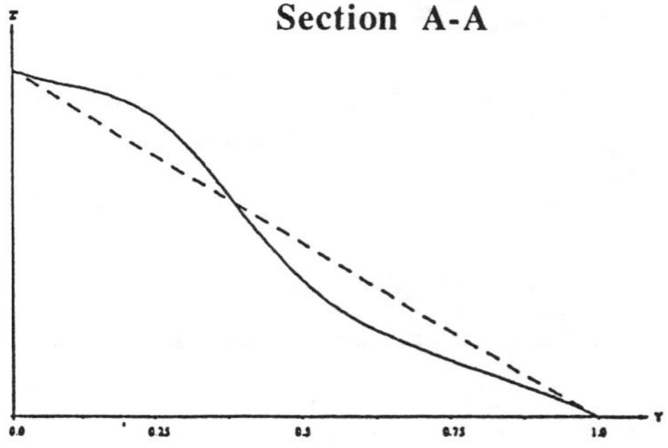

Section C-C

**Normalized Mode Characteristics:**

3rd Mode: $\omega$ = 765.0 rad/sec

Normal Displacement Posn (x,y,z): (3.01852,3.24760,0.80881)

| Position | | 0° | 7.5° | 15° | 22.5° | 30° | 37.5° | 45° | 52.5° | 60° |
|---|---|---|---|---|---|---|---|---|---|---|
| A-A | Ux | 0.000 | -0.328 | -0.772 | -0.685 | -0.095 | 0.451 | 0.515 | 0.204 | 0.000 |
| | Uy | 0.000 | -0.176 | -0.415 | -0.359 | 0.000 | 0.359 | 0.415 | 0.176 | 0.000 |
| | Uz | 0.000 | -0.046 | -0.147 | -0.125 | 0.165 | 0.531 | 0.594 | 0.262 | 0.000 |

| Position | | 0.0 | 0.125 | 0.25 | 0.375 | 0.5 | 0.625 | 0.75 | 0.875 | 1.0 |
|---|---|---|---|---|---|---|---|---|---|---|
| C-C | Ux | 0.000 | 0.679 | 0.784 | 0.008 | -0.772 | -1.000 | -0.742 | -0.322 | 0.000 |
| | Uy | 0.000 | 0.412 | 0.493 | 0.047 | -0.415 | -0.567 | -0.436 | -0.200 | 0.000 |
| | Uz | 0.000 | 0.186 | 0.222 | 0.028 | -0.149 | -0.182 | -0.117 | -0.041 | 0.000 |

Figure F.7d : Conoidal Shell: 3rd Mode.

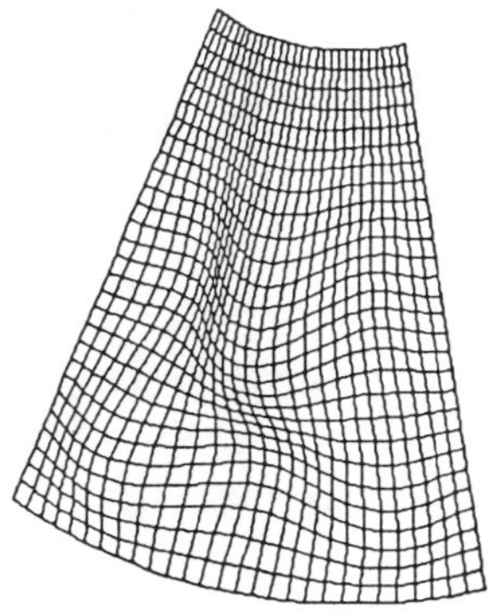

**Perspective**

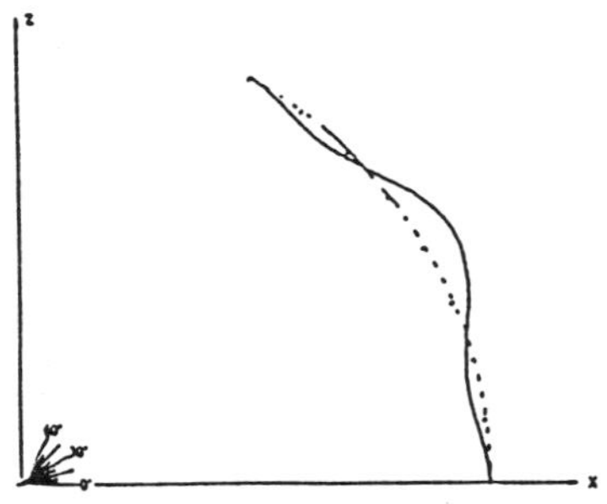


**Section A-A**

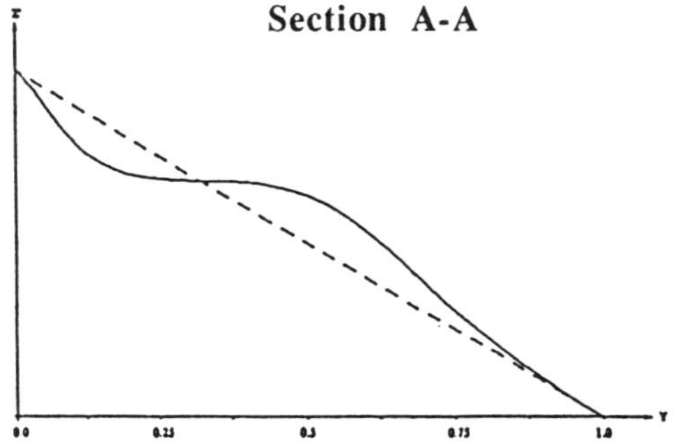


**Section B-B**

## Normalized Mode Characteristics:

4th Mode: $\omega$ = 863.0 rad/sec

Normal Displacement Posn (x,y,z): (3.03109,2.59808,1.75)

| Position | | 0° | 7.5° | 15° | 22.5° | 30° | 37.5° | 45° | 52.5° | 60° |
|---|---|---|---|---|---|---|---|---|---|---|
| A-A | Ux | 0.000 | -0.382 | -0.276 | 0.536 | 1.000 | 0.556 | -0.111 | -0.206 | 0.000 |
| | Uy | 0.000 | -0.206 | -0.137 | 0.359 | 0.658 | 0.359 | -0.137 | -0.206 | 0.000 |
| | Uz | 0.000 | -0.018 | 0.031 | 0.333 | 0.577 | 0.298 | -0.254 | -0.322 | 0.000 |

| Position | | 0.0 | 0.125 | 0.25 | 0.375 | 0.5 | 0.625 | 0.75 | 0.875 | 1.0 |
|---|---|---|---|---|---|---|---|---|---|---|
| B-B | Ux | 0.000 | -0.716 | -0.526 | 0.491 | 1.000 | 0.752 | 0.314 | 0.070 | 0.000 |
| | Uy | 0.000 | -0.481 | -0.369 | 0.303 | 0.658 | 0.513 | 0.226 | 0.056 | 0.000 |
| | Uz | 0.000 | -0.413 | -0.304 | 0.284 | 0.577 | 0.434 | 0.181 | 0.040 | 0.000 |

Figure F.7e : Conoidal Shell: 4th Mode.

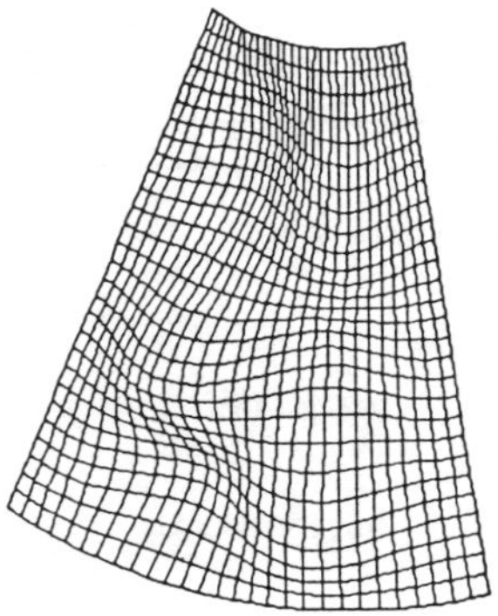

Perspective

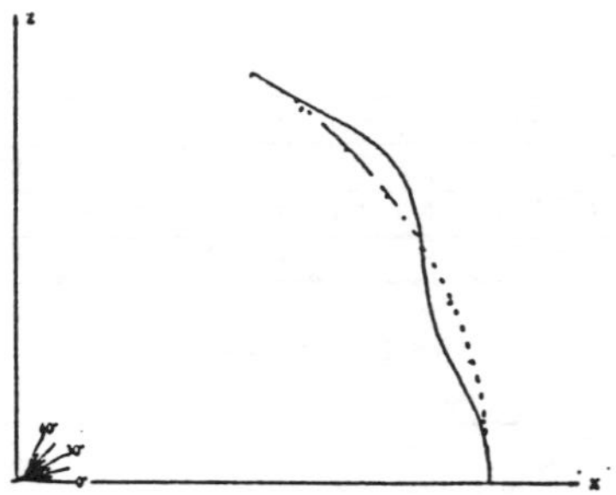

Section A-A

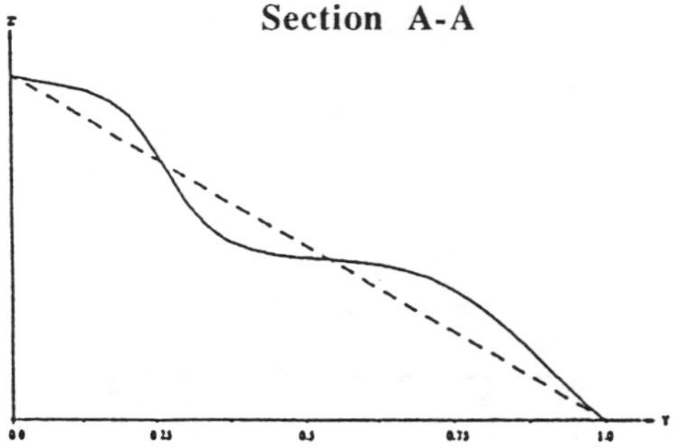

Section C-C

**Normalized Mode Characteristics:**

5th Mode: $\omega$ = 986.1 rad/sec

Normal Displacement Posn (x,y,z): (2.65630,3.89712,0.71175)

| Position | | 0° | 7.5° | 15° | 22.5° | 30° | 37.5° | 45° | 52.5° | 60° |
|---|---|---|---|---|---|---|---|---|---|---|
| A-A | Ux | 0.000 | -0.099 | -0.375 | -0.394 | -0.031 | 0.305 | 0.282 | 0.078 | 0.000 |
| | Uy | 0.000 | -0.070 | -0.256 | -0.265 | 0.000 | 0.265 | 0.256 | 0.070 | 0.000 |
| | Uz | 0.000 | -0.033 | -0.109 | -0.125 | 0.053 | 0.279 | 0.270 | 0.069 | 0.000 |

| Position | | 0.0 | 0.125 | 0.25 | 0.375 | 0.5 | 0.625 | 0.75 | 0.875 | 1.0 |
|---|---|---|---|---|---|---|---|---|---|---|
| C-C | Ux | 0.000 | 0.724 | 0.067 | -0.922 | -0.375 | 0.717 | 1.000 | 0.532 | 0.000 |
| | Uy | 0.000 | 0.442 | 0.047 | -0.560 | -0.256 | 0.387 | 0.574 | 0.320 | 0.000 |
| | Uz | 0.000 | 0.229 | 0.056 | -0.232 | -0.109 | 0.144 | 0.188 | 0.083 | 0.000 |

Figure F.7f : Conoidal Shell: 5th Mode.

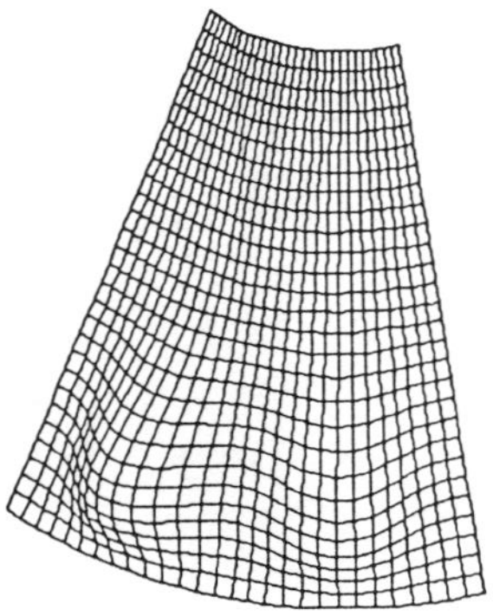

Perspective

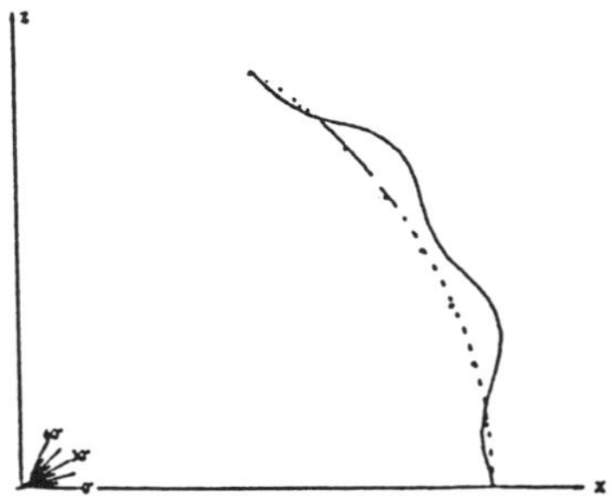

Section A-A

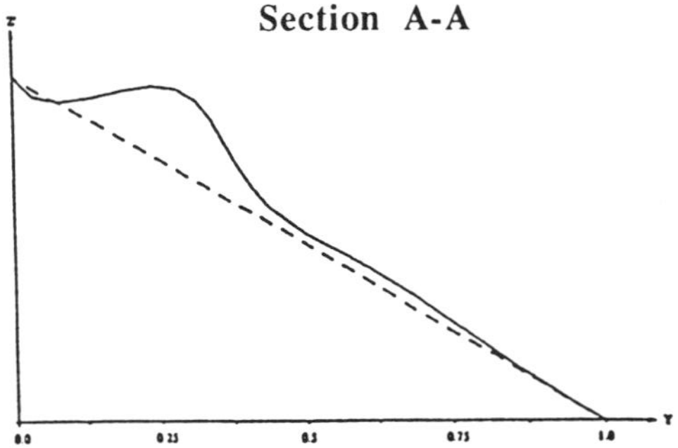

Section B-B

## Normalized Mode Characteristics:

6th Mode: $\omega$ = 999.9 rad/sec

Normal Displacement Posn (x,y,z): (3.68061,1.29904,2.125)

| Position | | 0° | 7.5° | 15° | 22.5° | 30° | 37.5° | 45° | 52.5° | 60° |
|---|---|---|---|---|---|---|---|---|---|---|
| A-A | Ux | 0.000 | -0.078 | 0.317 | 0.357 | 0.164 | 0.325 | 0.281 | -0.010 | 0.000 |
| | Uy | 0.000 | -0.083 | 0.161 | 0.164 | 0.019 | 0.164 | 0.161 | -0.083 | 0.000 |
| | Uz | 0.000 | 0.034 | 0.142 | 0.169 | 0.095 | 0.225 | 0.204 | -0.085 | 0.000 |

| Position | | 0.0 | 0.125 | 0.25 | 0.375 | 0.5 | 0.625 | 0.75 | 0.875 | 1.0 |
|---|---|---|---|---|---|---|---|---|---|---|
| B-B | Ux | 0.000 | 0.413 | 1.000 | 0.436 | 0.164 | 0.183 | 0.113 | 0.028 | 0.000 |
| | Uy | 0.000 | 0.384 | 0.662 | 0.188 | 0.019 | 0.072 | 0.052 | 0.010 | 0.000 |
| | Uz | 0.000 | 0.239 | 0.577 | 0.252 | 0.095 | 0.106 | 0.065 | 0.016 | 0.000 |

Figure F.7g : Conoidal Shell: 6th Mode.

## SUBJECT INDEX

# Lecture Notes in Engineering

Edited by C.A. Brebbia and S.A. Orszag

Vol. 1: J. C. F. Telles,
The Boundary Element Method
Applied to Inelastic Problems
IX, 243 pages. 1983.

Vol. 2: Bernard Amadei,
Rock Anisotropy and
the Theory of Stress Measurements
XVIII, 479 pages. 1983.

Vol. 3: Computational Aspects of
Penetration Mechanics
Proceedings of the Army Research
Office Workshop on Computational
Aspects of Penetration Mechanics
held at the Ballistic Research Laboratory
at Aberdeen Proving Ground, Maryland,
27–29 April, 1982
Edited by J. Chandra and J.E. Flaherty
VII, 221 pages. 1983.

Vol. 4: W.S. Venturini
Boundary Element Method in Geomechanics
VIII, 246 pages. 1983.

Vol. 5: Madassar Manzoor
Heat Flow Through Extended
Surface Heat Exchangers
VII, 286 pages. 1984.

Vol. 6: Myron B. Allen III
Collocation Techniques for Modeling
Compositional Flows in Oil Reservoirs
VI, 210 pages. 1984.

Vol. 7: Derek B. Ingham,
Mark A. Kelmanson
Boundary Integral Equation
Analyses of Singular, Potential,
and Biharmonic Problems
IV, 173 pages. 1984.

Vol. 8: Linda M. Abriola
Multiphase Migration of Organic
Compounds in a Porous Medium
A Mathematical Model
VIII, 232 pages. 1984.

Vol. 9: Theodore V. Hromadka II
The Complex Variable Boundary
Element Method
XI, 243 pages. 1984.

Vol. 10: C. A. Brebbia, H. Tottenham,
G. B. Warburton, J. M. Wilson, R. R. Wilson
Vibrations of Engineering Structures
VI, 300 pages. 1985.

Vol. 11: M. B. Beck
Water Quality Management:
A Review of the Development and
Application of Mathematical Models
VIII, 108 pages. 1985.

Vol. 12: G. Walker, J. R. Senft
Free Piston Stirling Engines
XIV, 286 pages. 1985.

Vol. 13: Nonlinear Dynamics
of Transcritical Flows
Proceedings of a DFVLR International
Colloquium, Bonn, Germany, March 26, 1984
VI, 203 pages. 1985.

Vol. 14: A. A. Bakr
The Boundary Integral
Equation Method in Axisymmetric
Stress Analysis Problems
XI, 213 pages. 1986.

Vol. 15: I. Kinnmark
The Shallow Water Wave
Equation: Formulation,
Analysis and Application
XXIII, 187 pages, 1986.

Vol. 16: G. J. Creus
Viscoelasticity – Basic
Theory and Applications
to Concrete Structures
VII, 161 pages. 1986.

Vol. 17: S. M. Baxter
C. L. Morfey
Angular Distribution
Analysis in Acoustics
VII, 202 pages. 1986.

Vol. 18: N. C. Markatos,
D. G. Tatchell, M. Cross, N. Rhodes
Numerical Simulation of Fluid Flow
and Heat/Mass Tranfer Processes
VIII, 482 pages. 1986.

Vol. 19: Finite Rotations
in Structural Mechanics
Proceedings of the Euromech
Colloquium 197, Jablonna 1985
VII, 385 pages. 1986.

Vol. 20: S. M. Niku
Finite Element Analysis
of Hyperbolic Cooling Towers
VIII , 216 pages. 1986.

# Lecture Notes in Engineering

Edited by C.A. Brebbia and S.A. Orszag

Vol. 21: B. F. Spencer, Jr.
Reliability of Randomly
Excited Hysteretic Structures
XIII, 138 pages. 1986.

Vol. 22: A. Gupta, R. P. Singh
Fatigue Behaviour
of Offshore Structures
XXI, 299 pages. 1986.

Vol. 23: P. Hagedorn, K. Kelkel,
J. Wallaschek
Vibrations and Impedances
of Rectangular Plates
with Free Boundaries
V, 152 pages. 1986.

Vol. 24: Supercomputers
and Fluid Dynamics
Proceedings of the First
Nobeyama Workshop
September 3-6, 1985
VIII, 200 pages. 1986.

Vol. 25: B. Hederson-Sellers
Modeling of Plume Rise
and Dispersion –
The University of Salford
Model: U. S. P. R.
VIII, 113 pages. 1987.

Vol. 26: Shell and Spatial Structures:
Computational Aspects
Proceeding of the International Symposium
July 1986, Leuven, Belgium
Edited by G. De Roeck, A. Samartin Quiroga,
M. Van Laethem and E. Backx
VII, 486 pages. 1987.

Vol. 27: Th. V. Hromadka, Ch.-Ch. Yen
G. F. Pinder
The Best Approximation Method
An Introduction
XIII, 168 pages. 1987.

Vol. 28: Refined Dynamical Theories
of Beams, Plates and Shells and
Their Applications
Proceedings of the Euromech-Colloquim 219
Edited by I. Elishakoff and H. Irretier
IX, 436 pages. 1987.

Vol. 29: G. Menges, N. Hövelmanns,
E. Baur (Eds.)
Expert Systems in Production Engineering
Proceedings of the International Workshop
Spa, Belgium, August 18-22, 1986
IV, 245 pages. 1987.

Vol. 30: R. Doležal
Simulation of Large State Variations
in Steam Power Plants
Dynamics of Large Scale Systems
X, 110 pages. 1987.

Vol. 31: Y. K. Lin, G. I. Schueller (Eds.)
Stochastic Structural Mechanics
U.S.-Austria Joint Seminar, May 4-5, 1987
Boca Raton, Florida, USA
XI, 507 pages. 1987.

Vol. 32: Y. K. Lin, R. Minai (Eds.)
Stochastic Approaches
in Earthquake Engineering
U.S.-Japan Joint Seminar, May 6-7, 1987
Boca Raton, Florida, USA
XI, 457 pages. 1987.

Vol. 33: P. Thoft-Christensen (Editor)
Reliability and Optimization
of Structural Systems
Proceedings of the First IFIP WG 7.5
Working Conference
Aalborg, Denmark, May 6-8, 1987
VIII, 458 pages. 1987.

Vol. 34: M. B. Allen III, G. A. Behie,
J. A. Trangenstein
Multiphase Flow in Porous Media
Mechanics, Mathematics, and Numerics
IV, 312 pages. 1988.

Vol. 35: W. Tang
A New Transformation
Approach in BEM
A Generalized Approach for
Transforming Domain Integrals
VI, 210 pages. 1988.

Vol. 36: R. H. Mendez, S. A. Orszag
Japanese Supercomputing
Architecture, Algorithms, and Applications
IV, 160 pages. 1988.

Vol. 37: J. N. Reddy, C. S. Krishnamoorthy,
K. N. Seetharamu (Eds.)
Finite Element Analysis
for Engineering Design
XIV, 869 pages. 1988.

Vol. 38: S. J. Dunnett, D. B. Ingham
The Mathematics of Blunt Body Sampling
VIII, 213 pages. 1988

Vol. 39: S.L. Koh, C.G. Speziale (Eds.)
Recent Advances in Engineering Science
A Symposium dedicated to A. Cemal Eringen
June 20–22, 1988, Berkeley, California
XVIII, 268 pages. 1989